Studies in Systems, Decision and Control

Volume 290

Series Editor

Janusz Kacprzyk, Systems Research Institute, Polish Academy of Sciences, Warsaw, Poland

The series "Studies in Systems, Decision and Control" (SSDC) covers both new developments and advances, as well as the state of the art, in the various areas of broadly perceived systems, decision making and control–quickly, up to date and with a high quality. The intent is to cover the theory, applications, and perspectives on the state of the art and future developments relevant to systems, decision making, control, complex processes and related areas, as embedded in the fields of engineering, computer science, physics, economics, social and life sciences, as well as the paradigms and methodologies behind them. The series contains monographs, textbooks, lecture notes and edited volumes in systems, decision making and control spanning the areas of Cyber-Physical Systems, Autonomous Systems, Sensor Networks, Control Systems, Energy Systems, Automotive Systems, Biological Systems, Vehicular Networking and Connected Vehicles, Aerospace Systems, Automation, Manufacturing, Smart Grids, Nonlinear Systems, Power Systems, Robotics, Social Systems, Economic Systems and other. Of particular value to both the contributors and the readership are the short publication timeframe and the world-wide distribution and exposure which enable both a wide and rapid dissemination of research output.

** Indexing: The books of this series are submitted to ISI, SCOPUS, DBLP, Ulrichs, MathSciNet, Current Mathematical Publications, Mathematical Reviews, Zentralblatt Math: MetaPress and Springerlink.

More information about this series at http://www.springer.com/series/13304

Ali Kaveh · Hossein Rahami ·
Iman Shojaei

Swift Analysis of Civil Engineering Structures Using Graph Theory Methods

Springer

Ali Kaveh
School of Civil Engineering
Iran University of Science and Technology
Tehran, Iran

Hossein Rahami
School of Engineering Science
College of Engineering
University of Tehran
Tehran, Iran

Iman Shojaei
Align Technology Inc.
San Jose, CA, USA

ISSN 2198-4182 ISSN 2198-4190 (electronic)
Studies in Systems, Decision and Control
ISBN 978-3-030-45548-4 ISBN 978-3-030-45549-1 (eBook)
https://doi.org/10.1007/978-3-030-45549-1

This Springer imprint is published by the registered company Springer Nature Switzerland AG
The registered company address is: Gewerbestrasse 11, 6330 Cham, Switzerland

Preface

The dual requirements of accuracy and speed in analyzing structural systems with ever-tighter design tolerances and larger numbers of elements have been relentlessly driving forward research into methods that can achieve the Swift Analysis of Structures at a reasonable computational cost.

The proviso of accuracy in the analysis is brought about by the need for demonstrating structural safety and material cost reductions. Unfortunately, more accurate analyses, with most conventional methods, lead to increased computational loads. Combined with the already heavy load of analyzing large and complex structures, conventional analysis methods turn into resource drains that are best avoided. In this book, we propose and validate a number of methods and shortcuts for frugal engineers by which they can shave off significant amounts from their computational budget in the analysis, reanalysis and thus structural design processes.

A major class of methods for saving computational resources make clever use of symmetries or regularities within structures and have been presented in an earlier book by the first author published by Springer, titled *Optimal Analysis of Structures by Concepts of Symmetry and Regularity*. In the present work, we will concentrate on a set of methods which instead apply to a larger and more commonly seen subset of structures in which regularity or symmetry are violated by small changes such as cutouts or additional non-regular members.

The value of the presented methods becomes more pronounced in situations where the analysis needs to be repeated hundreds or even thousands of times, as is the case with the optimal design of structures using different meta-heuristic algorithms.

This book is of most interest to researchers and engineers who engage in computer-aided analysis and design of structures and to software developers in this field. The methods are not only applicable to skeletal structures but by extension also to continuum models. Many of the underlying concepts can be further applied to non-structural systems, such as hydraulic and electrical networks. This work aims to serve as a complementary book to the previous books published on optimal analysis of large-scale structures utilizing concepts of symmetry and regularity.

Due to the novel application of graph-theoretical methods, we hope that mathematicians may also find some value and validation in our work.

This book attempts to explore a wide variety of methods and algorithms using concepts from a variety of fields. In order to ground this exploration, the early chapters will elucidate definitions and build a mathematical foundation to allow readers to make practical use of the knowledge gained. With the same goal in mind, we have also included solved examples in each chapter.

This book contains ten chapters as follows:

In Chap. 1, the required definitions from the field of graph theory are presented along with a discussion of graph products. In Chap. 2, further definitions and concepts relevant to the contents of the present book are provided. In Chap. 3, near-regular skeletal structures having additional members are discussed, and methods for their analysis and eigensolutions are developed. Chapter 4 is devoted to the analyses of near-regular structures having additional or missing nodes or members. Displacement and force methods as well as various eigensolutions for these structures are presented. In Chap. 5, finite element and mesh-free solutions for regular and near-regular structural and mechanical systems are developed. In Chap. 6, a dynamic analysis formulation is developed for single-degree-of-freedom systems and then generalized to multi-degree-of-freedom systems using modal analysis. In Chap. 7, we apply our swift analysis methods to speed up the optimal design of structures when using evolutionary methods. In Chap. 8, methods for static and dynamic analysis/reanalysis as well as eigensolutions including vibration analysis of such systems are developed by invoking the Kronecker products and matrix manipulations. In Chap. 9, a numerical method is presented for efficiently solving many classes of differential equations in arbitrary domains using geometrical transformations and Kronecker product rules. In Chap. 10, methods for numerically solving systems of linear equations are developed using the advantages of repetitive tridiagonal matrices.

We would like to thank our colleagues for using our joint papers and their help in various stages of writing this book: Dr. B. Barzgari and Dr. Mohammad Ardalan Asl. Our special thanks are due to Mr. Thomas Ditzinger, Editorial Director of the Applied Sciences of Springer, for his constructive comments, editing, and unfailing kindness in the course of the preparation of this book. My sincere appreciation is extended to our Springer colleagues Mrs. Sabine Schmitt, Ms. Saranya Kalidoss and Ms. Femina Joshi Arul Thas.

We would like to thank the publishers who permitted some of our papers to be utilized in the preparation of this book, consisting of Springer Verlag, Elsevier and Wiley.

Every effort has been made to render this book error-free. However, the authors would appreciate any remaining errors being brought to their attention through their email addresses: alikaveh@iust.ac.ir (Ali Kaveh), hrahami@ut.ac.ir (Hossein Rahami) and shojaei.iman@gmail.com (Iman Shojaei).

Tehran, Iran
February 2020

Ali Kaveh
Hossein Rahami
Iman Shojaei

Contents

Chapter 1
Definitions from Graph Theory and Graph Products

1.1 Introduction

Graph theory is a branch of mathematics started by Euler as early as 1736. Graph Theory has found many applications in engineering and science, such as chemical, electrical, civil and mechanical engineering, architecture, management and control, communication, operational research, sparse matrix technology, combinatorial optimisation, and computer science. Therefore many books have been published on applied graph theory.

In this chapter basic definitions and concepts of graph theory and graph products are presented; however, for proofs and details the reader may refer to textbooks on this subject [1–3].

1.2 Basic Definitions

There are many physical systems whose performance depends not only on the characteristics of their components, but also on their relative location. As an example, in a structure, if the properties of a member are altered, the overall behaviour of the structure will be changed. This indicates that the performance of a structure depends on the characteristics of its members. On the other hand, if the location of a member is changed, the properties of the structure will again be different. Therefore the connectivity (topology) of the structure influences the performance of the whole structure. Hence it is important to represent a system so that its topology can be understood clearly. The graph model of a system provides a powerful means for this purpose.

A. Kaveh et al., *Swift Analysis of Civil Engineering Structures Using Graph Theory Methods*, Studies in Systems, Decision and Control 290,
https://doi.org/10.1007/978-3-030-45549-1_1

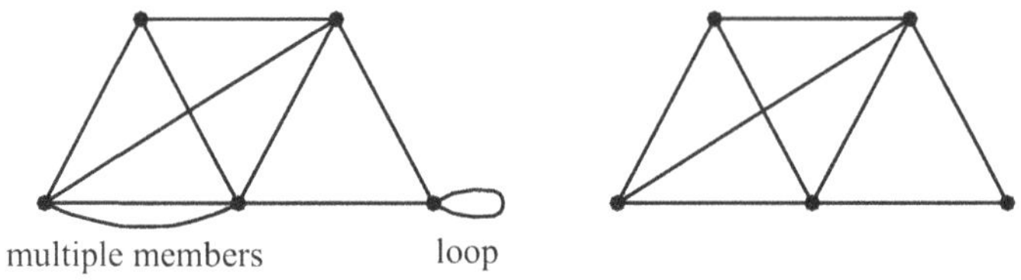

(a) A non-simple graph (b) A simple graph

Fig. 1.1 Non-simple and simple graphs

1.2.1 Definition of a Graph

A graph S consists of a set N(S) of elements called nodes (vertices or points) and a set M(S) of elements called members (edges or arcs) together with a relation of incidence which associates each member with a pair of nodes, called its ends.

Two or more members joining the same pair of nodes are known as a multiple member, and a member joining a node to itself is called a loop. A graph with no loops and multiple members is called a simple graph. If N(S) and M(S) are countable sets, then the corresponding graph S is finite.

The above definitions correspond to abstract graphs; however, a graph may be visualized as a set of points connected by line segments in Euclidean space; the points are identified with nodes, and the line segments without their end points are identified with members. Such a configuration is known as a topological graph. These definitions are illustrated in Fig. 1.1.

1.2.2 Adjacency and Incidence

Two nodes of a graph are called adjacent if these nodes are the end nodes of a member. A member is called incident with a node if it is an end node of that member. Two members are called incident if they have a common end node. The degree (valency) of a node n_i of a graph, denoted by deg (n_i), is the number of members incident with that node. Since each member has two end nodes, the sum of node-degrees of a graph is twice the number of its members (handshaking lemma - known as the first theorem of graph theory).

As an example, in Fig. 1.2 two nodes n_4 and n_5 are adjacent. Node n_3 is incident with member m_2 and m_6, and thus the deg (n_2) = 4.

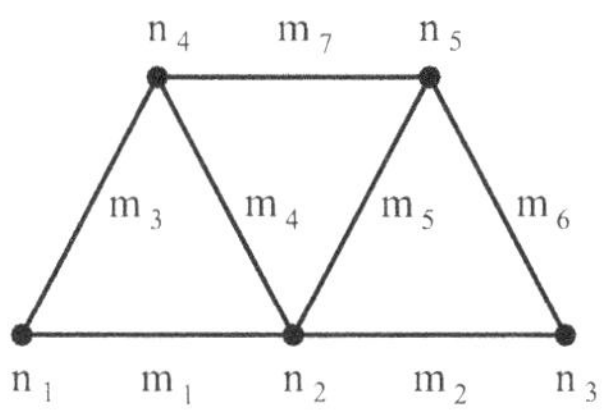

Fig. 1.2 A simple graph S

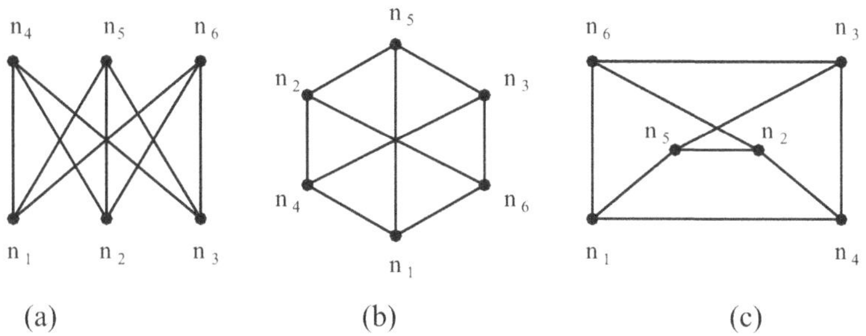

Fig. 1.3 Three isomorphic graphs

1.2.3 Isomorphic Graphs

Two graphs S_1 and S_2 are called isomorphic if there exists a one-to-one correspondence between their node sets and adjacency is preserved. As an example, the three graphs shown in Fig. 1.3 are isomorphic. The word isomorphic is derived from the Greek words same and form.

1.2.4 Graph Operations

A subgraph S_i of S is a graph for which $N(S_i) \subseteq N(S)$ and $M(S_i) \subseteq M(S)$, and each member of S_i has the same ends as in S.

The union of subgraphs $S_1, S_2, \ldots, S_k$ of S, denoted by $S^k = \cup_{i=1}^{k} S_i = S_1 \cup S_2 \cup \ldots \cup S_k$, is a subgraph of S with $N(S^k) = \cup_{i=1}^{k} N(S_i)$ and $M(S^k) = \cup_{i=1}^{k} M(S_i)$. The intersection of two subgraphs S_i and S_j is similarly defined using intersections of node-sets and member-sets of the two subgraphs. The ring sum of two subgraphs $S_i \oplus S_j = S_i \cup S_j - S_i \cap S_j$ is a subgraph which contains the nodes and members of S_i and S_j except those elements common to S_i and S_j. These definitions are illustrated in Fig. 1.4.

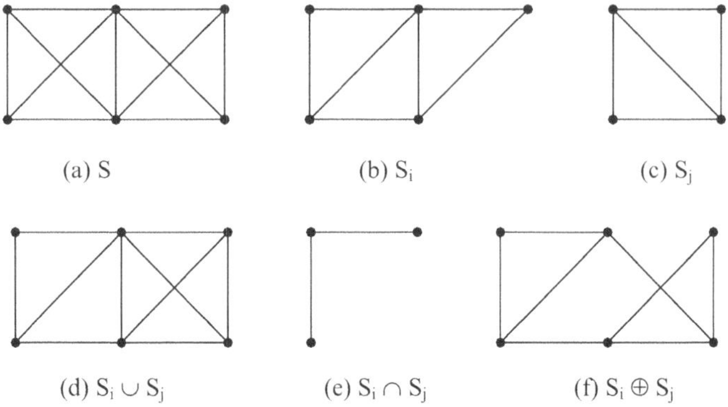

Fig. 1.4 A graph, two of its subgraphs, union, intersection and ring sum

There are many other useful operations, such as Cartesian product, strong Cartesian product, direct product and lexicographic product, and some other products successfully applied to structural engineering that is introduced in Sect. 1.5.

1.3 Different Types of Graphs

In order to simplify the study of properties of graphs, different types of graphs have been defined. Some important ones are as follows:

A null graph is a graph which contains no members. Thus N_k is a graph containing k isolated nodes.

A path graph is a graph consisting of a single path. Hence P_k is a path with k nodes and (k − 1) members.

A cycle graph is a graph consisting of a single cycle. Therefore C_k is a polygon with k members.

A star graph is a graph with one central node connected to some other nodes.

A wheel graph is a star graph with all the non-central nodes connected by a cycle.

A complete graph is a graph in which every two distinct nodes are connected by exactly one member, Fig. 1.5.

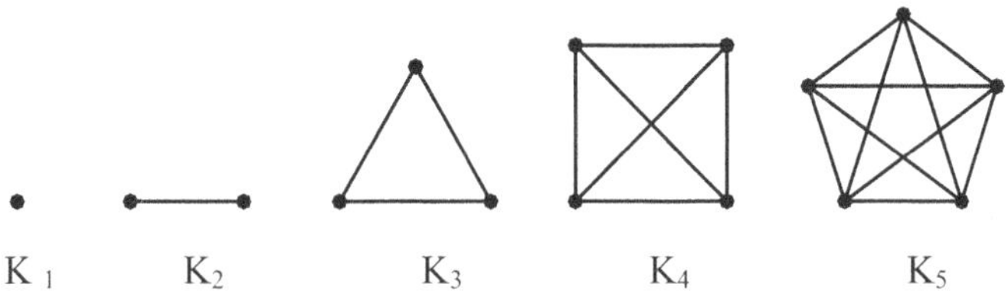

Fig. 1.5 Some complete graphs

A complete graph with N nodes is denoted by K_N. It is easy to prove that a complete graph with N nodes has N(N − 1)/2 members.

1.4 Matrix Representation of a Graph

Matrices play a dominant role in the theory of graphs and in particular in its applications to structural analysis. Some of these matrices conveniently describe the connectivity properties of a graph and others provide useful information about the patterns of the structural matrices, and some reveal additional information about transformations such as those of equilibrium and compatibility equations.

In this section various matrices are studied which reflect the properties of the corresponding graphs. For simplicity, all the graphs are assumed to be connected, since the generalization to non-connected graphs is trivial and consists of considering the direct sum of the matrices for their components.

A graph can be represented in various forms. Some of these representations are of theoretical importance, others are useful from the programming point of view when applied to realistic problems.

Node Adjacency Matrix: Let S be a graph with N nodes. The adjacency matrix **A** is an N × N matrix in which the entry in row i and column j is 1 if node n_i is adjacent to n_j, and is 0 otherwise. This matrix is symmetric and the row sums of **A** are the degrees of the nodes of S.

The adjacency matrix of the graph S, shown in Fig. 1.6, is a 5 × 5 matrix as:

$$\mathbf{A} = \begin{array}{c} \\ n_1 \\ n_2 \\ n_3 \\ n_4 \\ n_5 \end{array} \begin{array}{c} \begin{array}{ccccc} n_1 & n_2 & n_3 & n_4 & n_5 \end{array} \\ \left[\begin{array}{ccccc} 0 & 1 & 1 & 0 & 0 \\ 1 & 0 & 1 & 1 & 0 \\ 1 & 1 & 0 & 1 & 1 \\ 0 & 1 & 1 & 0 & 1 \\ 0 & 0 & 1 & 1 & 0 \end{array}\right] \end{array} \tag{1.1}$$

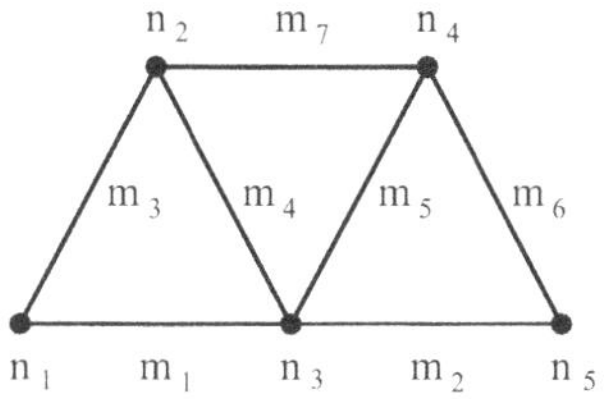

Fig. 1.6 A graph S

Node-Member Incidence Matrix: Let S be a graph with M members and N nodes. The node-member incidence matrix $\bar{\mathbf{B}}$ is an N × M matrix in which the entry in row i and column j is 1 if node n_i is incident with member m_j, and is 0 otherwise. As an example, the node-member incidence matrix of the graph in Fig. 1.6 is a 5 × 7 matrix of the form:

$$\bar{\mathbf{B}} = \begin{array}{c} \\ n_1 \\ n_2 \\ n_3 \\ n_4 \\ n_5 \end{array} \begin{array}{c} \begin{array}{ccccccc} m_1 & m_2 & m_3 & m_4 & m_5 & m_6 & m_7 \end{array} \\ \begin{bmatrix} 1 & 0 & 1 & 0 & 0 & 0 & 0 \\ 0 & 0 & 1 & 1 & 0 & 0 & 1 \\ 1 & 1 & 0 & 1 & 1 & 0 & 0 \\ 0 & 0 & 0 & 0 & 1 & 1 & 1 \\ 0 & 1 & 0 & 0 & 0 & 1 & 0 \end{bmatrix} \end{array} \tag{1.2}$$

Obviously, the pattern of an incidence matrix depends on the particular way that its nodes and members are labelled. One incidence matrix can be obtained from another by simply interchanging rows (corresponding to relabelling the nodes) and columns (corresponding to relabelling the members).

Member List: This type of representation is a common approach in structural mechanics. A member list consists of two rows (or columns) and M columns (or rows). Each column (or row) contains the labels of the two end nodes of each member, in which members are arranged sequentially. For example, the member list of S in Fig. 1.6 is:

$$\mathbf{ML} = \begin{array}{c} \\ n_i \\ n_j \end{array} \begin{array}{c} \begin{array}{ccccccc} m_1 & m_2 & m_3 & m_4 & m_5 & m_6 & m_7 \end{array} \\ \begin{bmatrix} 1 & 3 & 1 & 2 & 3 & 4 & 2 \\ 3 & 5 & 2 & 3 & 4 & 5 & 4 \end{bmatrix} \end{array}. \tag{1.3}$$

It should be noted that a member list can also represent orientations on members. The storage required for this representation is 2 × M. Some engineers prefer to add a third row containing the member's labels, for easy addressing. In this case the storage is increased to 3 × M.

A different way of preparing a member list is to use a vector containing the end nodes of members sequentially; e.g. for the previous example this vector becomes:

$$(1, 3; 3, 5; 1, 2; 2, 3; 3, 4; 4, 5; 2, 4). \tag{1.4}$$

This is a compact description of a graph; however, it is impractical because of the extra search required for its use in various algorithms.

Adjacency List: This list consists of N rows and D columns, where D is the maximum degree of the nodes of S. The i-th row contains the labels of the nodes adjacent to node i of S. For the graph S shown in Fig. 1.6, the adjacency list is:

$$\mathbf{AL} = \begin{matrix} n_1 \\ n_2 \\ n_3 \\ n_4 \\ n_5 \end{matrix} \begin{bmatrix} 2\ 3 & \\ 1\ 3\ 4 & \\ 1\ 2\ 4\ 5 & \\ 2\ 3\ 5 & \\ 3\ 4 & \end{bmatrix}_{N \times D} \tag{1.5}$$

The storage needed for an adjacency list is N × D.

1.5 Definitions of Graph Products

Before introducing graph products, it is required to define two important matrices associated with the graphs.

The adjacency matrix (**A**) of a graph is a square matrix with the dimension equal to the number of nodes of the graph. The ij-th entry of matrix **A** is zero if the nodes i and j are not connected in the graph and the entry is equal to 1 otherwise.

The Laplacian matrix is defined as $\mathbf{L} = \mathbf{D} - \mathbf{A}$, where **D** is a diagonal matrix in which the ii-th entry is the degree (i.e., the number of edges connected to a node) of node i.

A set of nodes connected together sequentially forms a path that is denoted by P_n. If the first and last nodes of a path coincide, it is called a cycle and is denoted by C_n.

We briefly introduce four types of graph products.

1.5.1 Cartesian Product

The Cartesian product of two graphs K and H is denoted by $S = K \times H$. For any two nodes $u = (u_1, u_2)$ and $v = (v_1, v_2)$ in $N(K) \times N(H)$, the edge uv exists in M(S) if

$$\begin{gathered} u_1 = v_1 \text{ and } u_2v_2 \in M(H) \\ \text{or} \\ u_2 = v_2 \text{ and } u_1v_1 \in M(K) \end{gathered} \tag{1.6}$$

This means two nodes in the Cartesian product are connected together if their first components are identical and the second components in the second graph are connected, and vice versa (Fig. 1.7).

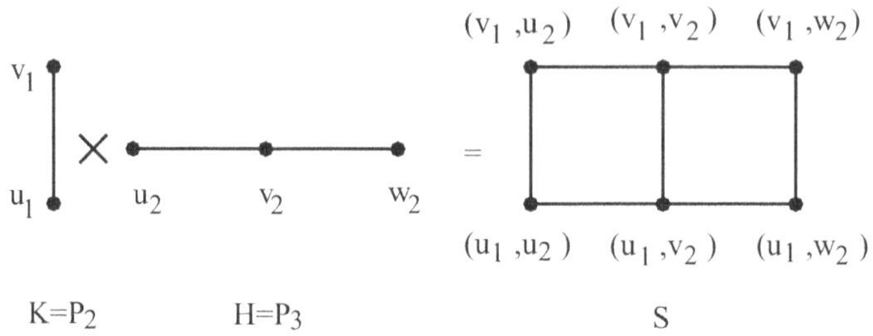

Fig. 1.7 The Cartesian product of two simple graphs: **a** generators; and **b** $S = K \times H$

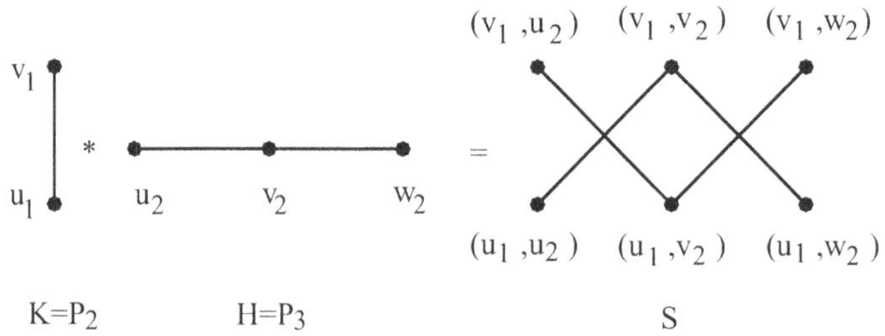

Fig. 1.8 The direct product of two simple graphs: **a** generators; and **b** $S = K{*}H$

1.5.2 Direct Product

The direct product of two graphs K and H is denoted by $S = K{*}H$. For any two nodes $u = (u_1, u_2)$ and $v = (v_1, v_2)$ in $N(K) \times N(H)$, the edge uv exists in M(S) if

$$u_1v_1 \in M(K) \text{ and } u_2v_2 \in M(H) \tag{1.7}$$

This means that two nodes in the direct product are connected together if their first components in the first graph and their second components in the second graph are connected (Fig. 1.8).

1.5.3 Strong Cartesian Product

The strong Cartesian product of two graphs K and H is denoted by $S = K \boxtimes H$. For any two nodes $u = (u_1, u_2)$ and $v = (v_1, v_2)$ in $N(K) \times N(H)$, the edge uv exists in M(S) if

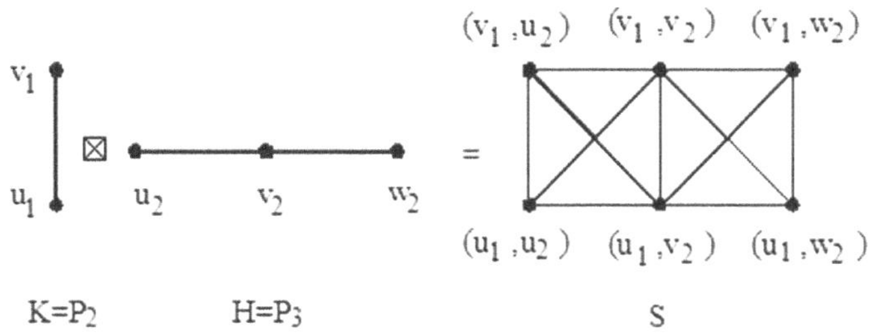

Fig. 1.9 The strong Cartesian product of two simple graphs: **a** generators; and **b** S = K ⊠ H

$$
\begin{gathered}
u_1 = v_1 \text{ and } {}_2v_2 \in M(H) \\
\text{or} \\
u_2 = v_2 \text{ and } u_1v_1 \in M(K) \\
\text{or} \\
u_1v_1 \in M(K) \text{ and } u_2v_2 \in M(H)
\end{gathered}
\tag{1.8}
$$

This product is the union of the previous two products (Fig. 1.9).

1.5.4 Lexicographic Product

The Lexicographic product of two graphs K and H is denoted by $S = K \circ H$. For any two nodes $u = (u_1, u_2)$ and $v = (v_1, v_2)$ in $N(K) \times N(H)$, the edge uv exists in M(S) if

$$
\begin{gathered}
u_1v_1 \in M(K) \\
\text{or} \\
u_1 = v_1 \text{ and } u_2v_2 \in M(H)
\end{gathered}
\tag{1.9}
$$

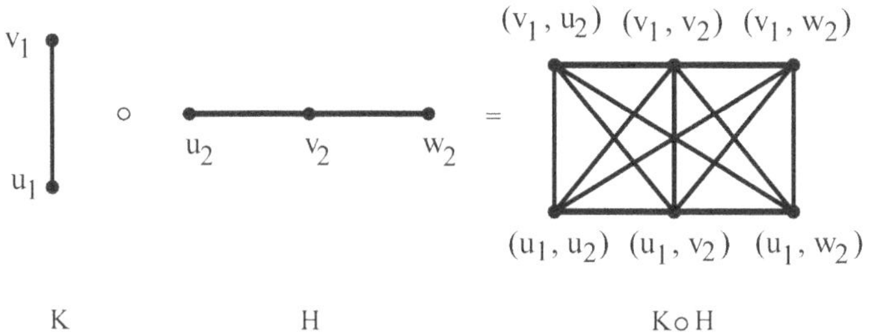

Fig. 1.10 The lexicographic product of two simple graphs: **a** generators; and **b** S = KoH

Graph H is copied on each node of K, and if two nodes i and j of the graph K are connected together, all nodes of the copied graph H in these two nodes are connected together (Fig. 1.10).

More general graph products can be found in the book by Kaveh [3]. For many applications of graph theory in structural engineering the interested reader may refer to Kaveh [1–3].

References

1. Kaveh, A.: Structural mechanics: graph and matrix methods, 3rd edn. Research Studies Press (John Wiley), Baldock, Hertfordshire, UK (2006). ISBN 0-86380-186-2/0-471-96028-4
2. Kaveh, A.: Optimal structural analysis, 2nd edn. Wiley, Chichester, UK (2006). https://doi.org/10.1002/9780470033326
3. Kaveh, A.: Optimal analysis of structures by concepts of symmetry and regularity. Springer, GmbH, Wien-NewYork (2013). ISBN 978-3-7091-1564-7

Chapter 2
Basic Concepts and Definitions of Symmetry and Regularity

2.1 Introduction

Concepts and definitions for different types of symmetry and regularity are already published in a book by Kaveh [1]. In this chapter definitions and concepts relevant to the contents of the present book are provided to make the present book self-content.

2.2 Kronecker Product and Different Matrix Patterns

The Kronecker product of two matrices $\boldsymbol{A}$ and $\boldsymbol{B}$ is a matrix obtained by replacing the *ij*-th entry of $\boldsymbol{A}$ by *aijB* for all *i*s and *j*s. As an example

$$\begin{bmatrix} 1 & -1 \\ 2 & 0 \end{bmatrix} \otimes \begin{bmatrix} a & b \\ c & d \end{bmatrix} = \begin{bmatrix} a & b & -a & -b \\ c & d & -c & -d \\ 2a & 2b & 0 & 0 \\ 2c & 2d & 0 & 0 \end{bmatrix} \tag{2.1}$$

Consider the 2-by-2 block matrix $\boldsymbol{M}$

$$\boldsymbol{M} = \begin{bmatrix} \boldsymbol{A} & \boldsymbol{B} \\ \boldsymbol{B} & \boldsymbol{A} \end{bmatrix} \tag{2.2}$$

Therefore, $\boldsymbol{M}$ can be written as

$$\boldsymbol{M} = \begin{bmatrix} 1 & 0 \\ 0 & 1 \end{bmatrix} \otimes \boldsymbol{A} + \begin{bmatrix} 0 & 1 \\ 1 & 0 \end{bmatrix} \otimes \boldsymbol{B} \tag{2.3}$$

A. Kaveh et al., *Swift Analysis of Civil Engineering Structures Using Graph Theory Methods*, Studies in Systems, Decision and Control 290,
https://doi.org/10.1007/978-3-030-45549-1_2

To generalize Eq. (2.3), form $\boldsymbol{F}$ is defined as follows:

$$\boldsymbol{F}_n(\boldsymbol{A}_m, \boldsymbol{B}_m, \boldsymbol{C}_m, \boldsymbol{D}_m) = \begin{bmatrix} \boldsymbol{A}_m & \boldsymbol{B}_m & \boldsymbol{D}_m & & & & \\ \boldsymbol{B}_m & \boldsymbol{C}_m & \boldsymbol{B}_m & \boldsymbol{D}_m & & & \\ \boldsymbol{D}_m & \boldsymbol{B}_m & \boldsymbol{C}_m & & \ddots & & \\ & \boldsymbol{D}_m & & \ddots & & \boldsymbol{D}_m & \\ & & \ddots & & \boldsymbol{C}_m & \boldsymbol{B}_m & \boldsymbol{D}_m \\ & & & \boldsymbol{D}_m & \boldsymbol{B}_m & \boldsymbol{C}_m & \boldsymbol{B}_m \\ & & & & \boldsymbol{D}_m & \boldsymbol{B}_m & \boldsymbol{A}_m \end{bmatrix}_n \tag{2.4}$$

Furthermore, form $\boldsymbol{G}$ is defined as a form similar to form $\boldsymbol{F}$ but including additional block matrices $\boldsymbol{B}_m$ in the corners:

$$\boldsymbol{G}_n(\boldsymbol{A}_m, \boldsymbol{B}_m, \boldsymbol{C}_m, \boldsymbol{D}_m) = \begin{bmatrix} \boldsymbol{A}_m & \boldsymbol{B}_m & \boldsymbol{D}_m & & & & \boldsymbol{B}_m \\ \boldsymbol{B}_m & \boldsymbol{C}_m & \boldsymbol{B}_m & \boldsymbol{D}_m & & & \\ \boldsymbol{D}_m & \boldsymbol{B}_m & \boldsymbol{C}_m & & \ddots & & \\ & \boldsymbol{D}_m & & \ddots & & \boldsymbol{D}_m & \\ & & \ddots & & \boldsymbol{C}_m & \boldsymbol{B}_m & \boldsymbol{D}_m \\ & & & \boldsymbol{D}_m & \boldsymbol{B}_m & \boldsymbol{C}_m & \boldsymbol{B}_m \\ \boldsymbol{B}_m & & & & \boldsymbol{D}_m & \boldsymbol{B}_m & \boldsymbol{A}_m \end{bmatrix}_n \tag{2.5}$$

Forms $\boldsymbol{F}$ and $\boldsymbol{G}$ hold Penta-block diagonal patterns and by setting $\boldsymbol{D}_m = 0$, Tri-block diagonal patterns are obtained. If this matrix is formed with three arguments, then $\boldsymbol{D}_m = 0$. The blocks might simply be numbers. The generalized pattern of $\boldsymbol{M}$ in Eq. (2.2) can be expressed as follows:

$$\boldsymbol{M} = \boldsymbol{F}_n(\boldsymbol{A}_m, \boldsymbol{B}_m, \boldsymbol{A}_m) = \boldsymbol{I}_n \otimes \boldsymbol{A}_m + \boldsymbol{F}_n(0, 1, 0) \otimes \boldsymbol{B}_m \tag{2.6}$$

The determinant of this matrix will be [2]

$$det(\boldsymbol{M}) = \sum_{n=0}^{[\frac{n}{2}]} (-1)^n \binom{n-i}{n} \boldsymbol{A}^{(n-2i)} \boldsymbol{B}^{(2i)} = \sum_{i=1}^{n} (\boldsymbol{A} + \alpha_i \boldsymbol{B}) \tag{2.7}$$

where

$$\alpha_i = 2\cos\left(\frac{i\pi}{n+1}\right); \quad i = 1 : n \tag{2.8}$$

2.3 Block Diagonalization of Compound Matrices

The eigenvalues of matrix $\boldsymbol{M}$ can efficiently be obtained using the following theorems. Assume matrix $\boldsymbol{M}$ can be expressed as the summation of two Kronecker products

$$\boldsymbol{M} = \boldsymbol{A}_1 \otimes \boldsymbol{B}_1 + \boldsymbol{A}_2 \otimes \boldsymbol{B}_2 \tag{2.9}$$

If $\boldsymbol{P}$ is a matrix that diagonalizes $\boldsymbol{A}_1$ and $\boldsymbol{A}_2$, we can show $\boldsymbol{U} = \boldsymbol{P} \otimes \boldsymbol{I}$ block diagonalizes $\boldsymbol{M}$, i.e., $\boldsymbol{U}^t\boldsymbol{M}\boldsymbol{U}$ is a block diagonal matrix. From algebra one can write:

$$(\boldsymbol{A} \otimes \boldsymbol{B})^t = \boldsymbol{A}^t \otimes \boldsymbol{B}^t \text{ and } (\boldsymbol{A} \otimes \boldsymbol{B})(\boldsymbol{C} \otimes \boldsymbol{D}) = \boldsymbol{AC} \otimes \boldsymbol{BD} \tag{2.10}$$

Therefore,

$$\begin{aligned}
\boldsymbol{U}^t\boldsymbol{M}\boldsymbol{U} &= \left(\boldsymbol{P}^t \otimes \boldsymbol{I}^t\right)(\boldsymbol{A}_1 \otimes \boldsymbol{B}_1 + \boldsymbol{A}_2 \otimes \boldsymbol{B}_2)(\boldsymbol{P} \otimes \boldsymbol{I}) \\
&= \left[\left(\boldsymbol{P}^t\boldsymbol{A}_1\right) \otimes (\boldsymbol{I}\boldsymbol{B}_1) + \left(\boldsymbol{P}^t\boldsymbol{A}_2\right) \otimes (\boldsymbol{I}\boldsymbol{B}_2)\right](\boldsymbol{P} \otimes \boldsymbol{I}) \\
&= \left(\boldsymbol{P}^t\boldsymbol{A}_1\boldsymbol{P}\right) \otimes (\boldsymbol{B}_1\boldsymbol{I}) + \left(\boldsymbol{P}^t\boldsymbol{A}_2\boldsymbol{P}\right) \otimes (\boldsymbol{B}_2\boldsymbol{I}) \\
&= \left(\boldsymbol{P}^t\boldsymbol{A}_1\boldsymbol{P}\right) \otimes \boldsymbol{B}_1 + \left(\boldsymbol{P}^t\boldsymbol{A}_2\boldsymbol{P}\right) \otimes \boldsymbol{B}_2 \\
&= \boldsymbol{D}_{A_1} \otimes \boldsymbol{B}_1 + \boldsymbol{D}_{A_2} \otimes \boldsymbol{B}_2
\end{aligned} \tag{2.11}$$

Regarding the assumption that $\boldsymbol{P}$ diagonalizes $\boldsymbol{A}_1$ and $\boldsymbol{A}_2$.

If $\boldsymbol{U}$ is orthogonal, the matrices $\boldsymbol{M}$ and $\boldsymbol{U}^t\boldsymbol{M}\boldsymbol{U}$ would be similar and the eigenvalues of $\boldsymbol{M}$ would be the union of the eigenvalues of the block matrices on the diagonal of matrix $\boldsymbol{U}^t\boldsymbol{M}\boldsymbol{U}$. Now, the question is whether one can find an orthogonal matrix like $\boldsymbol{P}$ that simultaneously diagonalizes matrices $\boldsymbol{A}_1$ and $\boldsymbol{A}_2$. The sufficient condition for a matrix like $\boldsymbol{A}_1$ to be diagonalizable is that the matrix is Hermitian. If the matrix is not Hermitian, it is sufficient to have independent eigenvectors. However, for two matrices like $\boldsymbol{A}_1$ and $\boldsymbol{A}_2$ to be diagonalizable simultaneously using matrix $\boldsymbol{P}$, the necessary and sufficient conditions are given in the following theorem:

Theorem 1 *The necessary and sufficient condition for simultaneous diagonalization of two Hermitian matrices* $\mathbf{A}_1$ *and* $\mathbf{A}_2$ *using an orthogonal matrix is*:

$$\boldsymbol{A}_1\boldsymbol{A}_2 = \boldsymbol{A}_2\boldsymbol{A}_1 \tag{2.12}$$

Proof Suppose there is an orthogonal matrix like $\boldsymbol{P}$ that diagonalizes $\boldsymbol{A}_1$ and $\boldsymbol{A}_2$ as:

$$\boldsymbol{D}_{A_1} = \boldsymbol{P}^t\boldsymbol{A}_1\boldsymbol{P} \text{ and } \boldsymbol{D}_{A_2} = \boldsymbol{P}^t\boldsymbol{A}_2\boldsymbol{P} \tag{2.13}$$

Since diagonal matrices are commutative, we can write

$$0 = \boldsymbol{D}_{A_1}\boldsymbol{D}_{A_2} - \boldsymbol{D}_{A_2}\boldsymbol{D}_{A_1} = \boldsymbol{P}^t(\boldsymbol{A}_1\boldsymbol{A}_2 - \boldsymbol{A}_2\boldsymbol{A}_1)\boldsymbol{P} \rightarrow \boldsymbol{A}_1\boldsymbol{A}_2 = \boldsymbol{A}_2\boldsymbol{A}_1 \tag{2.14}$$

The condition is also sufficient. Assume $\boldsymbol{A}_1$ and $\boldsymbol{A}_2$ are commutative and consider

$$\boldsymbol{P}^t\boldsymbol{A}_1\boldsymbol{P} = \boldsymbol{D} \text{ and } \boldsymbol{P}^t\boldsymbol{A}_2\boldsymbol{P} = \boldsymbol{T} \tag{2.15}$$

Suppose $\boldsymbol{P}$ diagonalizes $\boldsymbol{A}_1$. We should prove $\boldsymbol{P}$ can also diagonalize $\boldsymbol{A}_2$ that means $\boldsymbol{T}$ is diagonal. Since $\boldsymbol{A}_1$ and $\boldsymbol{A}_2$ are commutative, we can write $\boldsymbol{DT} = \boldsymbol{TD}$. For the *ij*-th element of the equation we will have $\boldsymbol{D}_{ii}\boldsymbol{T}_{ij} = \boldsymbol{T}_{ij}\boldsymbol{D}_{jj}$. Thus, for $\boldsymbol{D}_{ii} \neq \boldsymbol{D}_{jj}$ we will have $\boldsymbol{T}_{ij} = 0$, i.e. $\boldsymbol{T}$ is diagonal.

For $\boldsymbol{A}_1$ with identical eigenvalues for some diagonal entries, we will have $\boldsymbol{D}_{ii} = \boldsymbol{D}_{jj}$. In such a case $\boldsymbol{T}$ would be block diagonal and each block can be diagonalized since its corresponding block in $\boldsymbol{D}$ is a multiple of identity matrix. Therefore, $\boldsymbol{A}_1$ and $\boldsymbol{A}_2$ can be diagonalized simultaneously if and only if $\boldsymbol{A}_1\boldsymbol{A}_2 = \boldsymbol{A}_2\boldsymbol{A}_1$.

If this property holds, considering the similarity of $\boldsymbol{M}$ and $\boldsymbol{U}^t\boldsymbol{M}\boldsymbol{U}$ one can write:

$$\lambda_M = \bigcup_{i=1}^{n} eig(\boldsymbol{M}_i); \quad \boldsymbol{M}_i = \lambda_i(\boldsymbol{A}_1)\boldsymbol{B}_1 + \lambda_i(\boldsymbol{A}_2)\boldsymbol{B}_2 \tag{2.16}$$

where $\boldsymbol{A}_1$ and $\boldsymbol{A}_2$ are matrices of dimensions n and $\boldsymbol{B}_1$ and $\boldsymbol{B}_2$ are matrices of dimensions m.

$\lambda_i(\boldsymbol{A}_j)$ (here we have $j = 1, 2$) is a diagonal matrix containing the eigenvalues of $\boldsymbol{A}_j$, and n is the dimension of matrix $\boldsymbol{A}_i$. The order of eigenvalues in $\boldsymbol{A}_1$ and $\boldsymbol{A}_2$ should be the same as the order (appeared on the diagonal) obtained after simultaneous diagonalization of the two matrices.

As a specific case, if $\boldsymbol{A}_1 = \boldsymbol{I}_2$ and $\boldsymbol{A}_2 = \boldsymbol{F}_2(0, 1, 0)$, since $\boldsymbol{I}_2\boldsymbol{A}_2 = \boldsymbol{A}_2\boldsymbol{I}_2$ we will have

$$\lambda_M = \bigcup_{i=1}^{2} [eig(\boldsymbol{B}_1 + \lambda_i(\boldsymbol{A}_2)\boldsymbol{B}_2)]; \quad \lambda_i(\boldsymbol{A}_1) = 1; \quad \lambda_i(\boldsymbol{A}_2) = \{-1, 1\} \tag{2.17}$$

Therefore, it is only required to calculate the eigenvalues of $\boldsymbol{B}_1 + \boldsymbol{B}_2$ and $\boldsymbol{B}_1 - \boldsymbol{B}_2$

$$\lambda_M = \bigcup [eig(\boldsymbol{B}_1 + \boldsymbol{B}_2), eig(\boldsymbol{B}_1 - \boldsymbol{B}_2)] \tag{2.18}$$

To calculate eigenvectors of $\boldsymbol{M}$, assume the condition in Eq. (2.12) holds. Thus, matrices $\boldsymbol{A}_1$ and $\boldsymbol{A}_2$ can simultaneously be diagonalized using matrix $\boldsymbol{P}$. If $\boldsymbol{\mu}$ is the eigenvalue and $\boldsymbol{v}$ is the eigenvector of $\boldsymbol{M}_i$, we can write

$$(\lambda_i(\boldsymbol{A}_1)\boldsymbol{B}_1 + \lambda_i(\boldsymbol{A}_2)\boldsymbol{B}_2)\boldsymbol{v} = \boldsymbol{\mu v} \tag{2.19}$$

In the following we will show $\boldsymbol{u} \otimes \boldsymbol{v}$ is the eigenvector of $\boldsymbol{M}$. Since

$$(\boldsymbol{A} \otimes \boldsymbol{B})(\boldsymbol{C} \otimes \boldsymbol{D}) = \boldsymbol{AC} \otimes \boldsymbol{BD} \tag{2.20}$$

Therefore,

$$(\boldsymbol{A}_1 \otimes \boldsymbol{B}_1 + \boldsymbol{A}_2 \otimes \boldsymbol{B}_2)(\boldsymbol{u} \otimes \boldsymbol{v}) = (\boldsymbol{A}_1\boldsymbol{u}) \otimes (\boldsymbol{B}_1\boldsymbol{v}) + (\boldsymbol{A}_2\boldsymbol{u}) \otimes (\boldsymbol{B}_2\boldsymbol{v}) \tag{2.21}$$

If $\boldsymbol{u}$ is the generalized eigenvector of $\boldsymbol{A}_1$ and $\boldsymbol{A}_2$ that has been orthogonalized using Gram–Schmidt process (alternatively, we can say $\boldsymbol{u}$ is the eigenvector of $\boldsymbol{A}_1^{-1}\boldsymbol{A}_2$ or $\boldsymbol{Q}$ in $\boldsymbol{QZ}$ decomposition of $\boldsymbol{A}_1$ and $\boldsymbol{A}_2$, or eigenvector of a linear combination of $\boldsymbol{A}_1$ and $\boldsymbol{A}_2$), then

$$\boldsymbol{A}_1\boldsymbol{u} = \lambda_i(\boldsymbol{A}_1)\boldsymbol{u}; \quad \boldsymbol{A}_2\boldsymbol{u} = \lambda_i(\boldsymbol{A}_2)\boldsymbol{u} \tag{2.22}$$

and

$$\begin{aligned}(\boldsymbol{A}_1 \otimes \boldsymbol{B}_1 + \boldsymbol{A}_2 \otimes \boldsymbol{B}_2)(\boldsymbol{u} \otimes \boldsymbol{v}) &= \lambda_i(\boldsymbol{A}_1)\boldsymbol{u} \otimes (\boldsymbol{B}_1\boldsymbol{v}) + \lambda_i(\boldsymbol{A}_2)\boldsymbol{u} \otimes (\boldsymbol{B}_2\boldsymbol{v}) \\ &= \boldsymbol{u} \otimes (\lambda_i(\boldsymbol{A}_1)\boldsymbol{B}_1 + \lambda_i(\boldsymbol{A}_2)\boldsymbol{B}_2)\boldsymbol{v} = \boldsymbol{\mu}(\boldsymbol{u} \otimes \boldsymbol{v})\end{aligned} \tag{2.23}$$

Thus, $\boldsymbol{u} \otimes \boldsymbol{v}$ is an eigenvector of $\boldsymbol{M}$.

Now, assume $\boldsymbol{M}$ is the summation of three Kronecker products

$$\boldsymbol{M} = \sum_{i=1}^{3}(\boldsymbol{A}_i \otimes \boldsymbol{B}_i) \tag{2.24}$$

To block diagonalize $\boldsymbol{M}$, similar to what described for two matrices, one need to indicate that each pair of $\boldsymbol{A}_i$ can commute with respect to multiplication, i.e.

$$\boldsymbol{A}_i\boldsymbol{A}_j = \boldsymbol{A}_j\boldsymbol{A}_i; \quad i, j = 1:3 \ \ (i \neq j) \tag{2.25}$$

However, the question is how to recognize if the commutativity property with respect to multiplication holds. The following theorem provides helpful hints:

Theorem 2 *Two Penta-diagonal (Tri-diagonal as a special case) matrices of the form* $A_k = F(a_k, b_k, c_k, d_k)$ *satisfy* $A_iA_j = A_jA_i$ *if and only if* $\frac{a_k - c_k}{d_k}$ *and* $\frac{b_k}{d_k}$ *are identical for the two matrices.*

Proof Consider

$$A_i = F(a_i, b_i, c_i, d_i); \ A_j = F\left(a_j, b_j, c_j, d_j\right); \ C = A_iA_j - A_jA_i \tag{2.26}$$

Eight entries of matrix C will be non-zero

$$C_{13} = C_{n,n-2} = -C_{31} = -C_{n-2,n} = b_ib_j + a_id_j + d_ic_j - b_ib_j + a_jd_i + c_id_j$$

and

$$C_{12} = C_{n,n-1} = -C_{21} = -C_{n-1,n} = a_i b_j + b_i c_j + d_i b_j - a_j b_i + b_j c_i + d_j b_i \tag{2.27}$$

Setting the entries equal to zero results in

$$\frac{a_i - c_i}{d_i} = \frac{a_j - c_j}{d_j} \text{ and } \frac{a_i - c_i + d_i}{b_i} = \frac{a_j - c_j + d_j}{b_j} \tag{2.28}$$

By combining the two relationships, we will have

$$\frac{a_i - c_i}{d_i} = \frac{a_j - c_j}{d_j} \text{ and } \frac{d_i}{b_i} = \frac{d_j}{b_j} \tag{2.29}$$

Specifically, for Tri-diagonal matrices $\left(d_i = d_j = 0\right)$ the first relationship in Eq. (2.28) always holds and only the second relationship $\left(\frac{a_i - c_i}{b_i} = \frac{a_j - c_j}{b_j}\right)$ should be controlled.

Similarly, it can be shown that for matrices of the form $A_k = G(a_k, b_k, c_k, d_k)$, same relationships in Eq. (2.28) should be controlled.

2.3.1 *Inverse of a Block Matrix*

In this section solving a set of algebraic equations with available corresponding coefficient matrix in a block form is aimed. Consider the following equations:

$$(A_1 \otimes B_1 + A_2 \otimes B_2)x = C \text{ or } Mx = C \tag{2.30}$$

By calculating eigenvalues and eigenvectors of M, one can simply solve the equations. If λ_i and $\{\varphi\}_i$ are eigenvalues and eigenvectors of M for different values of i, by introducing $C_j = \{\varphi\}_j^t C$, one can write:

$$y_j = \frac{C_j}{\lambda_j} \to \{x\}_n = \sum_{i=1}^{n} \{\varphi\}_i y_i = \sum_{i=1}^{n} \{\varphi\}_i \frac{C_i}{\lambda_i} = \sum_{i=1}^{n} \frac{\{\varphi\}_i \{\varphi\}_i^t}{\lambda_i} C \tag{2.31}$$

If none of A_i's and B_i's have the commutativity property, then using *QZ* transformation the set of equations can be solved. This transformation and its applications are discussed in [3].

If finding M^{-1} using the eigenvalues (diagonal matrix D) and eigenvectors (matrix V) is aimed, regarding $M = VDV^t$ and the fact that eigenvectors of symmetric matrices are orthogonal, we can write:

$$M^{-1} = \left(VDV^t\right)^{-1} = VD^{-1}V^t = V\begin{bmatrix} 1/\lambda_1 & & 0 & & 0 \\ & 1/\lambda_2 & & \ddots & \\ 0 & & \ddots & & 0 \\ & \ddots & & \ddots & \\ 0 & & 0 & & 1/\lambda_{mn} \end{bmatrix}V^t \tag{2.32}$$

where D^{-1} is simply obtained by inverting the entries of the main diagonal of D, and the eigenvectors matrix M are $u \otimes v$.

2.4 Block Diagonalization of Adjacency and Laplacian Matrices

In this section the adjacency and Laplacian matrices of the graphs which are formed using graph products are studied using the theorems presented in Sect. 2.3, to transform them into block diagonal form. This simplifies the calculation of eigenvalues and eigenvectors, and ultimately simplifies the process of inverting matrices.

Obviously the nodal numbering of a graph affects the form of the adjacency and Laplacian matrices of the corresponding graph. Therefore first we will consider nodal numbering using rules from graph products. Here we will present methods for calculating the eigenvalues of paths and cycles which are often the generators of a graph generated using graph products.

Regarding the relationships in [4], eigenvalues of path (P) and cycle (C) graphs will be as follows:

For adjacency matrices $\lambda_m^P = 2\cos\left(\frac{k\pi}{m+1}\right)$ and $\lambda_m^C = 2\cos\left(\frac{2k\pi}{m}\right)$ for $k = 1:m$. It can be observed that $\lambda_m \in [-2, 2]$.

For Laplacian matrices $\lambda_m^P = 2 + 2\cos\left(\frac{k\pi}{m}\right)$ and $\lambda_m^C = 2 - 2\cos\left(\frac{2k\pi}{m}\right)$ for $k = 1:m$. It can be observed that $\lambda_m \in [0, 4]$.

The first eigenvalue of Laplacian matrix for both path and cycle are zero.

Using the above definitions and relationships as well as the earlier theorems, the eigenvalues and eigenvectors of the graph product can be obtained. Specifically, these relationships are important for calculating the second eigenvalues since they can be utilized in nodal ordering for profile reduction as well as bisection for parallel computations [5].

2.4.1 *Nodal Numbering of Graphs Generated by Graph Products*

In the following two sections (a and b) methods are presented for nodal numbering the graphs produced by graph products. In subsequent Sects. 2.4.2 and 2.4.3 suitable numbering resulting in matrices of special forms for the adjacency and Laplacian matrices leading to block diagonal matrices.

a. **First Approach**:

To achieve decomposable patterns for the adjacency and Laplacian matrices of the introduced graph products in Chap. 1, specific graph numbering should be performed. In the graph product of K and H, suppose the nodes of K and H are numbered through a_1 to a_m and b_1 to b_n, respectively. Now, the node (a_i, b_j) is numbered as $n(i-1)+j$ leading to a matrix of form F or G composed of m blocks of dimension n. As an example, the numbering of graph $K_3 \times H_2$ is shown in Fig. 2.1.

b. **Second Approach**:

Alternatively, the node (a_i, b_j) can be numbered as $m(j-1)+i$ that results in a canonical form composed of n blocks of dimension m. If so, the previous example would have the following pattern (Fig. 2.2)

In other words, in the first approach first all nodes of K located on the first node of H are numbered and then all nodes of K located on the second node of H and so forth. In the second approach the numbering is performed conversely.

For adjacency and Laplacian matrix of three types of graph products (e.g., Cartesian, strong Cartesian and direct products) both approaches of numbering are possible

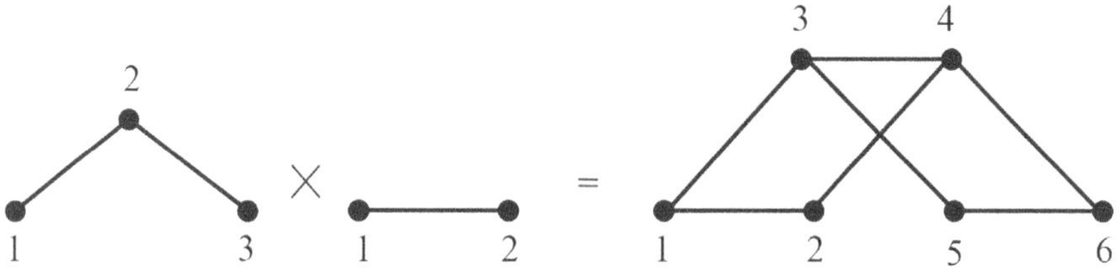

Fig. 2.1 $K_3 \times H_2 = F_3(A_2, B_2, C_2)$

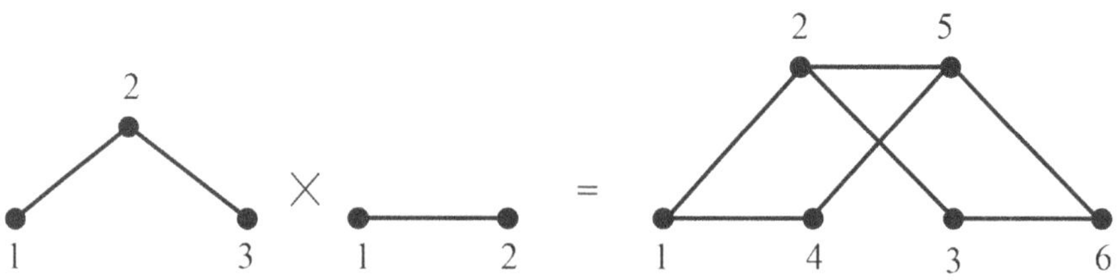

Fig. 2.2 $K_3 \times H_2 = F_2(A_3, B_3, C_3)$

and result in an adjacency or Laplacian matrix that satisfies the condition in Eq. (2.12). For Lexicographic product, however, the numbering through the first approach for adjacency matrix and through the second approach for Laplacian matrix lead to matrix patterns satisfying Eq. (2.12).

In the following the block diagonalization of the adjacency and Laplacian matrices of four types of graph products are discussed.

2.4.2 Block Diagonalization of Adjacency Matrices

For the adjacency matrix of four types of graph products (e.g., Cartesian, Strong Cartesian, Direct and Lexicographic products) using the first approach of numbering we will have

$$A_{mn} = I_n \otimes H_m + K_n \otimes B_m \tag{2.33}$$

where H_m is the adjucenty matrix of graph H, and K_n is the adjuctency matrix of graph K

$$B_m = \begin{cases} F_m(1, 0, 1) & \text{Cartesian} \\ F_m(0, 1, 0) & \text{Direct} \\ F_m(1, 1, 1) & \text{Strong Cartesian} \\ O_m & \text{Lexicographic} \end{cases} \tag{2.34}$$

where O_m is a matrix with unit entries.

Since Eq. (2.12) holds, M_{mn} can be diagonalized and we have

$$eig(A_{mn}) = \bigcup_{i=1}^{n} [eig(H_m + \lambda_i(K_n)B_m)]; \quad \lambda_i(I_n) = 1 \ (i = 1 : n) \tag{2.35}$$

Therefore, rather than calculating the eigenvalues of a matrix of dimension mn, we only need to calculate n times the eigenvalues of matrices of dimension m.

2.4.3 Block Diagonalization of Laplacian Matrices

Using the first approach of numbering, Laplacian matrix of Cartesian product holds the following form

$$L_{mn} = I_n \otimes H_m + K_n \otimes (-I_m) \tag{2.36}$$

where H_m is the Laplacian matrix of graph H, and K_n is the Laplacian of graph K.

This matrix is diagonalizable and we have

$$eig(L_{mn}) = \bigcup_{i=1}^{n} \{eig[H_m - \lambda_i(K_n)B_m]\} \tag{2.37}$$

For direct and strong Cartesian products of path generators, using the first approach of numbering Laplacian matrices can be written, respectively, as

$$L_{mn} = F_n(0,1,0) \otimes F_m(0,-1,0) + F_n(1,0,2) \otimes F_m(1,0,2) \tag{2.38}$$

and

$$L_{mn} = F_n(1,1,1) \otimes F_m(-1,-1,-1) + F_n(2,0,3) \otimes F_m(2,0,3) \tag{2.39}$$

However, since Eq. (2.12) does not hold, matrices L_{mn} cannot be block diagonalized. In [6] members are added to the edges of the graph such that A_2 and B_2 are converted to $2I$ in direct product and to $3I$ in strong Cartesian product, thus, decomposition becomes feasible. However, it can be shown that altering A_2 is sufficient for block diagonalization.

In these products if we add members only to the two opposite edges (let's say top and bottom) of the graph, rather than all around the graph, L_{mn} can be written as:

$$L_{mn} = I_n \otimes H_m + K_n \otimes (-B_m) \tag{2.40}$$

Since Eq. (2.12) holds, L_{mn} is diagonalizable (Eq. 2.37). Since only two edges of the graph are altered, the results are more accurate than the case where members were added to all edges of the graph. Using one iteration of Rayleigh quotient method the accuracy can be improved. In Sect. 2.5.1 a method for the decomposition of Laplacian matrix of direct and strong Cartesian products into two blocks will be presented. It should be noted that numbering Direct and Strong Cartesian products of path generators through the second approach does not lead to decomposable patterns as well.

To calculate the eigenvalues of the Laplacian matrix of Lexicographic product, it was mentioned that numbering through the second approach leads to matrix patterns satisfying Eq. (2.12). Alternatively, the following theorem can be used without changing the numbering.

Theorem 3 *two matrices* $M = \sum_{i=1}^{k}(A_i \otimes B_i)$ *and* $N = \sum_{i=1}^{k}(B_i \otimes A_i)$ *are similar.*

Proof It should be shown these matrices have the same eigenvalues. First, we assume $k = 1$ and try to show $A \otimes B$ and $B \otimes A$ are similar. Regarding the definition of the Kronecker product,

$$M([ib, ia], [jb, ja]) = A(ia, ja) \times A(ib, jb)$$
$$N([ia, ib], [ja, jb]) = B(ib, jb) \times A(ia, ja)$$

Matrix N can be obtained from matrix M through row and column operations. For instance, if A an B are matrices of dimension m and n, respectively, since the second entry of M is equal to $(m + 1)$ th entry of N, by interchanging the second and $(m + 1)$ th rows and columns of matrix M, matrix N is obtained. Similar operations for the rest of entries will transform matrix M to matrix N. Since performing simultaneous row and column operations, eigenvalues are not changed. Identifying the row and column that should be interchanged depends on the magnitude of m and n. Now, assume $k > 1$. Since all A_i s are of dimension m and all the B_i s are of dimension n, any operation that changes $A_1 \otimes B_1$ to $B_1 \otimes A_1$ will also alter $A_2 \otimes B_2$ to $B_2 \otimes A_2$ and so forth. Therefore, ultimately M will be transformed into N such that the matrices are similar and have the same set of eigenvalues.

In some graphs the commutative property of matrices A_i s may not hold; however, if B_i s commute the matrix can be block diagonalized regarding Theorem 3. According to Theorem 1, the sufficient condition for matrix M to be block diagonalizable is $A_i A_j = A_j A_i$. Considering Theorem 3, another sufficient condition is $B_i B_j = B_j B_i$ (assuming $A_i A_j \neq A_j A_i$). The difference between the two obtained block decomposition is matrix M will be decomposed to n blocks of dimension m whereas matrix N will be decomposed to m blocks of dimension n. The graph interpretation of such a difference is the numbering of the graph S that can be started with either nodes of K or nodes of H (e.g., first vs. second approach of numbering), leading to matrix M or N.

Laplacian matrix of Lexicographic product constructed using the first approach of numbering is written as

$$L_{mn} = I_n \otimes H_m + K_n \otimes O_m + DK_n \otimes R_m = \sum_{i=1}^{3} (A_i \otimes B_i) \tag{2.41}$$

where H_m is the Laplacian matrix of graph H, K_n is the Laplacian matrix of graph K, DK_n is a diagonal matrix including the entries on the diagonal of matrix K_n, and $R_m = mI_m - O_m$. In this relationship Eq. (2.12) for A_2 and A_3 does not hold. However, using the second approach of numbering or using Theorem 3, we will have

$$L_{nm} = H_m \otimes I_n + O_m \otimes K_n + R_m \otimes DK_n = \sum_{i=1}^{3} (B_i \otimes A_i) \tag{2.42}$$

Since $B_i B_j = B_j B_i$, we can write

$$eig(L_{nm}) = \bigcup_{i=1}^{m} \{eig[\lambda_i(H_m) I_n + \lambda_i(O_m) K_n + \lambda_i(R_m) DK_n]\} \tag{2.43}$$

By simultaneous diagonalization of H_m, O_m, and R_m, the eigenvalues of these matrices will appear on their main diagonal and we will have

$$eig(L_{nm}) = eig\left(\begin{bmatrix} 0 \\ \lambda_2 \\ \vdots \\ \lambda_m \end{bmatrix} I_n + \begin{bmatrix} m \\ 0 \\ \vdots \\ 0 \end{bmatrix} K_n + \begin{bmatrix} 0 \\ m \\ \vdots \\ m \end{bmatrix} DK_n\right) \tag{2.44}$$

or

$$eig(L_{nm}) = \bigcup_{i=1}^{m} \{eig[mDK_n + \lambda_i(H_m)I_n]\} \bigcup eig(mK_n) \tag{2.45}$$

Since matrix $mDK_n + \lambda_i(H_m)I_n$ is diagonal and the values on the diagonal are eigenvalues, rather than forming the matrix we only need to form the diagonal.

The steps for finding the eigenvalues of L_{nm} are summarized as follows:

1. Calculating eigenvalues of K_n and H_m, and removing the zero value from the eigenvalues of H_m.
2. Multiplying vector $d(K_n)$ by m and adding the result to the vector including the entries on the diagonal of $\lambda_i(H_m)I_n$. This would be the first set of answers.
3. Multiplying eigenvalues of K_n by m and calculating the second set of answers.

Example 1 Block diagonalization of Laplacian matrix of $P \boxtimes C$ is aimed:

$$L_{mn} = F_n(A_m, B_m, C_m)$$

where

$$A_m = G_m(5, -1, 5),\ \ B_m = G_m(-1, -1, -1) \text{ and } C_m = G_m(8, -1, 8)$$

Therefore,

$$A_m = 6I_m + B_m, C_m = 9I_m + B_m$$

$$L_{mn} = F_n(6I_m + B_m, B_m, 9I_m + B_m) = 3F_n(2, 0, 3) \otimes I_m + F_n(1, 1, 1) \otimes B_m$$

Since

$$eig(B_m) = -\left(1 + 2\cos\left(\frac{2k\pi}{m}\right)\right); \quad k = 1 : m$$

and

$$eig\left(\sum(A_i \otimes B_i)\right) = eig\left(\sum(B_i \otimes A_i)\right)$$

Therefore,

$$eig(L_{mn}) = eig\{I_m \otimes 3F_n(2, 0, 3) + B_m \otimes F_n(1, 1, 1)\}$$

Since I_m and B_m hold commutativity property, we will have

$$eig(L_{mn}) = \bigcup_{k=1}^{m} \left\{ eig\left[3F_n(2, 0, 3) - \left(1 + 2\cos\left(\frac{2k\pi}{m}\right)\right) F_n(1, 1, 1) \right] \right\}$$

Thus, Laplacian matrix becomes block diagonal.

2.5 Decomposition of the Block Matrices

In this section apart from those studied in Sect. 2.4, other forms of matrices are presented for which the Eq. (2.12) does not hold. Also the sufficient conditions for decomposability of a block diagonal matrix into numeric diagonal matrix are presented.

2.5.1 *Decomposition of the Matrices of the Form $F_n(A_m, B_m, C_m)$*

For the matrices of form $L_{mn} = F_n(A_m, B_m, C_m)$ the condition in Eq. (2.12), in general, does not hold. Such a form can be observed in Laplacian matrix of Strong Cartesian and Direct products. The form can, however, be decomposed into two blocks as follows:

A. If n is even (e.g., $n = 4$), rows and columns 3 and 4 are interchanged block wise:

$$L_{m\times 4} = \begin{bmatrix} A_m & B_m & 0 & 0 \\ B_m & C_m & 0 & B_m \\ 0 & 0 & A_m & B_m \\ 0 & B_m & B_m & C_m \end{bmatrix} = \begin{bmatrix} M_{m\times 2} & N_{m\times 2} \\ N_{m\times 2} & M_{m\times 2} \end{bmatrix} \tag{2.46}$$

This equation satisfies the condition in Eq. (2.12) and we will have

$$L_{m\times 4} = \begin{bmatrix} M_{m\times 2} & N_{m\times 2} \\ N_{m\times 2} & M_{m\times 2} \end{bmatrix} \equiv \begin{bmatrix} M_{m\times 2} + N_{m\times 2} & 0 \\ 0 & M_{m\times 2} - N_{m\times 2} \end{bmatrix} \tag{2.47}$$

And in a general ($n = 2k$) form we can write:

$$
\begin{aligned}
M_{mk} &= \begin{bmatrix} A_m & B_m & 0 & & & 0 \\ B_m & C_m & B_m & & & \\ 0 & B_m & & & \iddots & \\ & & & \ddots & & 0 \\ & & \ddots & & C_m & B_m \\ 0 & & & 0 & B_m & C_m \end{bmatrix}_k ; \\
N_{mk} &= \begin{bmatrix} 0 & 0 & 0 & & & 0 \\ 0 & 0 & 0 & & & \\ 0 & 0 & & & \iddots & \\ & & & \ddots & & 0 \\ & & \ddots & & 0 & 0 \\ 0 & & & 0 & 0 & B_m \end{bmatrix}_k
\end{aligned}
\tag{2.48}
$$

B. If n is odd (e.g., $n = 3$), one zero block row and one zero block column are added to the Laplacian matrix. After some row and column operations a matrix pattern similar to that of Eq. (2.47) is obtained:

$$
L_{m\times 3} = \begin{bmatrix} A_m & B_m & 0 \\ B_m & C_m & B_m \\ 0 & B_m & A_m \end{bmatrix} \rightarrow L_{m\times 4} = \begin{bmatrix} A_m & B_m & 0 & 0 \\ B_m & C_m & B_m & 0 \\ 0 & B_m & A_m & 0 \\ 0 & 0 & 0 & 0 \end{bmatrix}
$$

$$
\rightarrow \begin{vmatrix} A_m - \lambda I_m & B_m & 0 & 0 \\ B_m & C_m - \lambda I_m & B_m & 0 \\ 0 & B_m & A_m - \lambda I_m & 0 \\ 0 & 0 & 0 & 0 - \lambda I_m \end{vmatrix}
$$

$$
\xrightarrow{C_4 + C_2 \rightarrow C_4} \begin{vmatrix} A_m - \lambda I_m & B_m & 0 & B_m \\ B_m & C_m - \lambda I_m & B_m & C_m - \lambda I_m \\ 0 & B_m & A_m - \lambda I_m & B_m \\ 0 & 0 & 0 & -\lambda I_m \end{vmatrix}
$$

$$
\xrightarrow{R_4 + R_2 \rightarrow R_4} \begin{vmatrix} A_m - \lambda I_m & B_m & 0 & B_m \\ B_m & C_m - \lambda I_m & B_m & C_m - \lambda I_m \\ 0 & B_m & A_m - \lambda I_m & B_m \\ B_m & C_m - \lambda I_m & B_m & C_m - 2\lambda I_m \end{vmatrix}
$$

$$
\xrightarrow{C_2 - 0.5C_4 \rightarrow C_2} \begin{vmatrix} A_m - \lambda I_m & B_m/2 & 0 & B_m \\ B_m & (C_m - \lambda I_m)/2 & B_m & C_m - \lambda I_m \\ 0 & B_m/2 & A_m - \lambda I_m & B_m \\ B_m & C_m/2 & B_m & C_m - 2\lambda I_m \end{vmatrix}
$$

$$
\xrightarrow{R_2 - 0.5R_4 \rightarrow R_2} \begin{vmatrix} A_m - \lambda I_m & B_m/2 & 0 & B_m \\ B_m/2 & C_m/4 - \lambda I_m/2 & B_m/2 & C_m/2 \\ 0 & B_m/2 & A_m - \lambda I_m & B_m \\ B_m & C_m/2 & B_m & C_m - 2\lambda I_m \end{vmatrix}
$$

$$\overset{2C_2 \to C_2}{\longrightarrow} \begin{vmatrix} A_m - \lambda I_m & B_m & 0 & B_m \\ B_m/2 & C_m/2 - \lambda I_m & B_m/2 & C_m/2 \\ 0 & B_m & A_m - \lambda I_m & B_m \\ B_m & C_m & B_m & C_m - 2\lambda I_m \end{vmatrix}$$

$$\overset{0.5R_4 \to R_4}{\longrightarrow} \begin{vmatrix} A_m - \lambda I_m & B_m & 0 & B_m \\ B_m/2 & C_m/2 - \lambda I_m & B_m/2 & C_m/2 \\ 0 & B_m & A_m - \lambda I_m & B_m \\ B_m/2 & C_m/2 & B_m/2 & C_m/2 - \lambda I_m \end{vmatrix}$$

$$L_{m\times 4} = \begin{bmatrix} A_m & B_m & 0 & B_m \\ B_m/2 & C_m/2 & B_m/2 & C_m/2 \\ 0 & B_m & A_m & B_m \\ B_m/2 & C_m/2 & B_m/2 & C_m/2 \end{bmatrix} = \begin{bmatrix} M_{m\times 2} & N_{m\times 2} \\ N_{m\times 2} & M_{m\times 2} \end{bmatrix} \tag{2.49}$$

And in a general ($n = 2k - 1$) form we can write:

$$\begin{aligned} M_{mk} &= \begin{bmatrix} A_m & B_m & 0 & & & \\ B_m & C_m & B_m & \ddots & & \\ 0 & B_m & \ddots & \ddots & & 0 \\ & & \ddots & \ddots & C_m & B_m \\ & & & 0 & \frac{B_m}{2} & \frac{C_m}{2} \end{bmatrix}_k ; \\ N_{mk} &= \begin{bmatrix} 0 & 0 & 0 & & & \\ 0 & 0 & 0 & \ddots & & \\ 0 & 0 & & \ddots & & 0 \\ & & \ddots & & 0 & B_m \\ & & & 0 & \frac{B_m}{2} & \frac{C_m}{2} \end{bmatrix}_k \end{aligned} \tag{2.50}$$

After calculating eigenvalues of $M_{m\times 2} \pm N_{m\times 2}$, there will be m extra zeros corresponding to the added zero rows and columns.

There are specific cases wherein the form $F_n(A_m, B_m, C_m)$ can be fully block decomposed. For instance, if C can be written as $C_m = A_m + \alpha B_m$, we will have

$$M_{mn} = I_n \otimes A_m + F_n(0, 1, \alpha) \otimes B_m \tag{2.51}$$

Since $IF = FI$, the condition in Eq. (2.12) is satisfied.

2.5.2 Conditions for Full Decomposability of Block Matrices

In specific cases, matrix calculations can be simplified to numerical calculations (e.g., matrices of dimension 1). If both A_i s and B_i s satisfy the commutativity condition in Eq. (2.12), the eigenvalues of a summation of matrices will be equal to the summation of the matrices eigenvalues.

Theorem 4 *One sufficient condition for the relationship*

$$eig\left(\sum_{i=1}^{n}(A_i \otimes B_i)\right) = \sum_{i=1}^{n} eig(A_i \otimes B_i) \tag{2.52}$$

to hold is

$$A_i A_j = A_j A_i \text{ and } B_i B_j = B_j B_i;\ i, j = 1:n\ (i \neq j) \tag{2.53}$$

To calculate the eigenvectors, assume u and v are the generalized eigenvector of A_1 and A_2 and B_1 and B_2, respectively, that have been orthogonalized using Gram–Schmidt process, therefore

$$\begin{aligned} A_1 u &= \lambda_i(A_1)u;\ A_2 u = \lambda_i(A_2)u \\ B_1 v &= \lambda_i(B_1)v;\ B_2 v = \lambda_i(B_2)v \end{aligned} \tag{2.54}$$

and

$$\begin{aligned} (A_1 \otimes B_1 + A_2 \otimes B_2)(u \otimes v) &= \lambda_i(A_1)u \otimes \lambda_i(B_1)v + \lambda_i(A_2)u \otimes \lambda_i(B_2)v \\ &= \lambda_i(A_1)\lambda_i(B_1)(u \otimes v) + \lambda_i(A_2)\lambda_i(B_2)(u \otimes v) \\ &= \mu(u \otimes v) \end{aligned} \tag{2.55}$$

Therefore, μ the eigenvalue of $M = A_1 \otimes B_1 + A_2 \otimes B_2$ is the summation of the eigenvalues of $A_1 \otimes B_1$ (e.g., $\lambda_i(A_1)\lambda_i(B_1)$) and the eigenvalues of $A_2 \otimes B_2$ (e.g., $\lambda_i(A_2)\lambda_i(B_2)$).

Example 2 Calculating the eigenvalues of Laplacian matrix of $C \boxtimes C$ is aimed:

$$L_{mn} = G_n(A_m, B_m, C_m)$$

where

$$A_m = G_m(8, -1, 8);\ B_m = G_m(-1, -1, -1);\ C_m = A_m$$

Since $A_m = 9I_m + B_m$, One can write

$$L_{mn} = G_n(9I_m + B_m, B_m, 9I_m + B_m) = 9I_n \otimes I_m + G_n(1, 1, 1) \otimes B_m$$

Regarding that Eq. (2.53) is satisfied, using Theorem 4 we will have

$$eig(L_{mn}) = \bigcup_{k=1}^{m}\left\{eig\left[9I_n - \left(1 + 2\cos\left(\frac{2k\pi}{m}\right)\right)G_n(1, 1, 1)\right]\right\}$$

$$eig(L_{mn}) = 9 - \left(1 + 2\cos\left(\frac{2k\pi}{m}\right)\right)\left(1 + 2\cos\left(\frac{2k'\pi}{n}\right)\right); \quad \begin{matrix} k = 1:m \\ k' = 1:n \end{matrix}$$

A similar procedure can be repeated for Direct products.

2.5.3 Decomposition of the Matrices of Other Forms

Consider matrix M of the following form

$$M = \begin{bmatrix} A & B & P \\ B & A & P \\ P^t & P^t & R \end{bmatrix} \tag{2.56}$$

By adding one zero block row and one zero block column between the first and second rows and columns of matrix M and performing appropriate row and column operations, we will have

$$\begin{bmatrix} A & B & P \\ B & A & P \\ P^t & P^t & R \end{bmatrix} \rightarrow \begin{bmatrix} A & 0 & B & P \\ 0 & 0 & 0 & 0 \\ B & 0 & A & P \\ P^t & 0 & P^t & R \end{bmatrix}$$

$$\rightarrow \begin{vmatrix} A-\lambda I & 0 & B & P \\ 0 & 0-\lambda I & 0 & 0 \\ B & 0 & A-\lambda I & P \\ P^t & 0 & P^t & R-\lambda I \end{vmatrix}$$

$$\overset{C_4+C_2\rightarrow C_2}{\longrightarrow} \begin{vmatrix} A-\lambda I & P & B & P \\ 0 & -\lambda I & 0 & 0 \\ B & P & A-\lambda I & P \\ P^t & R-\lambda I & P^t & R-\lambda I \end{vmatrix}$$

$$\overset{R_4+R_2\rightarrow R_2}{\longrightarrow} \begin{vmatrix} A-\lambda I & P & B & P \\ P^t & R-2\lambda I & P^t & R-\lambda I \\ B & P & A-\lambda I & P \\ P^t & R-\lambda I & P^t & R-\lambda I \end{vmatrix}$$

$$\overset{C_4-0.5C_2\rightarrow C_4}{\longrightarrow} \begin{vmatrix} A-\lambda I & P & B & P/2 \\ P^t & R-2\lambda I & P^t & R/2 \\ B & P & A-\lambda I & P/2 \\ P^t & R-\lambda I & P^t & (R-\lambda I)/2 \end{vmatrix}$$

$$\overset{R_4-0.5R_2\to R_4}{\longrightarrow}\begin{vmatrix} A-\lambda I & P & B & P/2 \\ P^t & R-2\lambda I & P^t & R/2 \\ B & P & A-\lambda I & P/2 \\ P^t/2 & R/2 & P^t/2 & R/4-\lambda I/2 \end{vmatrix}$$

$$\overset{2C_4\to C_4}{\longrightarrow}\begin{vmatrix} A-\lambda I & P & B & P \\ P^t & R-2\lambda I & P^t & R \\ B & P & A-\lambda I & P \\ P^t/2 & R/2 & P^t/2 & R/2-\lambda I \end{vmatrix}$$

$$\overset{0.5R_2\to R_2}{\longrightarrow}\begin{vmatrix} A-\lambda I & P & B & P \\ P^t/2 & R/2-\lambda I & P^t/2 & R/2 \\ B & P & A-\lambda I & P \\ P^t/2 & R/2 & P^t/2 & R/2-\lambda I \end{vmatrix}$$

$$\begin{bmatrix} A & P & B & P \\ P^t/2 & R/2 & P^t/2 & R/2 \\ B & P & A & P \\ P^t/2 & R/2 & P^t/2 & R/2 \end{bmatrix} \to \begin{bmatrix} M & N \\ N & M \end{bmatrix} \tag{2.57}$$

The generalization of the matrix form in Eq. (2.56) would be as follows (for $n = 5$)

$$L_{5m+k} = \begin{bmatrix} A_m & B_m & Z_m & Z_m & B_m & P_{mk} \\ B_m & A_m & B_m & Z_m & Z_m & P_{mk} \\ Z_m & B_m & A_m & B_m & Z_m & P_{mk} \\ Z_m & Z_m & B_m & A_m & B_m & P_{mk} \\ B_m & Z_m & Z_m & B_m & A_m & P_{mk} \\ P^t_{mk} & P^t_{mk} & P^t_{mk} & P^t_{mk} & P^t_{mk} & R_k \end{bmatrix}_{5m+k} \tag{2.58}$$

where Z_m is a zero matrix of dimension m. Similar to the above operations the final matrix will have the following pattern:

$$\begin{bmatrix} M_{m+k} & N_{m+k} & W_{m+k} & W_{m+k} & N_{m+k} \\ N_{m+k} & M_{m+k} & N_{m+k} & W_{m+k} & W_{m+k} \\ W_{m+k} & N_{m+k} & M_{m+k} & N_{m+k} & W_{m+k} \\ W_{m+k} & W_{m+k} & N_{m+k} & M_{m+k} & N_{m+k} \\ N_{m+k} & W_{m+k} & W_{m+k} & N_{m+k} & M_{m+k} \end{bmatrix}_{5m+k} \tag{2.59}$$

where

$$M_{m+k} = \begin{bmatrix} A_m & P_{mk} \\ \frac{1}{n}P^t_{mk} & \frac{1}{n}R_k \end{bmatrix};\ N_{m+k} = \begin{bmatrix} B_m & P_{mk} \\ \frac{1}{n}P^t_{mk} & \frac{1}{n}R_k \end{bmatrix};\ W_{m+k} = \begin{bmatrix} Z_m & P_{mk} \\ \frac{1}{n}P^t_{mk} & \frac{1}{n}R_k \end{bmatrix} \tag{2.60}$$

The matrix in Eq. (2.59) is called a block circulant matrix that will be introduced later.

Laplacian matrix can be expressed as

$$L_{n(m+k)} = \sum_{i=1}^{3} (A_i \otimes B_i) = I_n \otimes M_{m+k} + G_n(0,1,0) \otimes N_{m+k} + K_n \otimes W_{m+k} \tag{2.61}$$

where

$$K_n = O_n - I_n - G_n(0,1,0); \quad O_n = ones(n) \tag{2.62}$$

Equation (2.12) holds and we will have:

$$\begin{aligned} \lambda_L &= \bigcup_{i=1}^{n} \{eig(L_i)\}; \; L_i = \sum_{j=1}^{3} \{\lambda_j(A_i) B_i\} \\ &= M_{m+k} + \lambda(G_n(0,1,0)) N_{m+k} + \lambda(K_n) W_{m+k} \end{aligned} \tag{2.63}$$

where

$$\lambda(G_n(0,1,0)) = 2\cos\left(\frac{2k\pi}{n}\right); \; k = 1:n \tag{2.64}$$

For calculating the eigenvalues of K_n we can write

$$K_n = O_n - I_n - G_n(0,1,0) = \sum_{i=1}^{3} (A_i \otimes B_i); \; B_i = 1 \tag{2.65}$$

Since

$$A_i A_j = A_j A_i; \; i, j = 1:3 \; (i \neq j) \tag{2.66}$$

We will have

$$\lambda(K_n) = \sum_{i=1}^{3} \{\lambda(A_i) B_i\} = \lambda(O_n) - \lambda(I_n) - \lambda(G_n(0,1,0)) \tag{2.67}$$

or

$$eig(K_n) = \begin{bmatrix} 0 \\ 0 \\ \vdots \\ 0 \\ n \end{bmatrix} - \begin{bmatrix} 1 \\ 1 \\ \vdots \\ 1 \\ 1 \end{bmatrix} - \begin{bmatrix} 2\cos(2\pi/n) \\ 2\cos(4\pi/n) \\ \vdots \\ 2\cos(2(n-1)\pi/n) \\ 2 \end{bmatrix} \tag{2.68}$$

Therefore,

$$eig(K_n) = \bigcup_{k=1}^{n-1} \left\{ -1 - 2\cos\left(\frac{2k\pi}{n}\right) \right\} \bigcup \{n-3\} \tag{2.69}$$

Finally,

$$\begin{aligned} \lambda_L = & \bigcup_{i=1}^{n-1} \left\{ eig\left(M_{m+k} + 2\cos\left(\frac{2i\pi}{n}\right)(N_{m+k} - W_{m+k}) - W_{m+k} \right) \right\} \\ & \bigcup \{ eig(M_{m+k} + 2N_{m+k} - (n-3)W_{m+k}) \} \end{aligned} \tag{2.70}$$

2.6 Circulant Block Matrices

A matrix of the following pattern is called a block circulant matrix where A_i s are matrices.

$$C = \begin{bmatrix} A_1 & A_2 & \cdots & A_{n-1} & A_n \\ A_n & A_1 & & A_{n-2} & A_{n-1} \\ \vdots & & \ddots & & \vdots \\ A_3 & A_4 & & A_1 & A_2 \\ A_2 & A_3 & \ldots & A_n & A_1 \end{bmatrix} \tag{2.71}$$

If entries (e.g., A_i s) are numbers, the matrix is called a circulant matrix. The following matrix is an example of a circulant matrix:

$$P_4 = \begin{bmatrix} 0 & 0 & 0 & 1 & 0 \\ 0 & 0 & 0 & 0 & 1 \\ 1 & 0 & 0 & 0 & 0 \\ 0 & 1 & 0 & 0 & 0 \\ 0 & 0 & 1 & 0 & 0 \end{bmatrix} \tag{2.72}$$

This matrix also fulfills the definition of a permutation matrix since each row and column contains only one entry as 1 and the remaining entries are zeros. The index 4 indicates that the non-zero entry (e.g., 1) first appears at column 4 of the first row.

Matrix $\boldsymbol{C}$ can be written as:

$$\boldsymbol{C} = \sum_{i=1}^{n} (\boldsymbol{P}_i \otimes \boldsymbol{A}_i) \tag{2.73}$$

Eigenvalues and eigenvectors of matrix $\boldsymbol{P}_i$ of the dimension n are as follows:

$$\begin{aligned} &eig(\boldsymbol{P}_i) = \omega^k;\ \omega = e^{\frac{2\pi i}{n}};\ i = \sqrt{-1};\ k = 0 : n-1 \\ &v_k = \left[1, \omega^k, \omega^{2k}, \ldots, \omega^{(n-1)k}\right]^t \end{aligned} \tag{2.74}$$

Defining the matrix function $\boldsymbol{H}(x)$ as

$$\begin{cases} \boldsymbol{H} : \boldsymbol{C} \to \boldsymbol{C}^2 \\ \boldsymbol{H}(x) = \sum_{i=1}^{n} \left(x^{i-1} \otimes \boldsymbol{A}_i\right) \end{cases} \tag{2.75}$$

Therefore, matrix C and the following matrix are similar

$$\boldsymbol{H} = \begin{bmatrix} \boldsymbol{H}(\omega^0) & 0 & \cdots & 0 & 0 \\ 0 & \boldsymbol{H}(\omega^1) & & 0 & 0 \\ \vdots & & \ddots & & \vdots \\ 0 & 0 & & \boldsymbol{H}(\omega^{n-2}) & 0 \\ 0 & 0 & \cdots & 0 & \boldsymbol{H}(\omega^{n-1}) \end{bmatrix} \tag{2.76}$$

Moreover, if u and v are the eigenvectors of $\boldsymbol{P}_i$ and $\boldsymbol{H}(\omega^i)$ respectively, $u \otimes v$ will be the eigenvector of $\boldsymbol{C}$. $\boldsymbol{G}_n(\boldsymbol{A}_m, \boldsymbol{B}_m, \boldsymbol{A}_m)$ is a specific case of $\boldsymbol{C}$. Since matrices $\boldsymbol{C}$ and $\boldsymbol{H}$ are similar, determinant of $\boldsymbol{C}$ is calculated as

$$\det(\boldsymbol{C}) = \prod_{i=0}^{n-1} det\left\{\boldsymbol{H}(\omega^i)\right\} \tag{2.77}$$

2.7 Complementary Examples

Example 3 Consider a simply supported a × a square plate under the loading condition in x direction (Fig. 2.3). Using finite difference method the critical load of the plate is calculated [2].

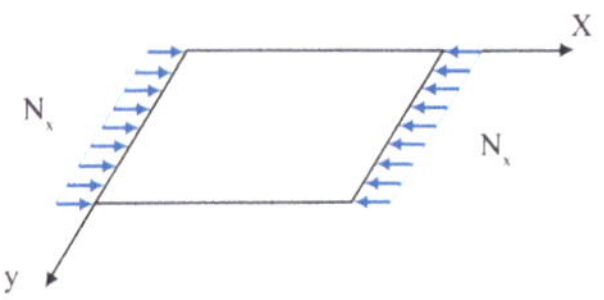

Fig. 2.3 A simply supported plate

First solution
Considering the governing differential equation as

$$\nabla^4 w + \frac{N_x}{D}\frac{\partial^2 w}{\partial x^2} = 0$$

where $D = \frac{Eh^3}{12(1-\mu^2)}$ is the bending stiffness per length unit of the plate. By dividing each side of the plate into n equal parts and generating a mesh, each element will be of dimension $h = a/n$. Formulation of finite difference method for a point like (j, k) will be as follows

$$\begin{aligned} &20w_{j,k} - 8\left(w_{j+h,k} + w_{j-h,k} + w_{j,k+h} + w_{j,k-h}\right) \\ &\quad + 2\left(w_{j+h,k+h} + w_{j+h,k-h} + w_{j-h,k+h} + w_{j-h,k-h}\right) \\ &\quad + \left(w_{j+2h,k} + w_{j,k-2h} + w_{j-2h,k} + w_{j,k+2h}\right) \\ &\quad + \frac{Nh^2}{D}\left(w_{j+h,k} - 2w_{j,k} + w_{j-h,k}\right) \\ &= 0 \end{aligned}$$

While for all points inside the domain this relationship is used, for nodes next to boundary nodes we will need additional information. For example, consider the boundary on the left side and assume we want to write the relationship for a point like (j, k) inside the domain right after the boundary. Since the point $(j - h, k)$ is located on the boundary, we will have $w_{j-h,k} = 0$. For the virtual point $(j - 2h, k)$, located outside of the plate, we will use the curvature boundary condition to define $w_{j-2h,k}$.

$$\frac{\partial^2 w}{\partial x^2} = 0 \rightarrow \frac{w_{j-2h,k} - 2w_{j-h,k} + w_{j,k}}{h^2} = 0 \rightarrow w_{j-2h,k} = -w_{j,k}$$

For simplicity of notation, let us assume $x = \frac{Nh^2}{D} = \frac{N(a/4)^2}{D}$. For the case of $n = 4$ (Fig. 2.4), we will solve the problem. The finite difference formulation for three representative points is written as

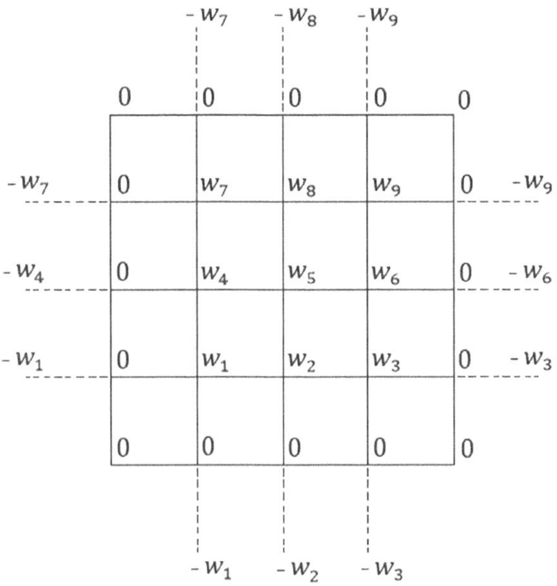

Fig. 2.4 Dividing each side of the plate into 4 equal parts

$$
\begin{aligned}
w_1 : (1, 1) &\rightarrow 20w_1 - 8(w_2 + w_4) + 2w_5 + (w_3 + w_7 - 2w_1) \\
&\quad + x(w_2 - 2w_1) = 0 \\
w_4 : (2, 1) &\rightarrow 20w_4 - 8(w_1 + w_5 + w_7) + 2(w_2 + w_8) + (w_6 - w_4) \\
&\quad + x(w_5 - 2w_4) = 0 \\
w_5 : (2, 2) &\rightarrow 20w_5 - 8(w_2 + w_4 + w_6 + w_8) + 2(w_1 + w_3 + w_7 + w_9) \\
&\quad + x(w_4 + w_6 - 2w_5) = 0
\end{aligned}
$$

Rearranging all equations in a matrix from results in:

$$
\begin{bmatrix}
18-2x & x-8 & 1 & -8 & 2 & 0 & 0 & 0 & 0 \\
x-8 & 19-2x & x-8 & 0 & -8 & 0 & 0 & 0 & 0 \\
1 & x-8 & 18-2x & 0 & 2 & -8 & 0 & 0 & 0 \\
-8 & 2 & 0 & 19-2x & x-8 & 1 & -8 & 2 & 0 \\
0 & -8 & 0 & x-8 & 20-2x & x-8 & 0 & -8 & 0 \\
0 & 2 & -8 & 1 & x-8 & 19-2x & 0 & 2 & -8 \\
0 & 0 & 0 & -8 & 2 & 0 & 18-2x & x-8 & 1 \\
0 & 0 & 0 & 0 & -8 & 0 & x-8 & 19-2x & x-8 \\
0 & 0 & 0 & 0 & 2 & -8 & 1 & x-8 & 18-2x
\end{bmatrix}
\begin{bmatrix} w_1 \\ w_2 \\ w_3 \\ w_4 \\ w_5 \\ w_6 \\ w_7 \\ w_8 \\ w_9 \end{bmatrix}
=
\begin{bmatrix} 0 \\ 0 \\ 0 \\ 0 \\ 0 \\ 0 \\ 0 \\ 0 \\ 0 \end{bmatrix}
$$

The matrix of coefficients, $\boldsymbol{M}$, can be written as

$$
\boldsymbol{M} = \begin{bmatrix} 1\,0\,0 \\ 0\,1\,0 \\ 0\,0\,1 \end{bmatrix} \otimes \begin{bmatrix} 18-2x & x-8 & 1 \\ x-8 & 19-2x & x-8 \\ 1 & x-8 & 18-2x \end{bmatrix}
$$

$$+\begin{bmatrix} 0 & 1 & 0 \\ 1 & 0 & 1 \\ 0 & 1 & 0 \end{bmatrix} \otimes \begin{bmatrix} -8 & 2 & 0 \\ 2 & -8 & 2 \\ 0 & 2 & -8 \end{bmatrix}$$

$$+\begin{bmatrix} 0 & 0 & 1 \\ 0 & 1 & 0 \\ 1 & 0 & 0 \end{bmatrix} \otimes \begin{bmatrix} 1 & 0 & 0 \\ 0 & 1 & 0 \\ 0 & 0 & 1 \end{bmatrix}$$

Since the determinant of the matrix should be equal to zero, one of the eigenvalues of the matrix is zero and, thus, x is calculated. Finally, through the relationship $N_{cr} = \frac{Dx}{(a/4)^2}$, the critical load is calculated.

If rather than the assumption $n = 4$, we solve the problem in a general form as a function of n, we will have

$$\boldsymbol{M} = \sum_{i=1}^{3} (\boldsymbol{A}_i \otimes \boldsymbol{B}_i)$$

where

$$\boldsymbol{A}_1 = I_n;\ \boldsymbol{A}_2 = \boldsymbol{F}_n(0, 1, 0);\ \boldsymbol{A}_3 = \boldsymbol{F}_n(0, 0, 1, 1)$$

$$\boldsymbol{B}_1 = \boldsymbol{F}_n(18 - 2x, x - 8, 19 - 2x, 1);\ \boldsymbol{B}_2 = \boldsymbol{F}_n(-8, 2, -8);\ \boldsymbol{B}_3 = I_n$$

Since Eq. (2.12) holds for each pair of $\boldsymbol{A}_i$ s, one can write

$$eig(\boldsymbol{M}) = \bigcup_{j=1}^{n} \left\{ eig\left(\sum_{i=1}^{3} \lambda_j(\boldsymbol{A}_i)\boldsymbol{B}_i \right) \right\}$$

$$eig(\boldsymbol{A}_1) = 1;\ eig(\boldsymbol{A}_2) = 2\cos\left(\frac{k\pi}{m}\right);\ eig(\boldsymbol{A}_3) = 1 + 2\cos\left(\frac{2k\pi}{m}\right);\ m = n + 1$$

$eig(\boldsymbol{M})$ can be written in the form of a diagonal matrix of form $\boldsymbol{F}_n$:

$$\Lambda = eig(\boldsymbol{M}) = \boldsymbol{F}_n(a, b, a + 1, 1) = \boldsymbol{I}_n \otimes a + \boldsymbol{F}_n(0, 1, 0) \otimes b + \boldsymbol{F}_n(0, 0, 1, 1) \otimes 1$$

$$a = 18 - 2x - 8 \times 2\cos\left(\frac{k\pi}{m}\right) + 1 + 2\cos\left(\frac{2k\pi}{m}\right);$$
$$b = x - 8 + 2 \times 2\cos\left(\frac{k\pi}{m}\right)$$

and, therefore, further decomposition is possible.

$$eig(\Lambda) = a + 2b\cos\left(\frac{k'\pi}{m}\right) + 1 + 2\cos\left(\frac{2k'\pi}{m}\right);\ k' = 1:n$$

$$\rightarrow 19 - 2x - 16\cos\left(\frac{k\pi}{m}\right) + 2\cos\left(\frac{2k\pi}{m}\right)$$

$$+ 2\cos\left(\frac{k'\pi}{m}\right)\left(x - 8 + 4\cos\left(\frac{k\pi}{m}\right)\right) + \left(1 + 2\cos\left(\frac{2k'\pi}{m}\right)\right) = 0$$

For the critical load ($k = k' = 1$), we will have

$$19 - 2x - 16\cos\left(\frac{\pi}{m}\right) + 2\cos\left(\frac{2\pi}{m}\right)$$

$$+ 2\cos\left(\frac{\pi}{m}\right)\left(x - 8 + 4\cos\left(\frac{\pi}{m}\right)\right) + \left(1 + 2\cos\left(\frac{2\pi}{m}\right)\right) = 0$$

$$\rightarrow x = 4\left(3 - 4\cos\left(\frac{k\pi}{m}\right) + \cos\left(\frac{2k\pi}{m}\right)\right)/\left(1 - \cos\left(\frac{k\pi}{m}\right)\right) = 8\left(1 - \cos\left(\frac{\pi}{m}\right)\right)$$

In this relationship if $m \rightarrow \infty\left(1 - \cos\alpha \cong \frac{\alpha^2}{2} \text{ when } \alpha \rightarrow 0\right)$, the critical load is obtained as

$$N_{cr} = \lim_{m\rightarrow\infty} \frac{Dx}{(a/m)^2} = \lim_{m\rightarrow\infty} \frac{D \times 8(1 - \cos(\pi/m))}{(a/m)^2} = \frac{D \times 4(\pi/m)^2}{(a/m)^2} = \frac{4\pi^2 D}{a^2}$$

Second solution

Using Theorem 2 we can see matrices $\boldsymbol{B}_i$ s are also commutative with respect to multiplication, therefore, the condition for full decomposability of block matrices holds:

$$\boldsymbol{A}_i\boldsymbol{A}_j = \boldsymbol{A}_j\boldsymbol{A}_i \text{ and } \boldsymbol{B}_i\boldsymbol{B}_j = \boldsymbol{B}_j\boldsymbol{B}_i;$$

$$i, j = 1:n\ (i \neq j) \rightarrow eig\left(\sum_{i=1}^{n}(\boldsymbol{A}_i \otimes \boldsymbol{B}_i)\right) = \sum_{i=1}^{n} eig(\boldsymbol{A}_i \otimes \boldsymbol{B}_i)$$

$$\boldsymbol{B}_2 = \boldsymbol{F}_n(-8, 2, -8) = -8\boldsymbol{I}_n + 2\boldsymbol{F}_n(0, 1, 0) \rightarrow eig(\boldsymbol{B}_2)$$

$$= -8 + 2 \times 2\cos\left(\frac{k\pi}{m}\right);\ m = n + 1$$

$$\boldsymbol{B}_1 = \boldsymbol{F}_n(18 - 2x, x - 8, 19 - 2x, 1)$$

$$= (18 - 2x)\boldsymbol{I}_n + (x - 8)\boldsymbol{F}_n(0, 1, 0) + \boldsymbol{F}_n(0, 0, 1, 1)$$

$$\rightarrow eig(\boldsymbol{B}_1) = 18 - 2x + (x - 8) \times 2\cos\left(\frac{k\pi}{m}\right) + 1 + 2\cos\left(\frac{2k\pi}{m}\right);\ eig(\boldsymbol{B}_3) = 1$$

$$
\begin{aligned}
eig\left(\sum_{i=1}^{n}(\boldsymbol{A}_i \otimes \boldsymbol{B}_i)\right) &= 1 \times \left(19 - 2x + (2x - 16)\cos\left(\frac{k\pi}{m}\right) + 2\cos\left(\frac{2k\pi}{m}\right)\right) \\
&\quad + 2\cos\left(\frac{k\pi}{m}\right)\left(-8 + 4\cos\left(\frac{k\pi}{m}\right)\right) \\
&\quad + \left(1 + 2\cos\left(\frac{2k\pi}{m}\right)\right) \times 1
\end{aligned}
$$

Since we need the determinant to be zero, we put the eigenvalue equal to zero and obtain the critical load:

$$
x = 8\left(1 - \cos\left(\frac{\pi}{m}\right)\right);\, k = 1 \rightarrow N_{cr} = \lim_{m\to\infty} \frac{Dx}{(a/m)^2} = \frac{4\pi^2 D}{a^2}
$$

Example 4 Consider the following structure that holds a P_2 by C_5 strong Cartesian product form (Fig. 2.5). The structure is composed of two equilateral 5-edges polygons where the distance of the external and internal nodes from polygons center are 3 and 1.5, respectively. The internal nodes are numbered as 2, 4, 6, 8 and 10 located 1.5 m above the plane of the external polygon. All the cross sectional areas are 5 cm^2 and the elastic modulus is taken as 200 kN/mm^2. The load P_1 at node 2 and the load P_2 at node 6 are equal to 30 and 20 kN, respectively [7].

Constructing the reduced stiffness matrix of the structure in the Cartesian coordinate system will not result in a decomposable pattern, whereas constructing the matrix in the cylindrical coordinate system leads to the following established pattern:

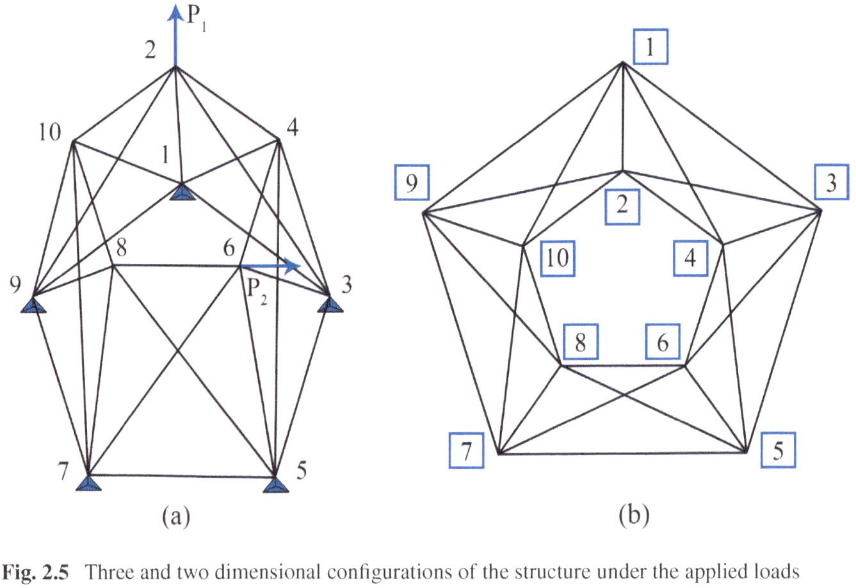

Fig. 2.5 Three and two dimensional configurations of the structure under the applied loads

$$\boldsymbol{K}_p = \boldsymbol{A}_1 \otimes \boldsymbol{B}_1 + \boldsymbol{A}_2 \otimes \boldsymbol{B}_2 + \boldsymbol{A}_3 \otimes \boldsymbol{B}_3$$

where $\boldsymbol{A}_i$ s and $\boldsymbol{B}_i$ s are matrices of dimension 5 and 3, respectively. To convert the stiffness matrix in the Cartesian coordinate system to that in the cylindrical coordinate system, the following relationships can be used:

$$\begin{aligned} \boldsymbol{F} &= \boldsymbol{K}\Delta;\ \ \boldsymbol{F}_p = \boldsymbol{R}\boldsymbol{F}; \\ \Delta_p &= \boldsymbol{R}\Delta \rightarrow \boldsymbol{R}^t\boldsymbol{F}_p = \boldsymbol{K}\boldsymbol{R}^t\Delta_p \rightarrow \boldsymbol{F}_p \\ &= \left(\boldsymbol{R}\boldsymbol{K}\boldsymbol{R}^t\right)\Delta_p \rightarrow \boldsymbol{F}_p = \boldsymbol{K}_p\Delta_p \end{aligned}$$

where $\boldsymbol{R}$ is the coordination conversion matrix from Cartesian to the Cylindrical coordinate system and $\boldsymbol{p}$ is an index representing the values (i.e., force or displacement) in the Cylindrical coordinate system. If the coordination conversion matrix for one node of the structure is indicated by $\boldsymbol{T}$, the coordination conversion matrix for the entire structure including n nodes would be block diagonal and can be written as $\boldsymbol{R} = \boldsymbol{I}_n \otimes \boldsymbol{T}$. The coordination conversion matrix $\boldsymbol{T}$ is constructed similar to the rotation matrix of a member but the angle α used in construction of matrix $\boldsymbol{T}$ is measured relative to the pole. The matrices $\boldsymbol{A}_i$ and $\boldsymbol{B}_i$ are introduced as follows:

$$\boldsymbol{A}_1 = \boldsymbol{I};\ \ \boldsymbol{B}_3 = \boldsymbol{B}_2^t;\ \ \boldsymbol{A}_3 = \boldsymbol{A}_2^t = \boldsymbol{P}_2$$

$$\boldsymbol{B}_1 = 10^5\begin{bmatrix} 0.6463 & 0 & -0.1867 \\ 0 & 1.2063 & 0 \\ -0.1867 & 0 & 0.3639 \end{bmatrix};$$

$$\boldsymbol{B}_2 = 10^4\begin{bmatrix} 1.9593 & 2.6967 & 0 \\ -2.6967 & -3.7117 & 0 \\ 0 & 0 & 0 \end{bmatrix}$$

Eigenvalues of $\boldsymbol{A}_2$ and $\boldsymbol{A}_3$ are calculated using Eq. (2.74) as follows:

$$\begin{aligned} eig(\boldsymbol{A}_2) = eig(\boldsymbol{A}_3) &= \left\{1, e^{\frac{2\pi i}{5}}, e^{\frac{4\pi i}{5}}, e^{\frac{6\pi i}{5}}, e^{\frac{8\pi i}{5}}\right\} \\ &= \{1, 0.3090 \pm 0.9511i, -0.8090 \pm 0.5878i\} \end{aligned}$$

Eigenvalues of the matrix $\boldsymbol{K}_p$ can be found as follows:

$$eig(\boldsymbol{K}_p) = \bigcup_{i=1}^{5}\left\{eig\left(B_1 + \lambda_i(A_2)B_2 + \lambda_i\left(A_2^t\right)B_2^t\right)\right\}$$

As an example the largest eigenvalue of matrix $\boldsymbol{K}_p$ would be $\lambda_{max} = 1.0864e5$.

Eigenvectors of matrix $\boldsymbol{K}_p$ will be $\boldsymbol{u} \otimes \boldsymbol{v}$ where $\boldsymbol{u}$ and $\boldsymbol{v}$ were already introduced. The eigenvector corresponding to λ_{max} is as

$$\{\varphi\} = [1, 1, 1, 1, 1]^t \otimes [0.4330, 0, -0.1119]^t$$

Since the governing equation of the structure $\boldsymbol{F}_p = \boldsymbol{K}_p \boldsymbol{\Delta}_p$ was expressed in the cylindrical coordinate system, the displacement values in the cylindrical coordinate system is obtained as:

$$\boldsymbol{\Delta}_p = \begin{matrix} \rho \\ \theta \\ z \end{matrix} \begin{bmatrix} 0.5330 & -0.2351 & 0.4221 & -0.2549 & -0.1770 \\ 0.0028 & 0.0924 & 0.2169 & 0.1621 & -0.1254 \\ 1.0978 & -0.1206 & 0.2166 & -0.1308 & -0.0908 \end{bmatrix}$$

The displacements will be obtained using $\boldsymbol{\Delta} = \boldsymbol{R}^t \boldsymbol{\Delta}_p$ in the Cartesian coordinate system

$$\boldsymbol{\Delta} = \begin{matrix} x \\ y \\ z \end{matrix} \begin{bmatrix} -0.0028 & -0.2521 & 0.4236 & 0.2810 & 0.2071 \\ 0.5330 & 0.0152 & -0.2140 & 0.1111 & 0.0646 \\ 1.0978 & -0.1206 & 0.2166 & -0.1308 & -0.0908 \end{bmatrix}$$

Columns 1 to 5 contain the displacements of the nodes 2, 4, 6, 8 and 10, respectively.

Example 5 Calculating frequencies and natural modes of the structure in Fig. 2.6 is aimed. The cross sectional area of all members is equal to 5 cm^2 and the elastic modulus is taken as 200 kN/mm^2 (MPa). Moreover, for all members $\rho = 78.5\ \text{kN/m}^3$ is considered [8].

Stiffness and mass matrices are constructed in the cylindrical coordinate system. Both matrices will have the following pattern:

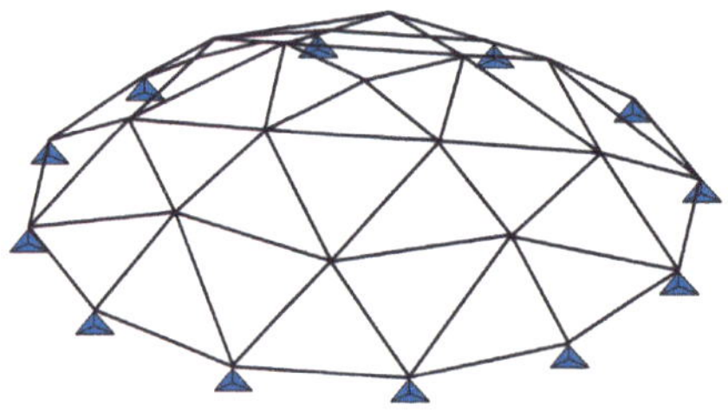

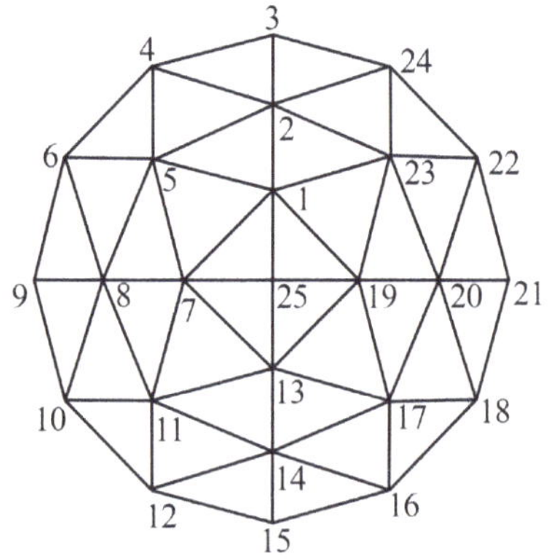

Fig. 2.6 A three-dimensional diamatic dome and the corresponding nodal numbering

$$
\begin{bmatrix}
\boldsymbol{A}_m & \boldsymbol{B}_m & \boldsymbol{Z}_m & \boldsymbol{B}^t_m & \boldsymbol{P}_{mk} \\
\boldsymbol{B}^t_m & \boldsymbol{A}_m & \boldsymbol{B}_m & \boldsymbol{Z}_m & \boldsymbol{P}_{mk} \\
\boldsymbol{Z}_m & \boldsymbol{B}^t_m & \boldsymbol{A}_m & \boldsymbol{B}_m & \boldsymbol{P}_{mk} \\
\boldsymbol{B}_m & \boldsymbol{Z}_m & \boldsymbol{B}^t_m & \boldsymbol{A}_m & \boldsymbol{P}_{mk} \\
\boldsymbol{P}^t_{mk} & \boldsymbol{P}^t_{mk} & \boldsymbol{P}^t_{mk} & \boldsymbol{P}^t_{mk} & \boldsymbol{R}_k
\end{bmatrix}
$$

where $m = 18$, $n = 4$, and $k = 3$. The last block column of the matrix includes three numerical columns, therefore, three zero columns and rows are added between block columns and rows of the matrix when conducting matrix operations to achieve a decomposable pattern similar to the one in Eq. (2.59). The pattern of the stiffness and mass matrices is shown in Fig. 2.7

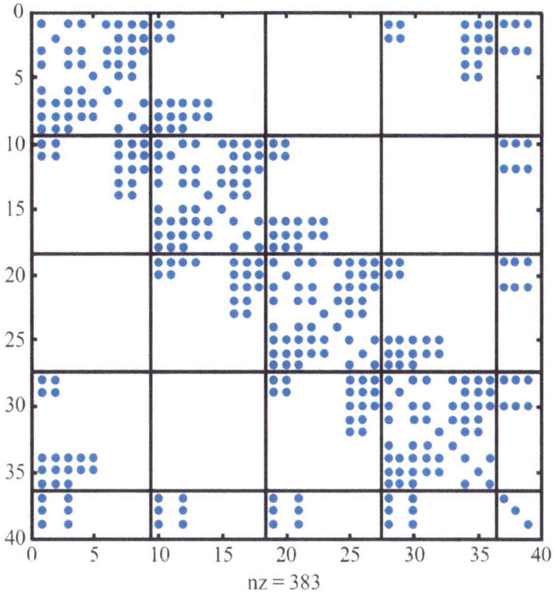

Fig. 2.7 The pattern of the stiffness ($\boldsymbol{K}$) and mass matrices ($\boldsymbol{M}$)

The matrix $\boldsymbol{K} - \boldsymbol{M}\Omega^2$ is constructed as follows:

$$\boldsymbol{K} - \boldsymbol{M}\Omega^2 = \begin{bmatrix} \boldsymbol{M}_K - \boldsymbol{M}_M\Omega^2 & \boldsymbol{N}_K - \boldsymbol{N}_M\Omega^2 & \boldsymbol{W}_K - \boldsymbol{W}_M\Omega^2 & \boldsymbol{S}_K - \boldsymbol{S}_M\Omega^2 \\ \boldsymbol{S}_K - \boldsymbol{S}_M\Omega^2 & \boldsymbol{M}_K - \boldsymbol{M}_M\Omega^2 & \boldsymbol{N}_K - \boldsymbol{N}_M\Omega^2 & \boldsymbol{W}_K - \boldsymbol{W}_M\Omega^2 \\ \boldsymbol{W}_K - \boldsymbol{W}_M\Omega^2 & \boldsymbol{S}_K - \boldsymbol{S}_M\Omega^2 & \boldsymbol{M}_K - \boldsymbol{M}_M\Omega^2 & \boldsymbol{N}_K - \boldsymbol{N}_M\Omega^2 \\ \boldsymbol{N}_K - \boldsymbol{N}_M\Omega^2 & \boldsymbol{W}_K - \boldsymbol{W}_M\Omega^2 & \boldsymbol{S}_{K-\boldsymbol{S}_M\Omega^2} & \boldsymbol{M}_{K-\boldsymbol{M}_M\Omega^2} \end{bmatrix}$$

The matrix holds the circulant pattern and $\boldsymbol{H}(x)$ can be written as

$$\boldsymbol{H}(x) = x^0 \otimes \left(\boldsymbol{M}_K - \boldsymbol{M}_M\Omega^2\right) + x^1 \otimes \left(\boldsymbol{N}_K - \boldsymbol{N}_M\Omega^2\right) + x^2 \otimes \left(\boldsymbol{W}_K - \boldsymbol{W}_M\Omega^2\right) + x^3 \otimes \left(\boldsymbol{S}_K - \boldsymbol{S}_M\Omega^2\right)$$

We also have

$$\omega^k = \left\{\left(e^{\frac{2\pi i}{4}}\right)^k\right\} = \{\pm 1, \pm i\};\; k = 0:3$$

For all values of x, the submatrices $\boldsymbol{H}(x)$ are obtained in terms of Ω, and using Eq. (2.77), the corresponding eigenvalues are calculated. The largest period of this structure is obtained as $T_1 = 0.0826$. For the construction of the eigenvectors, as we mentioned before, u and v should be calculated. Then we obtain the eigenvectors as

$$\omega = e^{\frac{2\pi i}{4}} = i \rightarrow v = \left\{1, \omega, \omega^2, \omega^3\right\}^t = \{1, i, -1, -i\}^t$$

For all values of u, the submatrices $\boldsymbol{H}\left(\omega^k\right)$ are calculated, and finally the vibrating modes $v \otimes u$ are obtained. As an example, for the 6th period ($T_6 = 0.0456$), the 6th natural mode is shown in Fig. 2.8.

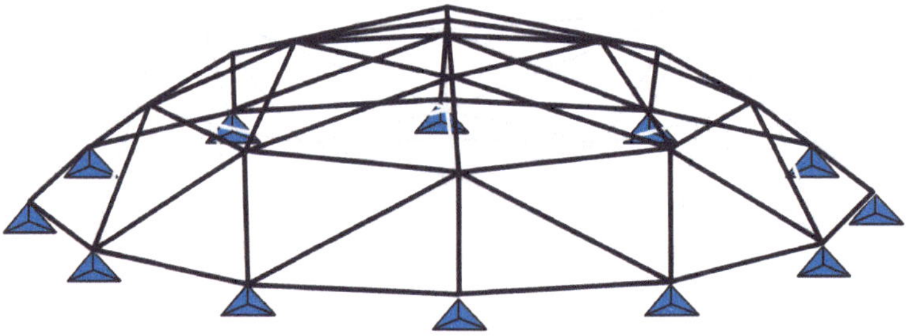

Fig. 2.8 The 6th natural mode of the structure in Example 5

References

1. Kaveh, A.: Optimal Analysis of Structures by Concepts of Symmetry and Regularity. Springer, GmbH, Wien-NewYork (2013). ISBN 978-3-7091-1564-7
2. Kaveh, A., Rahami, H.: Compound matrix block diagonalization for efficient solution of eigenproblems in structural mechanics. Acta Mech. **188**(3–4), 155–166 (2007)
3. Kaveh, A., Rahami, H.: Special decompositions for eigenproblems in structural mechanics. Commun. Numer. Methods Eng. **22**(9), 943–953 (2006)
4. Kaveh, A., Rahami, H.: An efficient method for decomposition of regular structures using graph products. Int. J. Numer. Methods Eng. **61**, 1797–1808 (2004)
5. Kaveh, A.: Structural Mechanics: Graph and Matrix Methods, 3rd edn. Research Studies Press (John Wiley), UK (2006)
6. Kaveh, A., Rahami, H.: Block diagonalization of adjacency and Laplacian matrices for graph product; applications in structural mechanics. Int. J. Numer. Methods Eng. **68**(1), 33–63 (2006)
7. Kaveh, A., Rahami, H.: An efficient analysis of repetitive structures generated by graph products. Int. J. Numer. Methods Eng. **84**(10), 108–126 (2010)
8. Kaveh, A., Rahami, H.: Block circulant matrices and applications in free vibration analysis of cyclically repetitive structures. Acta Mech. **217**(1–2), 51–62 (2011)

Chapter 3
Static Analysis of Near-Regular Skeletal Structures: Additional Members

3.1 Introduction

In Chap. 2, methods were presented for calculation of eigenpairs and inverse of decomposable (block) matrices. Specifically, stiffness matrix of a group of structures holds a pattern similar to that of decomposable matrices. Although there is no general rule for identification of structures with decomposable stiffness matrix, these structures have to hold symmetrical geometries. A structure with decomposable stiffness matrix is hereafter called a *regular structure*. Examples of regular structures were presented in Chap. 2.

In addition to regular structures, there is another group of structures for which the stiffness matrix cannot be directly decomposed. These structures might be solvable (i.e., eigensolution or inversion) using the relationships in Chap. 2 through manipulation of either the stiffness matrix or the structure. However, such manipulations have to be compensated for, which might be computationally demanding. A structure that can be solved with minimum manipulations is hereafter call a *near-regular structure*. In general, there are two types of near-regular structures solution of which requires different algorithms and techniques. In this chapter near-regular Type I structures are discussed and methods are developed for the analysis and eigensolution of these structures.

3.2 Analysis of Near-Regular Structures with Additional Members Using Displacement Methods

These near-regular structures are convertible to regular structures by adding or removing some members where the number of degrees of freedom is the same for the near-regular and the regular structure [1]. The regular structure is then solved and the effects of changes are compensated. Suppose after converting a near-regular structure to a regular one, the stiffness matrix of the regular structure is inverted using

A. Kaveh et al., *Swift Analysis of Civil Engineering Structures Using Graph Theory Methods*, Studies in Systems, Decision and Control 290,
https://doi.org/10.1007/978-3-030-45549-1_3

Eq. (2.32), and matrix $\boldsymbol{D}$ in Eq. (3.1) is obtained. Assume degrees of freedom m, n, … are the ones that are affected by member changes, whereas s, k, … are the unaffected degrees of freedom (the affected degrees of freedom are the degrees that are connected to changed members whereas the unaffected degrees are those to which the connected members are identical in both regular and near-regular structures).

$$\begin{bmatrix} \Delta_m \\ \Delta_n \\ \vdots \\ \cdots \\ \Delta_k \\ \Delta_s \\ \vdots \end{bmatrix} = \left[\begin{array}{ccc:ccc} \boldsymbol{D}_{mm} & \boldsymbol{D}_{mn} & & \boldsymbol{D}_{mk} & \boldsymbol{D}_{ms} & \\ & \boldsymbol{D}_{nn} & & \boldsymbol{D}_{nk} & \boldsymbol{D}_{ns} & \\ & & \ddots & & & \ddots \\ \cdots & \cdots & \cdots & \cdots & \cdots & \cdots \\ & & & \boldsymbol{D}_{kk} & \boldsymbol{D}_{ks} & \\ & & & & \boldsymbol{D}_{ss} & \\ Sym & & & & & \ddots \end{array}\right] \begin{bmatrix} \boldsymbol{F}_m \\ \boldsymbol{F}_n \\ \vdots \\ \cdots \\ \boldsymbol{F}_k \\ \boldsymbol{F}_s \\ \vdots \end{bmatrix} \tag{3.1}$$

or in a simpler form we will have

$$\begin{bmatrix} \boldsymbol{\Delta}_{I} \\ \cdots \\ \boldsymbol{\Delta}_{II} \end{bmatrix} = \left[\begin{array}{c:c} \boldsymbol{D}_{I,I} & \boldsymbol{D}_{I,II} \\ \cdots & \cdots \\ \boldsymbol{D}_{II,I} & \boldsymbol{D}_{II,II} \end{array}\right] \begin{bmatrix} \boldsymbol{F}_{I} \\ \cdots \\ \boldsymbol{F}_{II} \end{bmatrix} \tag{3.2}$$

If we construct the stiffness matrix of the near-regular structure, it will have the following form:

$$\begin{bmatrix} \boldsymbol{F}_m \\ \boldsymbol{F}_n \\ \vdots \\ \cdots \\ \boldsymbol{F}_k \\ \boldsymbol{F}_s \\ \vdots \end{bmatrix} = \left[\begin{array}{ccc:ccc} \boldsymbol{K}_{mm} + \boldsymbol{K}'_{mm} & \boldsymbol{K}_{mn} + \boldsymbol{K}'_{mn} & & \boldsymbol{K}_{mk} & \boldsymbol{K}_{ms} & \\ & \boldsymbol{K}_{nn} + \boldsymbol{K}'_{nn} & & \boldsymbol{K}_{nk} & \boldsymbol{K}_{ms} & \\ & & \ddots & & & \ddots \\ \cdots & \cdots & \cdots & \cdots & \cdots & \cdots \\ & & & \boldsymbol{K}_{kk} & \boldsymbol{K}_{ks} & \\ & & & & \boldsymbol{K}_{ss} & \\ Sym & & & & & \ddots \end{array}\right] \begin{bmatrix} \Delta'_{\boldsymbol{m}} \\ \Delta'_{\boldsymbol{n}} \\ \vdots \\ \cdots \\ \Delta'_{\boldsymbol{k}} \\ \Delta'_{\boldsymbol{s}} \\ \vdots \end{bmatrix} \tag{3.3}$$

where the components $\boldsymbol{K}$ belong to the regular structure and the components $\boldsymbol{K}'$ appear due to the applied changes. In a simpler form we can write

$$\begin{bmatrix} \boldsymbol{F}_I \\ \cdots \\ \boldsymbol{F}_{II} \end{bmatrix} = \begin{bmatrix} \boldsymbol{K}_{I,I} + \boldsymbol{K}'_{I,I} & \vdots & \boldsymbol{K}_{I,II} \\ \cdots & \vdots & \cdots \\ \boldsymbol{K}_{II,I} & \vdots & \boldsymbol{K}_{II,II} \end{bmatrix} \begin{bmatrix} \Delta'_I \\ \cdots \\ \Delta'_{II} \end{bmatrix} \tag{3.4}$$

where $\boldsymbol{K}'_{I,I}$ expresses the changes in the stiffness matrix caused by changes in the stiffness of members. Solving Eq. (3.4) results in

$$\begin{bmatrix} \Delta'_I \\ \cdots \\ \Delta'_{II} \end{bmatrix} = \begin{bmatrix} \boldsymbol{D}'_{I,I} & \vdots & \boldsymbol{D}'_{I,II} \\ \cdots & \vdots & \cdots \\ \boldsymbol{D}'_{II,I} & \vdots & \boldsymbol{D}'_{II,II} \end{bmatrix} \begin{bmatrix} \boldsymbol{F}_I \\ \cdots \\ \boldsymbol{F}_{II} \end{bmatrix} \tag{3.5}$$

Finding the solution of the near-regular structure using the inverse of the stiffness matrix of regular structure is aimed. From Eq. (3.4) we have

$$\begin{bmatrix} \boldsymbol{F}_I \\ \cdots \\ \boldsymbol{F}_{II} \end{bmatrix} - \begin{bmatrix} \boldsymbol{K}'_{I,I}\Delta'_I \\ \cdots \\ 0 \end{bmatrix} = \begin{bmatrix} \boldsymbol{K}_{I,I} & \vdots & \boldsymbol{K}_{I,II} \\ \cdots & \vdots & \cdots \\ \boldsymbol{K}_{II,I} & \vdots & \boldsymbol{K}_{II,II} \end{bmatrix} \begin{bmatrix} \Delta'_I \\ \cdots \\ \Delta'_{II} \end{bmatrix} \tag{3.6}$$

Left multiplication of the two sides of Eq. (3.6) by matrix $\boldsymbol{D}$ results in

$$\begin{bmatrix} \Delta'_I \\ \cdots \\ \Delta'_{II} \end{bmatrix} = \begin{bmatrix} \boldsymbol{D}_{I,I} & \vdots & \boldsymbol{D}_{I,II} \\ \cdots & \vdots & \cdots \\ \boldsymbol{D}_{II,I} & \vdots & \boldsymbol{D}_{II,II} \end{bmatrix} \begin{bmatrix} \boldsymbol{F}_I - \boldsymbol{K}'_{I,I}\Delta'_I \\ \cdots \\ \boldsymbol{F}_{II} \end{bmatrix} \tag{3.7}$$

Therefore,

$$\begin{aligned} \Delta'_I &= \left[\boldsymbol{I} + \boldsymbol{D}_{I,I}\boldsymbol{K}'_{I,I}\right]^{-1}\left[\boldsymbol{D}_{I,I}\boldsymbol{F}_I + \boldsymbol{D}_{I,II}\boldsymbol{F}_{II}\right] \\ \Delta'_{II} &= \boldsymbol{D}_{II,I}\left\{\boldsymbol{F}_I - \boldsymbol{K}'_{I,I}\left[\boldsymbol{I} + \boldsymbol{D}_{I,I}\boldsymbol{K}'_{I,I}\right]^{-1}\left[\boldsymbol{D}_{I,I}\boldsymbol{F}_I + \boldsymbol{D}_{I,II}\boldsymbol{F}_{I,I}\right]\right\} + \boldsymbol{D}_{II,II}\boldsymbol{F}_{II} \end{aligned} \tag{3.8}$$

Components $\boldsymbol{D}$ are obtained from Eq. (3.1) by inverting Eq. (2.32). Considering the following definition

$$\left[\boldsymbol{I} + \boldsymbol{D}_{I,I}\boldsymbol{K}'_{I,I}\right]^{-1} = \overline{\boldsymbol{D}}_{I,I} \tag{3.9}$$

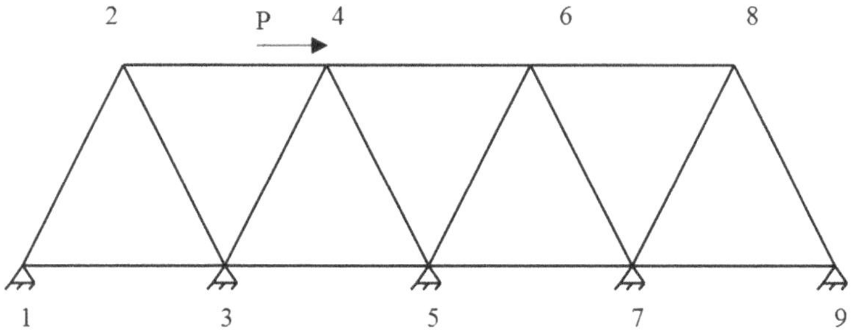

Fig. 3.1 A planar truss under lateral loading

And expanding Eq. (3.2) as

$$\begin{aligned}\Delta_I &= \boldsymbol{D}_{I,I}\boldsymbol{F}_I + \boldsymbol{D}_{I,II}\boldsymbol{F}_{II}\\ \Delta_{II} &= \boldsymbol{D}_{II,I}\boldsymbol{F}_I + \boldsymbol{D}_{II,II}\boldsymbol{F}_{II}\end{aligned} \tag{3.10}$$

We can re-write Eq. (3.8) as

$$\begin{aligned}\Delta'_I &= \overline{\boldsymbol{D}}_{I,I}\Delta_I\\ \Delta'_{II} &= -\boldsymbol{D}_{II,I}\boldsymbol{K}'_{I,I}\Delta'_I + \Delta_{II}\end{aligned} \tag{3.11}$$

Therefore, displacements of the near-regular structure are obtained. The solution process included inverting a matrix, of dimension equal to that of $\boldsymbol{K}'_{I,I}$, and some matrix operations. The required effort for the proposed solution depends on the number of affected degrees of freedom that defines the dimension of $\boldsymbol{K}'_{I,I}$. Substituting Eq. (3.10) in Eq. (3.11) results in the inverse of the stiffness matrix of the near-regular structure as follows:

$$\begin{bmatrix}\Delta'_I\\ \cdots\\ \Delta'_{II}\end{bmatrix} = \begin{bmatrix}\overline{\boldsymbol{D}}_{I,I}\boldsymbol{D}_{I,I} & \vdots & \overline{\boldsymbol{D}}_{I,I}\boldsymbol{D}_{I,II}\\ \cdots & \vdots & \cdots\\ -\boldsymbol{D}_{II,I}\boldsymbol{K}'_{I,I}\overline{\boldsymbol{D}}_{I,I}\boldsymbol{D}_{I,I} + \boldsymbol{D}_{II,I} & \vdots & -\boldsymbol{D}_{II,I}\boldsymbol{K}'_{I,I}\overline{\boldsymbol{D}}_{I,I}\boldsymbol{D}_{I,II} + \boldsymbol{D}_{II,II}\end{bmatrix} \begin{bmatrix}\boldsymbol{F}_I\\ \cdots\\ \boldsymbol{F}_{II}\end{bmatrix} \tag{3.12}$$

3.2.1 A Comprehensive Example

A truss is used to carry lateral loads (Fig. 3.1). For all members the length is equal to 5 m, the elastic modulus is taken as 210 kN/mm^2, and the cross sectional areas are 15 cm^2. The load $\boldsymbol{P}$ is equal to 35 kN. The analysis of the truss is aimed [2].

Method 1 Consider a second similar structure next to the main structure with the same loading condition (Fig. 3.2).

To achieve a decomposable matrix pattern, nodes 8 and 16 should be connected to nodes 10 and 2, respectively. Instead of direct connection of node 16 to node 2, the imaginary node 2_1 is used. The advantage of using such a node is the length of member 16-2_1 will be similar to the length of other horizontal members with no need to correct the stiffness of member. The relationship $\boldsymbol{F} = \boldsymbol{K}\Delta$ is written for the structure as

$$\begin{bmatrix} \boldsymbol{F}_2 \\ \boldsymbol{F}_4 \\ \boldsymbol{F}_6 \\ \vdots \\ \boldsymbol{F}_{12} \\ \boldsymbol{F}_{14} \\ \boldsymbol{F}_{16} \\ \cdots \\ \boldsymbol{F}_{2_1} \end{bmatrix} = \left[\begin{array}{ccccccccc} \boldsymbol{A} & \boldsymbol{B} & & & & & & \vdots & \\ \boldsymbol{B} & \boldsymbol{C} & \boldsymbol{B} & & & & & \vdots & \\ & \boldsymbol{B} & \boldsymbol{C} & \boldsymbol{B} & & & & \vdots & \\ & & \boldsymbol{B} & \ddots & \boldsymbol{B} & & & \vdots & \\ & & & \boldsymbol{B} & \boldsymbol{C} & \boldsymbol{B} & & \vdots & \\ & & & & \boldsymbol{B} & \boldsymbol{C} & \boldsymbol{B} & \vdots & \\ & & & & & \boldsymbol{B} & \boldsymbol{C} & \vdots & \boldsymbol{B} \\ \cdots & \cdots & \cdots & \cdots & \cdots & \cdots & \cdots & \cdots & \\ & & & & & & \boldsymbol{B} & & -\boldsymbol{B} \end{array}\right] \begin{bmatrix} \Delta_2 \\ \Delta_4 \\ \Delta_6 \\ \vdots \\ \Delta_{12} \\ \Delta_{14} \\ \Delta_{16} \\ \cdots \\ \Delta_{2_1} \end{bmatrix}$$

where $\boldsymbol{F}$s and Δs are the force and displacement vectors of dimension 2×1 in each degree of freedom and $\boldsymbol{A}$, $\boldsymbol{B}$ and $\boldsymbol{C}$ are the blocks of dimension 2.

$$\boldsymbol{A} = 10^7 \times \begin{bmatrix} 4.28 & 0 \\ 0 & 2.76 \end{bmatrix} \quad \boldsymbol{B} = 10^7 \times \begin{bmatrix} -1.53 & 0 \\ 0 & 0 \end{bmatrix} \quad \boldsymbol{C} = 10^7 \times \begin{bmatrix} 5.81 & 0 \\ 0 & 2.76 \end{bmatrix}$$

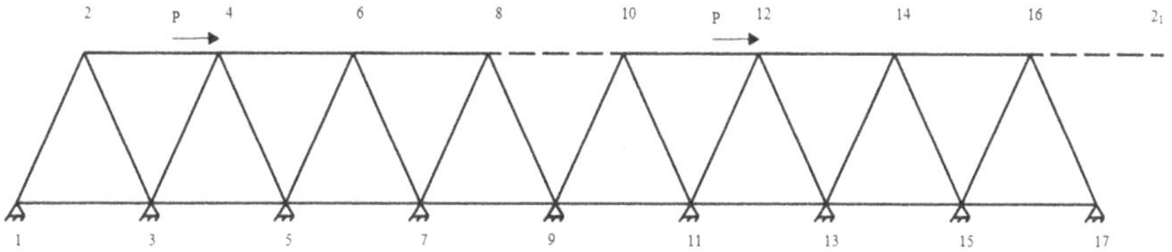

Fig. 3.2 The truss converted to cyclic truss using the concept of anti-symmetry

Since we assume nodes 2 and 2_1 are identical, they should have the same displacements ($\Delta_2 = \Delta_{2_1}$). Also, in reality these two nodes make one degree of freedom together, therefore

$$\boldsymbol{F}_{\text{real2}} = \boldsymbol{F}_2 + \boldsymbol{F}_{2_1}$$

And

$$\begin{bmatrix} \boldsymbol{F}_2 \\ \boldsymbol{F}_4 \\ \boldsymbol{F}_6 \\ \vdots \\ \boldsymbol{F}_{12} \\ \boldsymbol{F}_{14} \\ \boldsymbol{F}_{16} \\ \boldsymbol{F}_{2_1} \end{bmatrix} = \begin{bmatrix} \boldsymbol{A} & \boldsymbol{B} & & & & & \\ \boldsymbol{B} & \boldsymbol{C} & \boldsymbol{B} & & & & \\ & \boldsymbol{B} & \boldsymbol{C} & \boldsymbol{B} & & & \\ & & \boldsymbol{B} & \ddots & \ddots & & \\ & & & \ddots & \boldsymbol{C} & \boldsymbol{B} & \\ & & & & \boldsymbol{B} & \boldsymbol{C} & \boldsymbol{B} \\ \boldsymbol{B} & & & & & \boldsymbol{B} & \boldsymbol{C} \\ -\boldsymbol{B} & & & & & & \boldsymbol{B} \end{bmatrix} \begin{bmatrix} \Delta_2 \\ \Delta_4 \\ \Delta_6 \\ \vdots \\ \Delta_{12} \\ \Delta_{14} \\ \Delta_{16} \end{bmatrix}$$

$$\begin{bmatrix} \boldsymbol{F}_2 \\ \boldsymbol{F}_4 \\ \boldsymbol{F}_6 \\ \vdots \\ \boldsymbol{F}_{12} \\ \boldsymbol{F}_{14} \\ \boldsymbol{F}_{16} \end{bmatrix} = \begin{bmatrix} \boldsymbol{A}-\boldsymbol{B} & \boldsymbol{B} & & & & & \boldsymbol{B} \\ \boldsymbol{B} & \boldsymbol{C} & \boldsymbol{B} & & & & \\ & \boldsymbol{B} & \boldsymbol{C} & \boldsymbol{B} & & & \\ & & \boldsymbol{B} & \ddots & \ddots & & \\ & & & \ddots & \boldsymbol{C} & \boldsymbol{B} & \\ & & & & \boldsymbol{B} & \boldsymbol{C} & \boldsymbol{B} \\ \boldsymbol{B} & & & & & \boldsymbol{B} & \boldsymbol{C} \end{bmatrix} \begin{bmatrix} \Delta_2 \\ \Delta_4 \\ \Delta_6 \\ \vdots \\ \Delta_{12} \\ \Delta_{14} \\ \Delta_{16} \end{bmatrix}$$

Since $\boldsymbol{A}-\boldsymbol{B} = \boldsymbol{C}$, the stiffness matrix holds the form $\boldsymbol{G}_n(\boldsymbol{C}_m, \boldsymbol{B}_m, \boldsymbol{C}_m)$ for which eigenpairs and then the inverse of the matrix can be calculated.

To obtain the actual answers, the applied changes must be compensated. However, the added members between nodes 8 and 10 as well as nodes 2 and 16 have zero internal forces (anti-symmetrical loading condition). Therefore, the initial and the changed structures are equal with no need for compensation.

Method 2 Node 8 is connected to node 2 (Fig. 3.3).

In graph theory there is no difference between this edge and other edges. However, when it comes to structures the stiffness of this member should be chosen appropriately. Because the length of the member is four times the length of the horizontal members at the top, its area should be choose four times bigger too. The relationship $\boldsymbol{F} = \boldsymbol{K}\Delta$ can be written as:

$$\begin{bmatrix} \boldsymbol{F}_2 \\ \boldsymbol{F}_4 \\ \boldsymbol{F}_6 \\ \boldsymbol{F}_8 \end{bmatrix} = \begin{bmatrix} \boldsymbol{C} & \boldsymbol{B} & & \boldsymbol{B} \\ \boldsymbol{B} & \boldsymbol{C} & \boldsymbol{B} & \\ & \boldsymbol{B} & \boldsymbol{C} & \boldsymbol{B} \\ \boldsymbol{B} & & \boldsymbol{B} & \boldsymbol{C} \end{bmatrix} \begin{bmatrix} \Delta_2 \\ \Delta_4 \\ \Delta_6 \\ \Delta_8 \end{bmatrix}$$

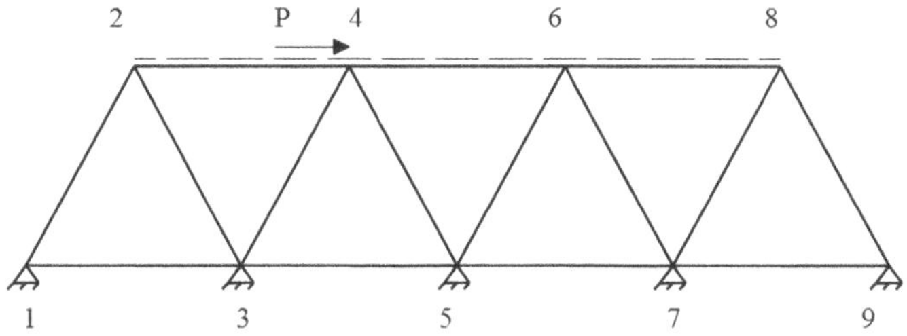

Fig. 3.3 Changed truss by adding a member

The inverse of this matrix can be defined using its eigenpairs.

Now the initial structure is solved using Eq. (3.11). The added member changes nodes 2 and 8, therefore, $\boldsymbol{K}'_{I,I}$ is a matrix of dimension $2 \times 2 = 4$. Thus, instead of inverting a matrix of dimension 8, a matrix of dimension 4 should be inverted. By increasing the number of spans, the matrix of dimension 8 will increase while the dimension of matrix $\boldsymbol{K}'_{I,I}$ is always 4 because we only connect the first and the last nodes.

Discussion

In the two methods above the initial truss was manipulated and then solved. If the stiffness matrix of the initial truss is constructed, we will have

$$\begin{bmatrix} \boldsymbol{F}_2 \\ \boldsymbol{F}_4 \\ \boldsymbol{F}_6 \\ \boldsymbol{F}_8 \end{bmatrix} = \begin{bmatrix} \boldsymbol{A} & \boldsymbol{B} & & \\ \boldsymbol{B} & \boldsymbol{C} & \boldsymbol{B} & \\ & \boldsymbol{B} & \boldsymbol{C} & \boldsymbol{B} \\ & & \boldsymbol{B} & \boldsymbol{A} \end{bmatrix} \begin{bmatrix} \Delta_2 \\ \Delta_4 \\ \Delta_6 \\ \Delta_8 \end{bmatrix}$$

The form $\boldsymbol{F}(\boldsymbol{A}, \boldsymbol{B}, \boldsymbol{C})$ with the condition $\boldsymbol{A} - \boldsymbol{B} = \boldsymbol{C}$ is decomposable as

$$\boldsymbol{F}(\boldsymbol{A}, \boldsymbol{B}, \boldsymbol{C}) = \begin{bmatrix} \boldsymbol{A} & \boldsymbol{B} & & & \\ \boldsymbol{B} & \boldsymbol{C} & \boldsymbol{B} & & \\ & \boldsymbol{B} & \ddots & \ddots & \\ & & \ddots & \boldsymbol{C} & \boldsymbol{B} \\ & & & \boldsymbol{B} & \boldsymbol{A} \end{bmatrix} = \boldsymbol{I} \otimes (\boldsymbol{A} + \boldsymbol{B}) + \boldsymbol{T} \otimes (-\boldsymbol{B})$$

where $\boldsymbol{T} = \boldsymbol{F}(1, -1, 2)$. Since $\boldsymbol{IT} = \boldsymbol{TI}$, the matrix $\boldsymbol{F}$ can be block diagonalized and eigenvalues can be obtained.

However, since only a few structures possess such a property, in most cases it is necessary to add or remove members to obtain decomposable patterns.

3.2.2 Complementary Examples

Example 1 The displacements of the truss shown in Fig. 3.4 are calculated. For all members the length is equal to 5 m, the elastic modulus is taken as 210 kN/mm^2, and the cross sectional areas are 10 cm^2. The load $\boldsymbol{P}$ is equal to 25 kN.

The form of the stiffness matrix for this structure is as follows:

$$\boldsymbol{F}(\boldsymbol{A},\boldsymbol{B},\boldsymbol{C}) = \begin{bmatrix} \boldsymbol{A} & \boldsymbol{B} & & & \\ \boldsymbol{B} & \boldsymbol{C} & \boldsymbol{B} & & \\ & \boldsymbol{B} & \ddots & \ddots & \\ & & \ddots & \boldsymbol{C} & \boldsymbol{B} \\ & & & \boldsymbol{B} & \boldsymbol{A} \end{bmatrix}$$

In which $\boldsymbol{A}-\boldsymbol{B} \neq \boldsymbol{C}$. By adding members to the truss (Fig. 3.5) the stiffness matrix will have the decomposable form $\boldsymbol{G}(\boldsymbol{C},\boldsymbol{B},\boldsymbol{C})$ as follows:

$$\boldsymbol{K} = \begin{bmatrix} \boldsymbol{C} & \boldsymbol{B} & & & \boldsymbol{B} \\ \boldsymbol{B} & \boldsymbol{C} & \boldsymbol{B} & & \\ & \boldsymbol{B} & \boldsymbol{C} & \boldsymbol{B} & \\ & & \boldsymbol{B} & \boldsymbol{C} & \boldsymbol{B} \\ \boldsymbol{B} & & & \boldsymbol{B} & \boldsymbol{C} \end{bmatrix}$$

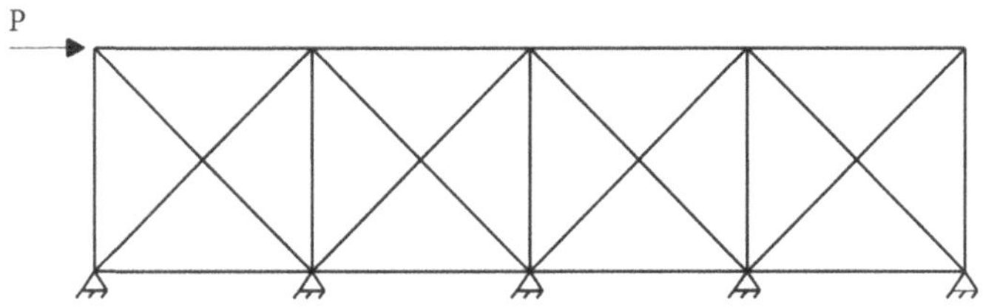

Fig. 3.4 The truss under study

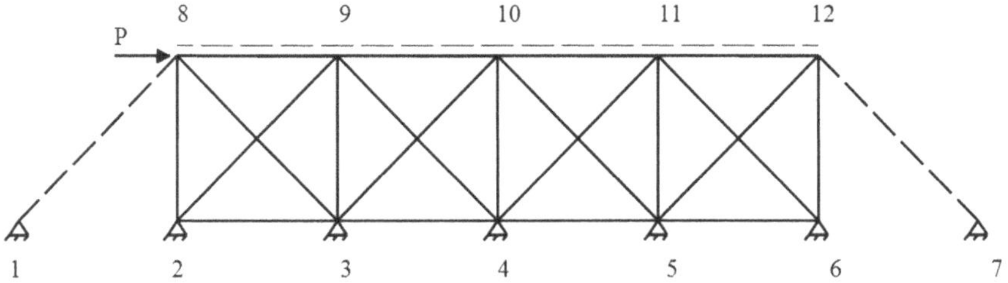

Fig. 3.5 Changed truss by addition members

where the matrices $\boldsymbol{B}$ and $\boldsymbol{C}$ are

$$\boldsymbol{B} = 10^7 \times \begin{bmatrix} -2.1 & 0 \\ 0 & 0 \end{bmatrix} \quad \boldsymbol{C} = 10^7 \times \begin{bmatrix} 5.68 & 0 \\ 0 & 3.58 \end{bmatrix}$$

The eigenvalues and eigenvectors are obtained as:

$$\boldsymbol{\lambda} = 10^7 \times \begin{bmatrix} 1.48 &&&&&&&&& \\ & 3.58 &&&&&&&& \\ && 3.58 &&&&&&& \\ &&& 3.58 &&&&&& \\ &&&& 3.58 &&&&& \\ &&&&& 3.58 &&&& \\ &&&&&& 4.38 &&& \\ &&&&&&& 4.38 && \\ &&&&&&&& 9.08 & \\ &&&&&&&&& 9.08 \end{bmatrix}$$

$$\boldsymbol{V} = \begin{bmatrix} 0.44 & 0 & 0 & 0 & 0 & 0 & 0.038 & -0.63 & -0.35 & 0.52 \\ 0 & 0 & 0 & -1 & 0 & 0 & 0 & 0 & 0 & 0 \\ 0.44 & 0 & 0 & 0 & 0 & 0 & 0.61 & -0.16 & 0.59 & -0.22 \\ 0 & 0.185 & -0.2 & 0 & 0 & 0.96 & 0 & 0 & 0 & 0 \\ 0.44 & 0 & 0 & 0 & 0 & 0 & 0.34 & 0.53 & -0.61 & -0.17 \\ 0 & -0.57 & -0.82 & 0 & 0 & -0.06 & 0 & 0 & 0 & 0 \\ 0.44 & 0 & 0 & 0 & 0 & 0 & -0.4 & 0.48 & 0.39 & 0.49 \\ 0 & -0.8 & 0.53 & 0 & 0 & 0.26 & 0 & 0 & 0 & 0 \\ 0.44 & 0 & 0 & 0 & 0 & 0 & -0.59 & -0.23 & -0.03 & -0.63 \\ 0 & 0 & 0 & 0 & 1 & 0 & 0 & 0 & 0 & 0 \end{bmatrix}$$

And

$$\boldsymbol{D} = \boldsymbol{K}^{-1} = \boldsymbol{V}\boldsymbol{\lambda}^{-1}\boldsymbol{V}^T$$

$$\boldsymbol{D} = 10^{-7} \times \begin{bmatrix} 0.27 & 0 & 0.13 & 0 & 0.07 & 0 & 0.07 & 0 & 0.13 & 0 \\ 0 & 0.28 & 0 & 0 & 0 & 0 & 0 & 0 & 0 & 0 \\ 0.13 & 0 & 0.27 & 0 & 0.13 & 0 & 0.07 & 0 & 0.07 & 0 \\ 0 & 0 & 0 & 0.28 & 0 & 0 & 0 & 0 & 0 & 0 \\ 0.07 & 0 & 0.13 & 0 & 0.27 & 0 & 0.13 & 0 & 0.07 & 0 \\ 0 & 0 & 0 & 0 & 0 & 0.28 & 0 & 0 & 0 & 0 \\ 0.07 & 0 & 0.07 & 0 & 0.13 & 0 & 0.27 & 0 & 0.13 & 0 \\ 0 & 0 & 0 & 0 & 0 & 0 & 0 & 0.28 & 0 & 0 \\ 0.13 & 0 & 0.07 & 0 & 0.07 & 0 & 0.13 & 0 & 0.27 & 0 \\ 0 & 0 & 0 & 0 & 0 & 0 & 0 & 0 & 0 & 0.28 \end{bmatrix}$$

According to Eq. (3.2), the degrees of freedom 8 and 12 are included in one block

$$\begin{bmatrix} \Delta_8 \\ \Delta_{12} \\ \cdots \\ \Delta_9 \\ \Delta_{10} \\ \Delta_{11} \end{bmatrix} = \begin{bmatrix} \boldsymbol{D}_{88} & \boldsymbol{D}_{812} & \vdots & \boldsymbol{D}_{89} & \boldsymbol{D}_{810} & \boldsymbol{D}_{811} \\ & \boldsymbol{D}_{1212} & \vdots & \boldsymbol{D}_{912} & \boldsymbol{D}_{1012} & \boldsymbol{D}_{1112} \\ \cdots & \cdots & \vdots & \cdots & \cdots & \cdots \\ & & \vdots & \boldsymbol{D}_{99} & \boldsymbol{D}_{910} & \boldsymbol{D}_{911} \\ & & \vdots & & \boldsymbol{D}_{1010} & \boldsymbol{D}_{1011} \\ Sym & & \vdots & & & \boldsymbol{D}_{1111} \end{bmatrix} \begin{bmatrix} \boldsymbol{F}_8 \\ \boldsymbol{F}_{12} \\ \cdots \\ \boldsymbol{F}_9 \\ \boldsymbol{F}_{10} \\ \boldsymbol{F}_{11} \end{bmatrix}$$

Considering Eq. (3.2), we will have

$$\boldsymbol{D}_{I,I} = \begin{bmatrix} 0.27 & 0 & 0.13 & 0 \\ 0 & 0.28 & 0 & 0 \\ 0.13 & 0 & 0.27 & 0 \\ 0 & 0 & 0 & 0.28 \end{bmatrix} \quad \boldsymbol{D}_{I,II} = \begin{bmatrix} 0.13 & 0 & 0.07 & 0 & 0.07 & 0 \\ 0 & 0 & 0 & 0 & 0 & 0 \\ 0.07 & 0 & 0.07 & 0 & 0.13 & 0 \\ 0 & 0 & 0 & 0 & 0 & 0 \end{bmatrix}$$

$$\boldsymbol{D}_{II,I} = \begin{bmatrix} 0.13 & 0 & 0.07 & 0 \\ 0 & 0 & 0 & 0 \\ 0.07 & 0 & 0.07 & 0 \\ 0 & 0 & 0 & 0 \\ 0.07 & 0 & 0.13 & 0 \\ 0 & 0 & 0 & 0 \end{bmatrix} \quad \boldsymbol{D}_{II,II} = \begin{bmatrix} 0.27 & 0 & 0.13 & 0 & 0.07 & 0 \\ 0 & 0.28 & 0 & 0 & 0 & 0 \\ 0.13 & 0 & 0.27 & 0 & 0.13 & 0 \\ 0 & 0 & 0 & 0.28 & 0 & 0 \\ 0.07 & 0 & 0.13 & 0 & 0.27 & 0 \\ 0 & 0 & 0 & 0 & 0 & 0.28 \end{bmatrix}$$

According to Eqs. (3.3) and (3.4), we obtain

$$\boldsymbol{K}'_{I,I} = 10^7 \times \begin{bmatrix} -2.84 & -0.74 & 2.1 & 0 \\ -0.74 & -0.74 & 0 & 0 \\ 2.1 & 0 & -2.84 & 0.74 \\ 0 & 0 & 0.74 & -0.74 \end{bmatrix}$$

$$\boldsymbol{K}_{I,I} = 10^7 \times \begin{bmatrix} 5.68 & 0 & -2.1 & 0 \\ 0 & 3.58 & 0 & 0 \\ -2.1 & 0 & 5.68 & 0 \\ 0 & 0 & 0 & 3.58 \end{bmatrix}$$

$$\boldsymbol{K}_{I,II} = 10^7 \times \begin{bmatrix} -2.1 & 0 & 0 & 0 & 0 & 0 \\ 0 & 0 & 0 & 0 & 0 & 0 \\ 0 & 0 & 0 & 0 & -2.1 & 0 \\ 0 & 0 & 0 & 0 & 0 & 0 \end{bmatrix} \quad \boldsymbol{K}_{II,I} = 10^7 \times \begin{bmatrix} -2.1 & 0 & 0 & 0 \\ 0 & 0 & 0 & 0 \\ 0 & 0 & 0 & 0 \\ 0 & 0 & 0 & 0 \\ 0 & 0 & -2.1 & 0 \\ 0 & 0 & 0 & 0 \end{bmatrix}$$

$$
\boldsymbol{K}_{II,II} = 10^7 \times \begin{bmatrix} 5.68 & 0 & -2.1 & 0 & 0 & 0 \\ 0 & 3.58 & 0 & 0 & 0 & 0 \\ -2.1 & 0 & 5.68 & 0 & -2.1 & 0 \\ 0 & 0 & 0 & 3.58 & 0 & 0 \\ 0 & 0 & -2.1 & 0 & 5.68 & 0 \\ 0 & 0 & 0 & 0 & 0 & 3.58 \end{bmatrix}
$$

Using Eq. (3.11) displacements of the initial structure are obtained

$$
\begin{bmatrix} \Delta_8 \\ \Delta_9 \\ \Delta_{10} \\ \Delta_{11} \\ \Delta_{12} \end{bmatrix} = \begin{bmatrix} \overset{x}{0.24} & \overset{y}{-0.018} \\ 0.21 & 0.14 \\ 0.17 & -0.016 \\ 0.13 & 0.013 \\ 0.10 & -0.15 \end{bmatrix}
$$

While initially inverting a matrix of dimension 10 was expected, using the changed structure inverting a matrix of dimension 4 is required.

Comment It was mentioned that the original structure (Fig. 3.4) does not hold the condition $\boldsymbol{A} - \boldsymbol{B} \neq \boldsymbol{C}$. It can be shown that by adding only the lateral members 1–8 and 7–12 to the truss, we can generate the form $\boldsymbol{F}(\boldsymbol{A}, \boldsymbol{B}, \boldsymbol{C})$ with the condition $\boldsymbol{A} - \boldsymbol{B} = \boldsymbol{C}$ which is decomposable. But how to understand we can generate such a form by adding the two members before construction of the stiffness matrix? In fact, after adding the lateral members to the truss the difference between, let's say, nodes 8 and 9 (corresponding to blocks $\boldsymbol{A}$ and $\boldsymbol{C}$, respectively) in terms of connected members would be only the member 8–9. Furthermore, this member is corresponding to block $\boldsymbol{B}$ in stiffness matrix as it is connecting the two nodes. Therefore, after adding stiffness of member 8–9 (corresponding to $-\boldsymbol{B}$) to block $\boldsymbol{A}$, we will have $\boldsymbol{A} - \boldsymbol{B} = \boldsymbol{C}$.

Example 2 For the structure shown in the Fig. 3.6, the displacements of the roof are calculated. The elastic modulus for all members is taken as 210 kN/mm^2, and the cross sectional areas are 10 cm^2. The load $\boldsymbol{P}$ is equal to 45 kN. The length of each span is 5 m and the height of each story is 3 m.

The first six stories of the structure are of the same type whereas the last three stories are similar (Fig. 3.6). The stiffness matrix of the structure follows no special pattern. Nevertheless, by converting the structure to the one indicated in Fig. 3.7, the stiffness matrix can be written as

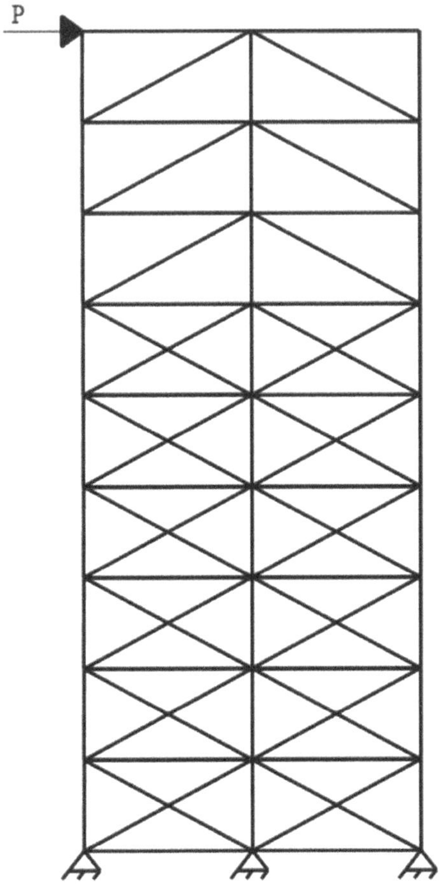

Fig. 3.6 A nine-story structure under lateral load P

$$\boldsymbol{K} = \begin{bmatrix} \boldsymbol{A} & \boldsymbol{B} & & & & \\ \boldsymbol{B} & \boldsymbol{A} & \boldsymbol{B} & & & \\ & \boldsymbol{B} & \boldsymbol{A} & \ddots & & \\ & & \ddots & \ddots & \boldsymbol{B} & \\ & & & \boldsymbol{B} & \boldsymbol{A} & \boldsymbol{B} \\ & & & & \boldsymbol{B} & \boldsymbol{A} \end{bmatrix}$$

Matrix $\boldsymbol{K}$ is a matrix of dimension 54 and the blocks $\boldsymbol{A}$ and $\boldsymbol{B}$ are of dimension 6.

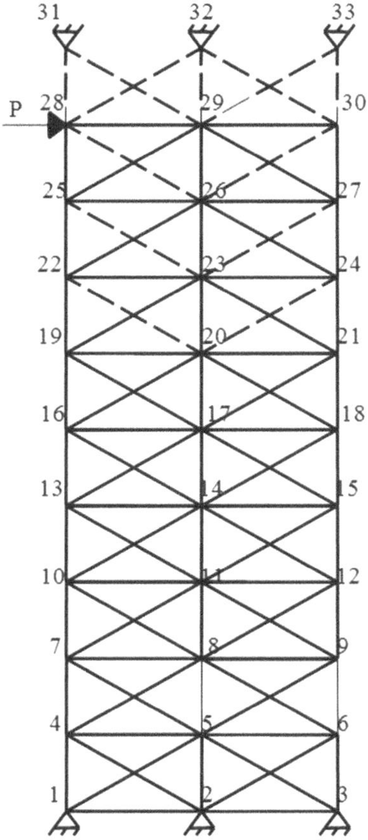

Fig. 3.7 The changed structure obtained by adding members

$$A = 10^7 \begin{bmatrix} 3.58 & 0 & -2.1 & 0 & 0 & 0 \\ 0 & 5.68 & 0 & 0 & 0 & 0 \\ -2.1 & 0 & 7.17 & 0 & -2.1 & 0 \\ 0 & 0 & 0 & 7.17 & 0 & 0 \\ 0 & 0 & -2.1 & 0 & 3.58 & 0 \\ 0 & 0 & 0 & 0 & 0 & 5.68 \end{bmatrix}$$

$$B = 10^7 \begin{bmatrix} 0 & 0 & -0.74 & 0.74 & 0 & 0 \\ 0 & -2.1 & 0.74 & -0.74 & 0 & 0 \\ -0.74 & -0.74 & 0 & 0 & -0.74 & 0.74 \\ -0.74 & -0.74 & 0 & -2.1 & 0.74 & -0.74 \\ 0 & 0 & -0.74 & -0.74 & 0 & 0 \\ 0 & 0 & -0.74 & -0.74 & 0 & -2.1 \end{bmatrix}$$

Furthermore, Matrices $\boldsymbol{A}$ and $\boldsymbol{B}$ follow the Form III pattern of Ref. [3], and then in finding their eigenvalues, the dimension will be half. Thus, to calculate the eigenvalues of the changed structure, one should solve matrices of dimension 3. Since $\boldsymbol{K}'_{1,1}$ is a matrix of dimension 20, the initial structure is solved by inverting matrices of dimension 20 and 3 (rather than inverting a matrix of dimension 54). Displacements of the roof in the initial structure are obtained using Eq. (3.11).

$$\begin{bmatrix} \Delta_{28} \\ \Delta_{29} \\ \Delta_{30} \end{bmatrix} = \begin{bmatrix} \overset{x}{0.98} & \overset{y}{-0.023} \\ 0.84 & -0.033 \\ 0.76 & -0.038 \end{bmatrix}$$

Example 3 The structure in Fig. 3.8 follows the pattern of the strong Cartesian product of C_5 and P_2. Cartesian product of path C_n and cycle P_m can be considered as repeated C_n cycles that are connected together by the means of P_m paths at the location of nodes of the C_n.

Two equilateral polygons with 5 edges form the structure according to Fig. 3.9. The distances of the external and internal nodes from the center are 3 m and 1.5 m,

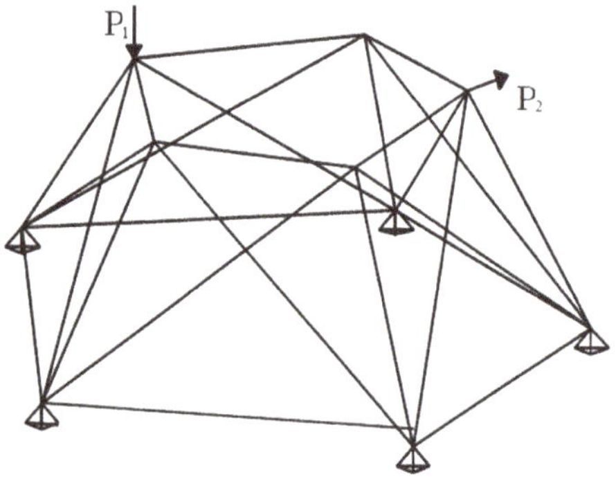

Fig. 3.8 A 3-dimensional truss structure

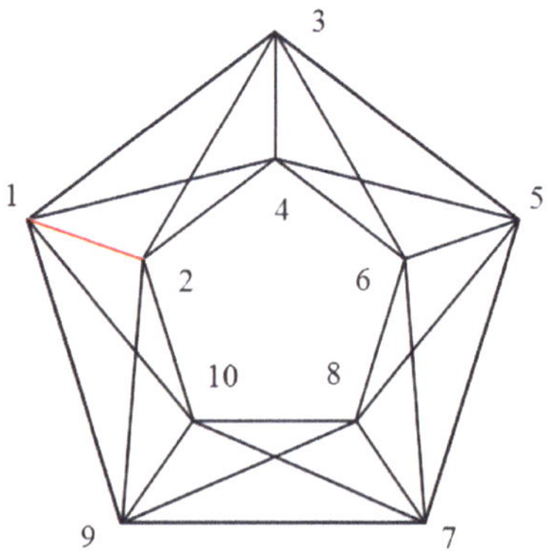

Fig. 3.9 The top view of the truss

respectively. The nodes 2, 4, 6, 8 and 10 are located at the height of 1.5 m relative to the external nodes. The cross sectional area for all members is taken as 5 cm^2, except for the member between nodes 1 and 2 for which the cross sectional area is 2.5 cm^2. For all members the elastic modulus is taken as 200 kN/mm^2. The load $\boldsymbol{P}_1$ at node 2 is equal to 30 kN and the load $\boldsymbol{P}_2$ at the node 6 is equal to 20 kN.

The cross sectional area for the member between nodes 1 and 2 is taken similar to the other members. Therefore, the stiffness matrix, constructed in a cylindrical coordinate system, will have a decomposable form:

$$\boldsymbol{K} = \boldsymbol{A}_1 \otimes \boldsymbol{B}_1 + \boldsymbol{A}_2 \otimes \boldsymbol{B}_2 + \boldsymbol{A}_3 \otimes \boldsymbol{B}_3$$

where

$$\boldsymbol{A}_1 = \begin{bmatrix} 1 & & & & \\ & 1 & & & \\ & & 1 & & \\ & & & 1 & \\ & & & & 1 \end{bmatrix} \quad \boldsymbol{A}_2 = \boldsymbol{A}_3^T = \begin{bmatrix} 0 & 0 & & & 1 \\ 1 & 0 & 0 & & \\ & 1 & 0 & 0 & \\ & & 1 & 0 & 0 \\ 0 & & & 1 & 0 \end{bmatrix}$$

$$\boldsymbol{B}_1 = 10^5 \begin{bmatrix} 0.64 & 0 & -0.18 \\ 0 & 1.2 & 0 \\ -0.18 & 0 & 0.36 \end{bmatrix} \quad \boldsymbol{B}_2 = \boldsymbol{B}_3^T = 10^4 \begin{bmatrix} 1.95 & 2.69 & 0 \\ -2.69 & -3.71 & 0 \\ 0 & 0 & 0 \end{bmatrix}$$

Eigenvalues of matrix $\boldsymbol{A}_2$ will be

$$\begin{aligned} eig(\boldsymbol{A}_2) &= \left\{1, e^{\frac{2\pi i}{5}}, e^{\frac{4\pi i}{5}}, e^{\frac{6\pi i}{5}}, e^{\frac{8\pi i}{5}}\right\} \\ &= \{1, 0.31 + 0.95i, -0.81 + 0.58i, -0.81 - 0.58i, 0.31 - 0.95i\} \end{aligned}$$

The eigenvalues of the matrix $\boldsymbol{K}$ can be written as

$$eig(\boldsymbol{K}) = \bigcup_{i=1}^{5} \left[eig\left\{\boldsymbol{B}_1 + \lambda_i(\boldsymbol{A}_2)\boldsymbol{B}_2 + \lambda_i\left(\boldsymbol{A}_2^T\right)\boldsymbol{B}_2^T\right\}_3\right]$$

And we will have

$$\begin{aligned} \lambda_K = {} & 10^5\{0.12, 0.12, 0.21, 0.21, 0.32, 0.46, 0.49, 0.49, 0.50, 0.50, 1.08, \\ & 1.41, 1.41, 1.87, 1.87\} \end{aligned}$$

Eigenvectors of matrix $\boldsymbol{K}$ are calculated through $\boldsymbol{u} \otimes \boldsymbol{v}$ where $\boldsymbol{v}$ and $\boldsymbol{u}$ were already introduced in Chap. 2.

Stiffness matrix of the changed structure is inverted through $\boldsymbol{D} = \boldsymbol{K}^{-1} = \boldsymbol{V}\boldsymbol{\lambda}^{-1}\boldsymbol{V}^T$. Constructing the stiffness matrix of the initial structure in the cylindrical coordinate system, matrix $\boldsymbol{K}'_{I,I}$ is obtained as:

$$\boldsymbol{K}'_{I,I} = 10^5 \begin{bmatrix} -11.76 & 0.02 & 11.81 \\ 0.02 & 0 & -0.02 \\ 11.81 & -0.02 & -11.86 \end{bmatrix}$$

Using Eq. (3.11) displacements of the structure in the Cylindrical coordinate system is obtained:

$$\begin{bmatrix} \Delta_8 \\ \Delta_9 \\ \Delta_{10} \\ \Delta_{11} \\ \Delta_{12} \end{bmatrix} = \begin{bmatrix} \overset{\rho}{0.57} & \overset{\theta}{0.01} & \overset{z}{1.12} \\ -0.26 & 0.11 & -0.14 \\ 0.46 & 0.25 & 0.25 \\ -0.28 & 0.18 & -0.15 \\ -0.21 & -0.16 & -0.12 \end{bmatrix}$$

By transforming the displacements from cylindrical coordinate system to the Cartesian coordinate system, we will have:

$$\begin{bmatrix} \Delta_8 \\ \Delta_9 \\ \Delta_{10} \\ \Delta_{11} \\ \Delta_{12} \end{bmatrix} = \begin{bmatrix} \overset{x}{-0.01} & \overset{y}{-0.57} & \overset{z}{1.12} \\ -0.31 & 0.02 & -0.14 \\ 0.44 & -0.25 & 0.25 \\ 0.32 & 0.15 & -0.15 \\ 0.23 & 0.07 & -0.12 \end{bmatrix}$$

Example 4 Consider the 29-story telecommunication tower indicated in Fig. 3.10.

The tower has 1392 degrees of freedom and includes 4 similar sides (Fig. 3.11). Sections of the first 15 stories (Fig. 3.12) and sections of the last 14 stories (Fig. 3.13) are similar.

Since the sections of the last 14 stories are not symmetric, the stiffness matrix of the structure is not decomposable. By adding proper members to the sections of the last 14 stories (Fig. 3.14) the pattern of the stiffness matrix becomes as $\boldsymbol{G}_n(\boldsymbol{C}_m, \boldsymbol{B}_m, \boldsymbol{C}_m)$ wherein n = 4 and m = 348.

Therefore, instead of a matrix of dimension 1392, matrices of dimension 348 should be inverted. Matrix $\boldsymbol{K}'_{I,I}$, related to the effects of added members, is of dimension $14 \times 2 \times 3 = 84$ where 14 is the number of stories, 2 is number of the changed nodes in each story, and 3 is number of degrees of freedom at each node.

3.3 Analysis of Near-Regular Structures with Additional Members Using Force Methods

In this section a simplified analysis is presented for those non-regular structures which are obtained by addition or removal of some members to regular structural models.

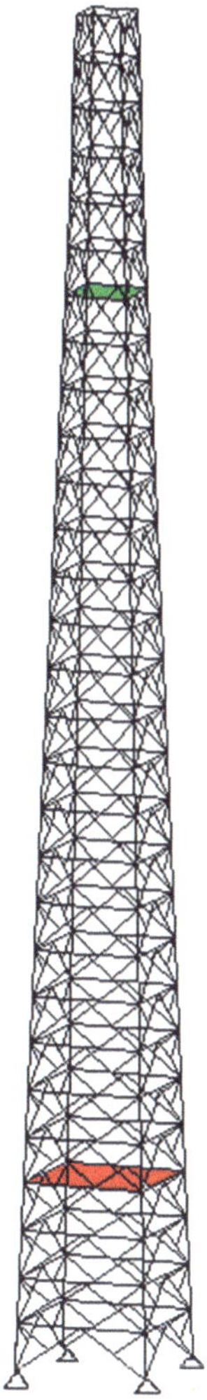

Fig. 3.10 A 29-story communication tower

we use the force method, however, instead of selecting a statically determinate basic structure we employ the regular part of the structure as the basic structure. Here an irregular structure is divided into two sets, namely "the regular part of the structure" and "the excessive members". Regular part refers to the structure for which the inverse of the stiffness matrix can be obtained by the previously developed simplified methods, and excessive members refer to those which cause the non-regularity of the regular structure [4].

To demonstrate this problem, consider the truss shown in Fig. 3.15a. This structure consists of a regular part $P_4 \boxtimes P_{10}$ as shown in Fig. 3.15b and has become irregular because of having additional 10 bars. The main aim is to decompose these two parts

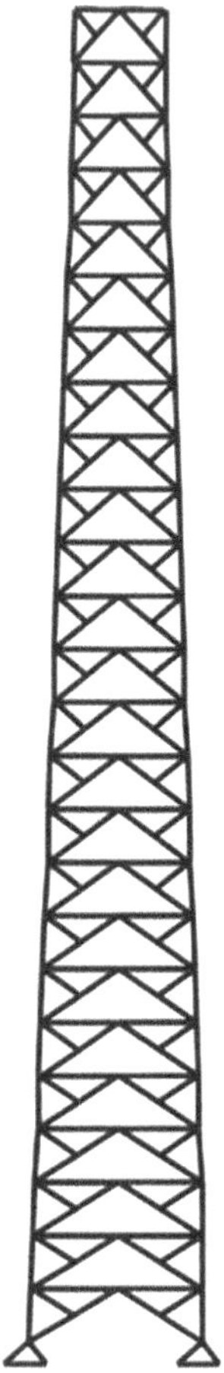

Fig. 3.11 Front view of the communication tower

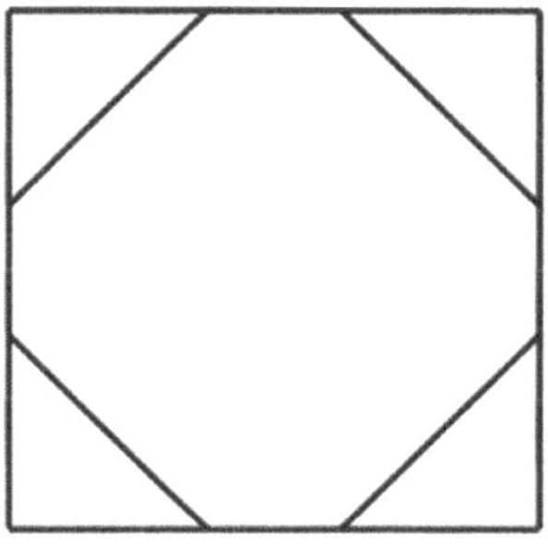

Fig. 3.12 Section of the first 15 stories. The section is in red in Fig. 3.10

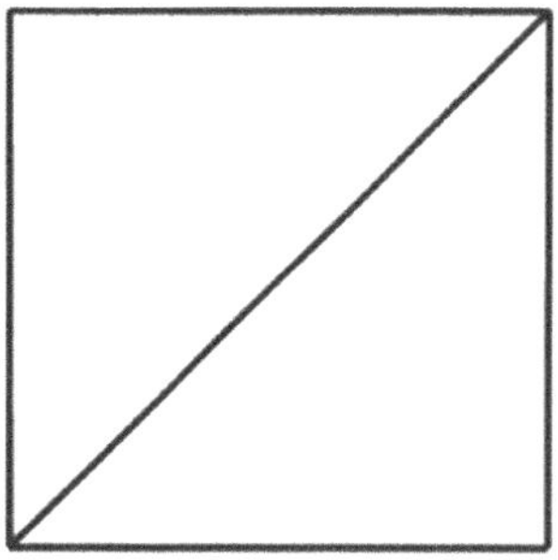

Fig. 3.13 Section of the second 14 stories. The section is in green in Fig. 3.10

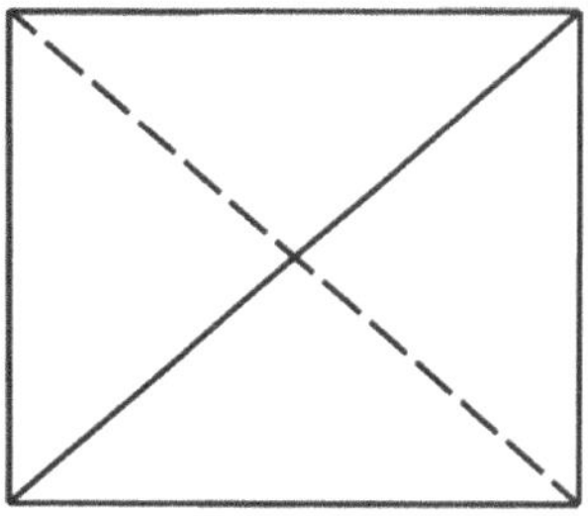

Fig. 3.14 Section of the last 14 stories after addition of a member

in order to arrive at the analysis of the irregular structure using the results of the analysis of the regular part. In Fig. 3.15c the positions of the excessive members are highlighted, where the regular part is shown in broken lines.

It should be mentioned that for some regular structures the stiffness matrices can be formed in special block forms, known as the canonical forms [5]. Here we assume that only the members cause irregularity and no additional nodes are present except those of the regular part, i.e. the nodes of the two ends of each excessive members are in the regular part of the structure.

At the beginning, methods suggested are presented for the formation of the matrices required in the force method. Obviously one can also obtain these matrices by other approaches.

The present method consists of two groups of structures as described in the following:

The first group is related to the analysis of those structures in which the excessive members have caused the irregularity. The second group is about those structures which require addition of some members to alter the irregular structure to a regular one. In this case, by assuming pairs of members with two identical modulus of elasticity having positive and negative signs are added to those places where we need to have members to make the irregular structure into regular one. In this case, the members with negative sign will be treated as the excessive members.

In the above force method, the internal forces of the excessive members will be considered as redundants, and the corresponding forces will be applied at the regular part of the structure as external loads to incorporate the effect of such members. Thus

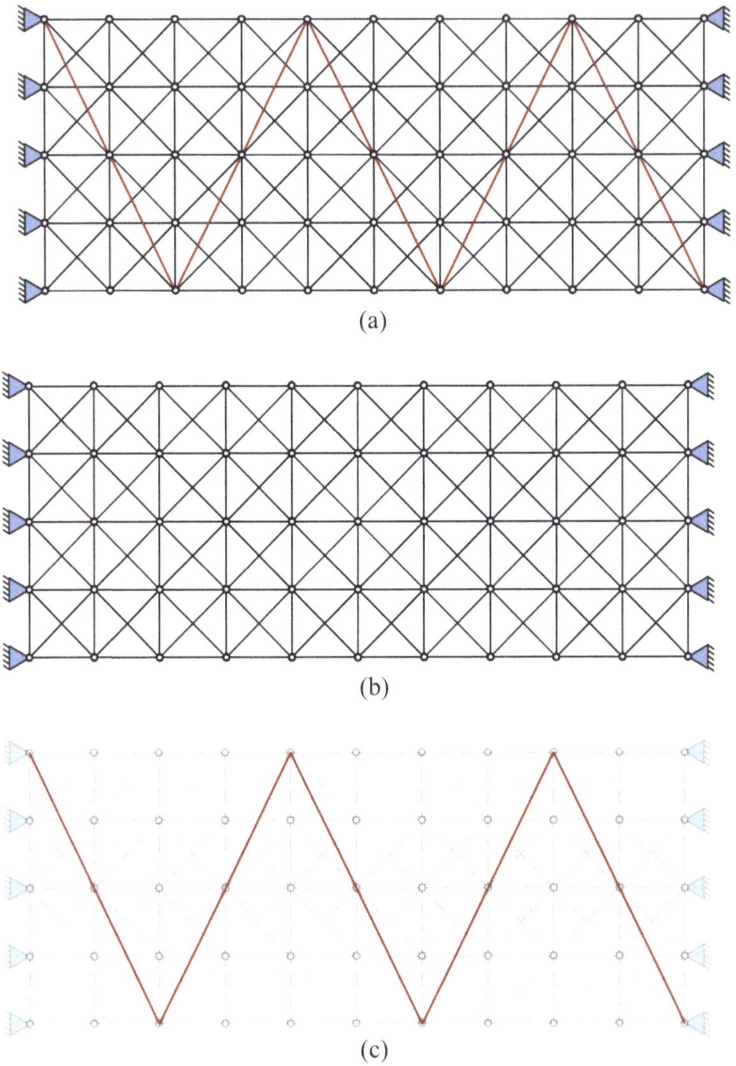

Fig. 3.15 **a** An irregular truss, **b** the regular part as the strong Cartesian product $P_4 \boxtimes P_{10}$, **c** the excessive members being highlighted

the regular part will be the main structure to be analyzed. This means we analyze the irregular structure by considering the regular part and adding the effect of the internal forces of the complementary members as external loads.

Here we assume that the removal of the excessive members will leave the structure geometrically stable, and considering the topology of the regular structures this assumption is quite logical.

In each remaining section first the formulation will be presented, and then through a simple example the process of analysis will be described in a step by step manner. Then by some practical examples, the efficiency of the method will be demonstrated.

First the formation of the flexibility matrix is described. Again it should be mentioned that this matrix can also be formed using some other methods. In Sect. 3.3.5 a method is presented which can be considered as the concise form of the next section. In this method the relations are simplified, and a simple formation of the flexibility matrix has become feasible.

3.3.1 Formulation of the Flexibility Matrix

In this section a method is presented for the formation of the flexibility matrix $\boldsymbol{B}$ in the following form [6]:

$$\boldsymbol{B} = \left[\, \boldsymbol{B}_0 \; \boldsymbol{B}_1 \,\right] \tag{3.13}$$

where the ith column of $\boldsymbol{B}_0$ is a vector of internal force of the structure under a unit value of a load applied at the ith DOF of the structure ($\boldsymbol{P}_i = 1$), and the ith column of $\boldsymbol{B}_1$ is a vector containing the internal forces of the structure under the unit load applied at the position of the ith redundant of the ($\boldsymbol{X}_i = 1$) structure.

According to the above definitions for the formation a matrix $\boldsymbol{B}$ we are looking for a method by means of which having the externally applied loads of the structure we find the internal forces of the members. In the following a method is presented for this problem using the equilibrium matrix, though one can also find this employing the existing traditional method.

In order to calculate the internal forces of the regular structure under the external loading we proceed as the following:

In the global coordinate system we have

$$\boldsymbol{S}.\boldsymbol{\Delta} = \boldsymbol{P} \quad \Rightarrow \quad \boldsymbol{\Delta} = \boldsymbol{S}^{-1}.\boldsymbol{P} \tag{3.14}$$

where $\boldsymbol{S}^{-1}$ is the inverse of the stiffness matrix of the DOFs of the regular part of the structure. Using the theorems previously developed for the block matrices, $\boldsymbol{S}^{-1}$ can be formed using the blocks constituting $\boldsymbol{S}$. This matrix can be obtained using some concepts of graph products or employing concepts from group theory.

According to the definition of equilibrium matrices of the members of the structure and Eq. (3.14), in general, the following form can be written for the deformation of the local coordinate systems of the members of the regular structure:

$$\boldsymbol{\delta} = \boldsymbol{A}^t.\boldsymbol{\Delta} = \boldsymbol{A}^t.\boldsymbol{S}^{-1}.\boldsymbol{P} \tag{3.15}$$

where $\boldsymbol{A}$ is the equilibrium matrix of the regular structure. Considering the equilibrium equations in the local coordinate system and Eq. (3.15), in general the vector of internal forces of the members of the regular structure under the action of an imaginary external unit load an be obtained as:

$$\boldsymbol{Q}_0 = s.\delta = s.\left(\boldsymbol{A}^t.S^{-1}.\boldsymbol{P}\right) = \left(s.\boldsymbol{A}^t.S^{-1}\right).\boldsymbol{P} \tag{3.16}$$

Here $\boldsymbol{s}$ is the block diagonal matrix containing the stiffness of the members of the regular part of the structure. Therefore the vector of internal forces of a regular structure can be obtained having the external forces in the following form:

$$\boldsymbol{Q}_0 = \boldsymbol{R}.\boldsymbol{P}; \quad \boldsymbol{R} = s.\boldsymbol{A}^t.S^{-1} \tag{3.17}$$

If $\overline{\boldsymbol{X}}$ contains the internal forces of the members in the global coordinate system and $\boldsymbol{P}$ is the external force vector of the structure, then the internal forces of the members of the regular structure when part of it is irregular can be obtained as:

$$\boldsymbol{Q}_1 = \boldsymbol{R}.\boldsymbol{P} + \boldsymbol{R}.\overline{\boldsymbol{X}} \tag{3.18}$$

Thus for the analysis of irregular structure discussed in this section, $\boldsymbol{Q}_1$ is the vector of internal forces of the regular structure. The vectors $\boldsymbol{P}$ and $\boldsymbol{X}$ can be expressed as:

$$\overline{\boldsymbol{X}} = \boldsymbol{N}.\boldsymbol{X}; \quad \boldsymbol{P} = \boldsymbol{I}.\boldsymbol{P} \tag{3.19}$$

$\boldsymbol{I}$ is a unit matrix and $\boldsymbol{N}$ is a matrix for transforming the local coordinate system to the global coordinate system. $\boldsymbol{X}$ is the internal force vector of excessive members.

Here the method for the formation of $\boldsymbol{A}$ and $\boldsymbol{N}$ is explained. If $\boldsymbol{A}_T$ is the equilibrium matrix of irregular structure, then by partitioning according to the numbers of internal forces of the excessive members, the matrices $\boldsymbol{A}$ and $\boldsymbol{N}$ can be formed as:

$$\boldsymbol{A}_T = \left[\boldsymbol{A}\ \boldsymbol{N}\right] \tag{3.20}$$

where $\boldsymbol{A}$ is the equilibrium matrix of the regular structure. In Eqs. (3.6) and (3.7) by taking the common factor and extracting the vector of the assumed forces we have:

$$\boldsymbol{Q}_1 = \boldsymbol{R}.\left[\boldsymbol{I}\ \boldsymbol{N}\right].\begin{bmatrix} \boldsymbol{P} \\ \boldsymbol{X} \end{bmatrix} \tag{3.21}$$

In general case the internal forces of the irregular structure will be as follows:

$$\boldsymbol{Q} = \begin{bmatrix} \boldsymbol{Q}_1 \\ \boldsymbol{X} \end{bmatrix} \tag{3.22}$$

Therefore adding $\boldsymbol{X}$ to Eq. (3.21) the matrix $\boldsymbol{Q}$ can be written as

$$\boldsymbol{Q} = \begin{bmatrix} \boldsymbol{Q}_1 \\ \boldsymbol{X} \end{bmatrix} = \begin{bmatrix} \boldsymbol{R} & \boldsymbol{R.N} \\ \boldsymbol{Z} & \boldsymbol{I} \end{bmatrix} . \begin{bmatrix} \boldsymbol{P} \\ \boldsymbol{X} \end{bmatrix} \tag{3.23}$$

Now the matrices $\boldsymbol{B}_0$ and $\boldsymbol{B}_1$ can be formed by partitioning the above matrix according to the numbering of the internal forces of the excessive members of the structure in the following form:

$$\boldsymbol{B}_0 = \begin{bmatrix} \boldsymbol{R} \\ \boldsymbol{Z} \end{bmatrix}; \quad \boldsymbol{B}_1 = \begin{bmatrix} \boldsymbol{R.N} \\ \boldsymbol{I} \end{bmatrix} \tag{3.24}$$

In the above equations $\boldsymbol{Z}$ is a matrix of zeros with dimension $t \times k$ and $\boldsymbol{I}$ is a unit matrix of dimension $t \times t$. The matrices $\boldsymbol{B}_0$ and $\boldsymbol{B}_1$ have dimensions $(e + t) \times k$ and $(e + t) \times t$, respectively. t is the total number of internal forces of the excessive members, k is the DOFs of the irregular structure in global coordinate system and e is the number of internal forces of the regular structure.

In this method we need to form the matrix $\boldsymbol{A}_T$ and in the subsequent section a simple method will be presented for this formation.

The formation of the matrix $\boldsymbol{B}$ can be summarized as follows:

Step 1: Form the matrices $\boldsymbol{S}^{-1}$ and s for the regular structure.

Step 2: Form the matrix $\boldsymbol{A}_T$ for all the members of the structure consisting of regular and excessive members.

Step 3: Partition $\boldsymbol{A}_T$ using Eq. (3.20) and form the matrices $\boldsymbol{A}$ and $\boldsymbol{N}$.

Step 4: Calculate the matrix $\boldsymbol{R}$ using Eq. (3.17).

Step 5: Calculate the matrices $\boldsymbol{B}_0$ and $\boldsymbol{B}_1$ using Eq. (3.24).

A simple method for the formation of the matrix $\boldsymbol{A}_T$

A general method for the formation of the equilibrium matrix consists of writing equilibrium of the forces at the nodes of the structure. For a quick calculation of the matrix $\boldsymbol{A}_T$ one can assemble the rotation matrices of the members of the irregular structure. Then it can be partitioned using the relationship presented in the previous section. In the following the approach for positioning the rotation matrices of the members in each column of the equilibrium matrix is illustrated. For the formation of the equilibrium matrix $\boldsymbol{A}_T$ of the irregular matrix we perform the following process:

If we consider i as the nodal DOFs of the assumed member j in the local coordinate system, and r are the nodal DOFs of the assumed member j in the global coordinate system, then the columns corresponding to i in the matrix $\boldsymbol{A}_T$ will be as follows:

$$\boldsymbol{A}_T(r, i) = \boldsymbol{T'}^{t}_{j} \tag{3.25}$$

The remaining rows of these columns are zero. We repeat this process for all the members of the irregular structure. $\boldsymbol{T}_j$ is the modified rotation matrix of the jth member. This matrix can be represented as follows:

$$\text{Space truss member}$$
$$\boldsymbol{T}_j = \left[\boldsymbol{T}_1\ -\boldsymbol{T}_1\right],\quad \boldsymbol{T}_1 = \left[\cos\alpha\ \cos\beta\ \cos\gamma\right]$$
$$\text{Planar frame member}$$
$$\boldsymbol{T}_j = \left[\boldsymbol{T}_1\ \boldsymbol{T}_3\right],\quad \boldsymbol{T}_3 = s_1^{-1}.s_2.\boldsymbol{T}_2,\quad s_j = \begin{bmatrix} s_1 & s_2 \\ s_2 & s_1 \end{bmatrix} \tag{3.26}$$
$$\boldsymbol{T}_1 = \begin{bmatrix} \cos\alpha & \sin\alpha & 0 \\ -\sin\alpha & \cos\alpha & 0 \\ 0 & 0 & 1 \end{bmatrix},\quad \boldsymbol{T}_2 = \begin{bmatrix} -\cos\alpha & -\sin\alpha & 0 \\ \sin\alpha & -\cos\alpha & 0 \\ 0 & 0 & 1 \end{bmatrix}$$

s_j is the jth block of the stiffness matrix s. α, β and γ are the angles with the x, y and z axis, respectively.

Similar to Eq. (3.20) the above matrix can be transformed to $\boldsymbol{A}$ and $\boldsymbol{N}$ by partitioning and numbering the internal forces of the excessive members which are separated from the structure.

The algorithm can be summarized as:

Step 1: Formation of the matrices $\boldsymbol{T}_j$ for members of the irregular structure.

Step 2: Formation of the equilibrium matrix $\boldsymbol{A}_T$ by assembling the rotation matrices of the irregular matrix using Eq. (3.25).

Step 3: Formation of the matrices $\boldsymbol{A}$ and $\boldsymbol{N}$ by partitioning of the matrix $\boldsymbol{A}_T$.

3.3.2 Analysis of Regular Structures with Excessive Members

In this section the analysis of those structures for which the irregularity is produced by excessive members is studied. Here the force method is used for the analysis, with the only difference that instead of removing member to obtain a statically determinate structure, members are removed to transform the structure into a regular one. Here the relationships required for the force method are presented. Base on the concepts of the force method, the internal forces of the members of the irregular structure can be expressed as:

$$\left[\boldsymbol{Q}\right] = [\boldsymbol{B}].\begin{bmatrix} \boldsymbol{P} \\ \boldsymbol{X} \end{bmatrix} \tag{3.27}$$

After the formation of the matrix $\boldsymbol{B}$ which was described in Sect. 3.3.1, one can calculate the internal forces of the excessive members using the following relationships:

$$\boldsymbol{D}_2 = \boldsymbol{B}_1^t.\boldsymbol{F}.\boldsymbol{B}_1;\quad \boldsymbol{D}_1 = \boldsymbol{B}_1^t.\boldsymbol{F}.\boldsymbol{B}_0 \tag{3.28}$$

$$\boldsymbol{X} = -\boldsymbol{D}_2^{-1}.\boldsymbol{D}_1.\boldsymbol{P} \tag{3.29}$$

Here $\boldsymbol{F}$ is a block matrix of dimension $(e+t) \times (e+t)$ and contains all the flexibility matrices of the members of the irregular structure. The matrix $\boldsymbol{D}_1$ is of dimension $t \times k$ and the matrix $\boldsymbol{D}_2$ is of dimension $t \times t$. This means that for calculating the internal forces of the excessive members, only the inverse of a matrix of dimension t is needed.

At the end, the forces of $\boldsymbol{X}$ are added to the external force vector $\boldsymbol{P}$ denoted by $\boldsymbol{P}^*$ which is defined as the equivalent external load of the regular structure. According to this, the displacements of the structure can be obtained by the inverse of the stiffness matrix of the regular structure as follows:

$$\boldsymbol{P}^* = \boldsymbol{P} + \boldsymbol{N}.\boldsymbol{X} \tag{3.30}$$

$$\boldsymbol{\Delta} = \boldsymbol{S}^{-1}.\boldsymbol{P}^* \tag{3.31}$$

where $\boldsymbol{S}^{-1}$ is the inverse of the stiffness matrix of the DOFs of the regular structure and can be obtained using the existing methods [5, 7]. The vector $\boldsymbol{\Delta}$ contains the displacements of the irregular structure.

The matrix $\boldsymbol{N}$ is the transformation matrix of the internal forces in excessive members from local to global coordinate systems. A simple method for the formation of this matrix is given in Eq. (3.34) of Sect. 3.3.5.

Summary of the algorithm

Step 1: Numbering the DOFs, nodes and members of the irregular structure and formation of the external force vector $\boldsymbol{P}$.

Step 2: Formation of the matrices $\boldsymbol{S}^{-1}$ and $\boldsymbol{s}$.

Step 3: Formation of the equilibrium matrix of the irregular structure using Eq. (3.25).

Step 4: Calculation of the $\boldsymbol{B}_0$ and $\boldsymbol{B}_1$ matrices using Eq. (3.24). The steps 3 and 4 can also be performed simultaneously using Eqs. (3.34) and (3.35) of the Sect. 3.3.5.

Step 5: Formation of the flexibility matrix F in a block diagonal form for all the members of the irregular structure.

Step 6: Calculation of the matrices D_1 and D_2 using Eq. (3.28).

Step 7: Calculation of the vector X using Eq. (3.29).

Step 8: Calculation of the equivalent external load of the regular structure using Eq. (3.30).

Step 9: Calculation of the nodal displacements of the irregular structure using Eq. (3.31).

The above explanations are further explained through the following simple example.

Investigation of a simple example

For the 10-bar truss shown in Fig. 3.16, deleting member 10, the structures become regular. Here using the force method, the internal force of the member 10 is calculated and as an additional force it is added to the external forces. Then the regular structure is analyzed with the new loads.

It should be noted that in the standard force method the basic structure is selected for a redundant structure is often statically determinate. For the structure of this

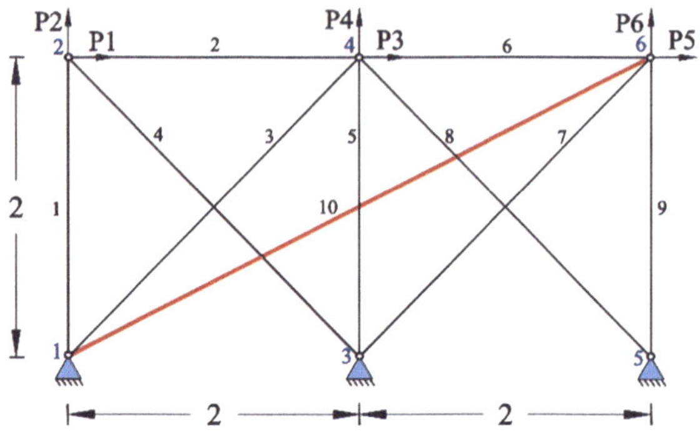

Fig. 3.16 A 10-bar truss transformable to a regular structure

example we have 4 statical indeterminacy and 4 redundants should be chosen. However, in our approach the basic structure is selected as a regular structure which is not necessarily statically determinate.

In this example, EA is assumed to be unit for all the members and the external load vector is as follows:

$$\boldsymbol{P} = \begin{bmatrix} 10\ 0\ 0\ 20\ 0\ 0 \end{bmatrix}^t$$

One of the methods for the formation of the equilibrium matrix is to use the equilibrium equations of forces at the nodes and formation of the matrix of the coefficients of the forces. In this example the equilibrium matrix of the irregular structure is calculated using this approach. The obtained equilibrium matrix is partitioned into A and N using Eq. (3.20).

The force equilibrium equations will be as follows:

$$\begin{aligned} &P_1 = -Q_2 - \sqrt{2}/2Q_4, && P_2 = Q_1 + \sqrt{2}/2Q_4 \\ &P_3 = Q_2 + \sqrt{2}/2Q_3 - Q_6 - \sqrt{2}/2Q_8, && P_4 = \sqrt{2}/2Q_3 + Q_5 + \sqrt{2}/2Q_8 \\ &P_5 = Q_6 + \sqrt{2}/2Q_7 + 2/\sqrt{5}Q_{10}, && P_6 = \sqrt{2}/2Q_7 + Q_9 + 1/\sqrt{5}Q_{10} \end{aligned}$$

The relation between the equilibrium matrix $\boldsymbol{A}$ and the vector of external and internal forces of the structure can be written as:

$$\boldsymbol{P} = \boldsymbol{A}.\boldsymbol{Q} \tag{3.32}$$

In this way the matrix $\boldsymbol{A}$ and the partitioning considering the excessive member 10 will be as follows:

$$
A_T = \begin{bmatrix} A & N \end{bmatrix} = \begin{bmatrix} 0 & -1 & 0 & -0.7071 & 0 & 0 & 0 & 0 & 0 & 0 \\ 1 & 0 & 0 & 0.7071 & 0 & 0 & 0 & 0 & 0 & 0 \\ 0 & 1 & 0.7071 & 0 & 0 & -1 & 0 & -0.7071 & 0 & 0 \\ 0 & 0 & 0.7071 & 0 & 1 & 0 & 0 & 0.7071 & 0 & 0 \\ 0 & 0 & 0 & 0 & 0 & 1 & 0.7071 & 0 & 0 & 0.9844 \\ 0 & 0 & 0 & 0 & 0 & 0 & 0.7071 & 0 & 1 & 0.4472 \end{bmatrix}
$$

Thus the matrix N will be as

$$
N = \begin{bmatrix} 0\ 0\ 0\ 0\ 0.9844\ 0.4472 \end{bmatrix}^t
$$

The stiffness matrix of the regular structure by elimination of the member 10 will become:

$$
S = \begin{bmatrix} 0.6767 & -0.1767 & -0.5 & 0 & 0 & 0 \\ -0.1767 & 0.6767 & 0 & 0 & 0 & 0 \\ -0.5 & 0 & 1.3535 & 0 & -0.5 & 0 \\ 0 & 0 & 0 & 0.8535 & 0 & 0 \\ 0 & 0 & -0.5 & 0 & 0.6767 & 0.1767 \\ 0 & 0 & 0 & 0 & 0.1767 & 0.6767 \end{bmatrix}
$$

Replacing two columns 5 and 6 with columns 3 and 4, and also their corresponding rows, This matrix will get Form III pattern and calculating its eigenvalues leads to the formation of S^{-1}. Here, s is a block diagonal matrix having the stiffness of the members of the regular structure. Since the structure is a truss, therefore this matrix becomes a diagonal one.

$$
s = diag\{0.5, 0.5, 0.3535, 0.3535, 0.5, 0.5, 0.3535, 0.3535, 0.5\}
$$

In this relation $diag$ represents a block diagonal matrix.

Substituting the above matrices in Eq. (3.24) leads to the formation of B_0 and B_1 matrices.

$$
B_0 = \begin{bmatrix} 0.3535 & 0.8311 & 0.1846 & 0 & 0.1464 & -0.0382 \\ -0.6464 & -0.1688 & 0.1846 & 0 & 0.1464 & -0.0382 \\ 0.3535 & 0.0923 & 0.4459 & 0.2928 & 0.3535 & -0.0923 \\ -0.5 & 0.2387 & -0.2612 & 0 & -0.2071 & 0.0540 \\ 0 & 0 & 0 & 0.5857 & 0 & 0 \\ -0.1464 & -0.0382 & -0.1847 & 0 & 0.6464 & -0.1688 \\ 0.2071 & 0.0540 & 0.2612 & 0 & 0.5 & 0.2387 \\ -0.3535 & 0.0923 & -0.4459 & 0.2928 & -0.3535 & 0.0923 \\ -0.1464 & -0.0382 & -0.1847 & 0 & -0.3535 & 0.8311 \\ 0 & 0 & 0 & 0 & 0 & 0 \end{bmatrix}
$$

$$B_1 = \begin{bmatrix} -0.1138 & -0.1138 & -0.2749 & 0.1610 & 0 & -0.5026 & -0.5540 & 0.2749 & -0.0557 & 1 \end{bmatrix}^t$$

The flexibility matrix of the irregular structure F in general is a block diagonal matrix. Since the considered structure is a truss, thus this matrix has numerical values in its diagonal.

$$F = diag\left\{2, 2, 2\sqrt{2}, 2\sqrt{2}, 2, 2, 2\sqrt{2}, 2\sqrt{2}, 2, 2\sqrt{5}\right\}$$

Using Eq. (3.28) the matrices D_1 and D_2 are formed as follows:

$$D_1 = \begin{bmatrix} -0.8719 & -0.2277 & -1.0997 & 0 & -2.1050 & -0.1109 \end{bmatrix}; \quad D_2 = [6.4045]$$

Now employing Eq. (3.29), the internal forces of the excessive members are calculated as:

$$X = -D_2^{-1}.D_1.P = -\frac{1}{6.4045} \times \begin{bmatrix} \overset{1}{-0.8719} \ldots \overset{4}{0} \ldots \end{bmatrix} . \begin{bmatrix} 10 & 0 & 0 & 20 & 0 & 0 \end{bmatrix}^t$$
$$= -1.3624$$

Substituting the values of X in Eq. (3.30), the vector P^* can be obtained. Now according to Eq. (3.31), multiplying this vector with S^{-1}, the displacement vector of the nodal forces of the irregular structure can be obtained.

$$P^* = \{10, 0, 0, 20, 1.2177, 0.6089\}^t$$
$$\Delta = \left\{ 25.8839 \; 6.7609 \; 12.6449 \; 23.4314 \; 8.3473 \; -3.0400 \right\}^t$$

It should be noted that the aim of this example was the explanation of the method by means of a simple example and for showing the capabilities of the presented method is not sufficient. The reduction in dimensions of the matrices achieved by the present method and the speed of calculation will be illustrated in Sect. 3.3.4.

3.3.3 Analysis of Regular Structures with Some Missing Members

In this section we consider those structures which need addition of some members to become a regular one. Obviously the method presented in the previous section can not be applied directly for these structures. However, for transforming this case to the previous one, a pair of members with equal modulus of elasticity having different signs, are added where we have lack of members for regularity. In the next step the members with negative modulus of elasticity are considered as excessive members and separated from the structure. The remaining process of the analysis is the same as the previous case. The internal forces of the members with negative modulus of elasticity are calculated and added to the external forces. Then the regular structure with the external loads together with the internal forces of the excessive members which are applied as the additional external forces, is analyzed. In the following a simple example is considered for further explanation.

Investigation of a simple example

In this section a simple example is used to describe the process of the algorithm. In a 4-bar structure shown in Fig. 3.17a, it is obvious that if we add a member between the nodes 2 and 3, the structure will be transformed into Form II and one can easily calculate its inverse. Now we add two members 5 and 6 of identical properties have modulus of elasticity of different signs between the two nodes 2 and 3. This is logical assumption because the property of one member can be nullified by the other member.

Member 6 has negative modulus of elasticity and we consider it as a member separation of which transforms the structure into a regular one as shown in Fig. 3.17c. The structure obtained in this way is equivalent to the basic structure of the force method. From here onward all the previous steps can be employed. Figure 3.17d shows the excessive bar with negative modulus of elasticity which is separated from the truss shown in Fig. 3.17c.

The external force vector will be as follows:

$$\boldsymbol{P} = \left[0\,0\,0\,-10\right]^{t}$$

For the formation of the equilibrium matrix of the structure shown in Fig. 3.17b, one can either use the equations corresponding to the equilibrium of the forces at the nodes, or alternatively use Eq. (3.25).

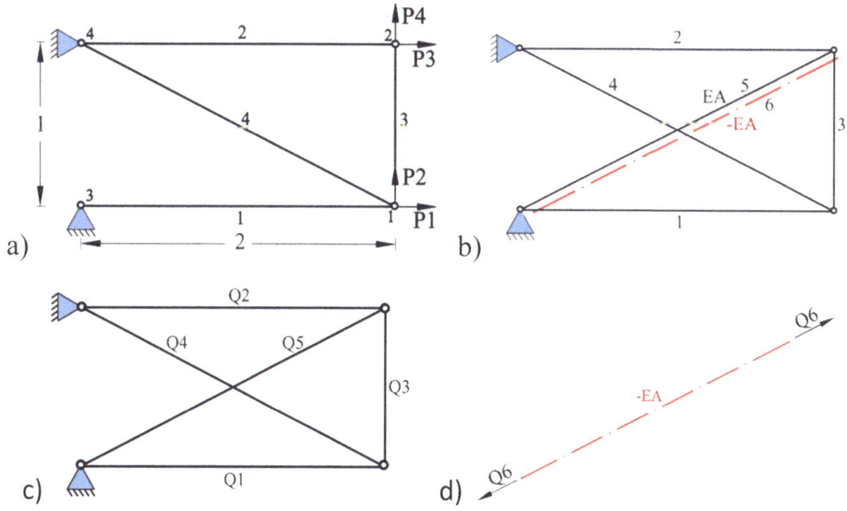

Fig. 3.17 **a** An irregular structure, **b** the irregular structure with a pair of members being added, **c** representation of the internal forces in the regular structure, **d** the added member with negative modulus

$$A_T = \begin{bmatrix} 1 & 0 & 0 & 0.8944 & 0 & 0 \\ 0 & 0 & -1 & -0.4472 & 0 & 0 \\ 0 & 1 & 0 & 0 & 0.8944 & 0.8944 \\ 0 & 0 & 1 & 0 & 0.4472 & 0.4472 \end{bmatrix}$$

The matrices A and N can be obtained by partitioning the above matrix according to Eq. (3.20) and numbering the internal force of the separated member.

$$A = \begin{bmatrix} 1 & 0 & 0 & 0.8944 & 0 \\ 0 & 0 & -1 & -0.4472 & 0 \\ 0 & 1 & 0 & 0 & 0.8944 \\ 0 & 0 & 1 & 0 & 0.4472 \end{bmatrix}; \quad N = \begin{bmatrix} 0 \\ 0 \\ 0.8944 \\ 0.4472 \end{bmatrix}$$

The matrix $\boldsymbol{S}^{-1}$ corresponding to the regular structure shown in Fig. 3.17c, and the matrix $\boldsymbol{s}$ are as follows:

$$\boldsymbol{S}^{-1} = \begin{bmatrix} 1.5935 & 2.0508 & -0.4065 & 1.9491 \\ 2.0508 & 9.8338 & -1.9491 & 9.3465 \\ -0.4065 & -1.9491 & 1.5935 & -2.0508 \\ 1.9491 & 9.3465 & -2.0508 & 9.8338 \end{bmatrix}$$

$$s = diag\{0.5 \; 0.5 \; 1 \; 0.4472 \; 0.4472\}$$

It can be seen that the matrix S can be transformed into Form II by multiplying the row and column 2 by -1. Using the above matrices and Eq. (3.17), the matrix $\boldsymbol{R}$ is obtained as:

$$\boldsymbol{R} = \begin{bmatrix} 0.7967 & 1.0254 & -0.2032 & 0.9745 \\ -0.2032 & -0.9745 & 0.7967 & -1.0254 \\ -0.1016 & -0.4873 & -0.1016 & 0.4872 \\ 0.2272 & -1.1464 & 0.2272 & -1.0896 \\ 0.2272 & 1.0896 & 0.2272 & 1.1464 \end{bmatrix}$$

Using Eq. (3.24), the matrices $\boldsymbol{B}_0$ and $\boldsymbol{B}_1$ are obtained as:

$$\boldsymbol{B}_0 = \begin{bmatrix} 0.7967 & 1.0254 & -0.2032 & 0.9745 \\ -0.2032 & -0.9745 & 0.7967 & -1.0254 \\ -1.1016 & -0.4873 & -0.1016 & 0.4872 \\ 0.2272 & -1.1464 & 0.2272 & -1.0896 \\ 0.2272 & 1.0896 & 0.2272 & 1.1464 \\ 0 & 0 & 0 & 0 \end{bmatrix}; \quad \boldsymbol{B}_1 = \begin{bmatrix} 0.2540 \\ 0.2540 \\ 0.1270 \\ -0.2840 \\ 0.7159 \\ 1 \end{bmatrix}$$

The flexibility matrix $\boldsymbol{F}$ for the truss shown in Fig. 3.17b will be as follows:

$$\boldsymbol{F} = diag\{2\ 2\ 1\ 2.2360\ 2.2360\ -2.2360\}$$

Using Eq. (3.28) the following matrices are obtained:

$$\boldsymbol{D}_1 = [0.5081\ 2.4364\ 0.5081\ 2.5635]; \quad \boldsymbol{D}_2 = [-0.6351]$$

The matrix $\boldsymbol{X}$ is calculated from Eq. (3.29) as

$$\boldsymbol{X} = -\boldsymbol{D}_2{}^{-1}.\boldsymbol{D}_1.\boldsymbol{P} = \frac{1}{0.6351} \times [0.5081\ 2.4364\ 0.5081\ 2.5635].[0\ 0\ 0\ -10]^t$$
$$= -40.3607$$

Adding $\boldsymbol{X}$ to the vector of external loads according to Eq. (3.30) and multiplying the matrix $\boldsymbol{S}^{-1}$ employing Eq. (3.31) we will have:

$$\boldsymbol{P}^* = [0\ 0\ -36.0997\ -28.0498]^t$$
$$\boldsymbol{\Delta} = [-40\ -191.803\ 0\ -201.803]^t$$

Finally, using X and Eq. (3.23) one can find the internal forces of the structure shown in Fig. 3.17b as follows:

$$\boldsymbol{Q} = [-20\ 0\ -10\ 22.3607\ -40.3607\ -40.3607]^t$$

One can recognize the equality of the internal forces of the entries 5 and 6.

3.3.4 Complementary Examples

Here four examples are presented. The first two examples correspond to Sect. 3.3.2 and the third example belongs to Sect. 3.3.3. The fourth example corresponds to the combination of the methods presented in Sects. 3.3.2 and 3.3.3. The latter example is chosen as a frame structure to showing the applicability of the presented method to other skeletal structures other than trusses.

Example 1 A truss with 47 members is considered in the form of a single layer rotational dome, having two members 26 and 27 making the truss an irregular one, Fig. 3.18. If we remove these two members then the remaining regular structure can easily be solved using the method of Kaveh and Rahami [8]. The value of EA = 1 N is assumed to be identical for all the members and the force $P_z^2 = 10$ N and $P_y^8 = 20$ N are applied at nodes 2 and 8, in z direction and y direction, respectively.

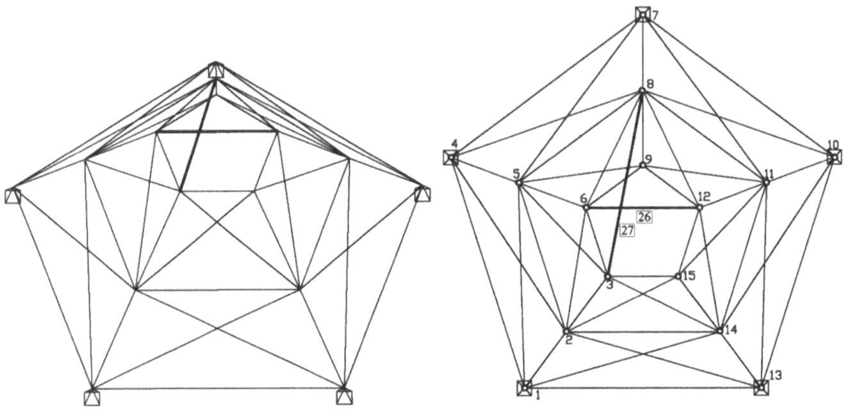

Fig. 3.18 The space dome of the Example 1 with 47 members

For solution of this problem the members 26 and 27 are considered as excessive members. For the above irregular structure the equilibrium matrix $\boldsymbol{A}_T$ has dimension 30×47 and by partitioning using Eq. (3.20), the matrices $\boldsymbol{A}$ and $\boldsymbol{N}$ with dimensions 30×45 and 30×2 are obtained. The matrix S of the regular structure with dimension 30×30 is formed by deleting the excessive members, shown in Fig. 3.19a, as follows:

$$\boldsymbol{S} = \sum_{i=1}^{5} (\boldsymbol{P}_i \otimes \boldsymbol{A}_i)$$

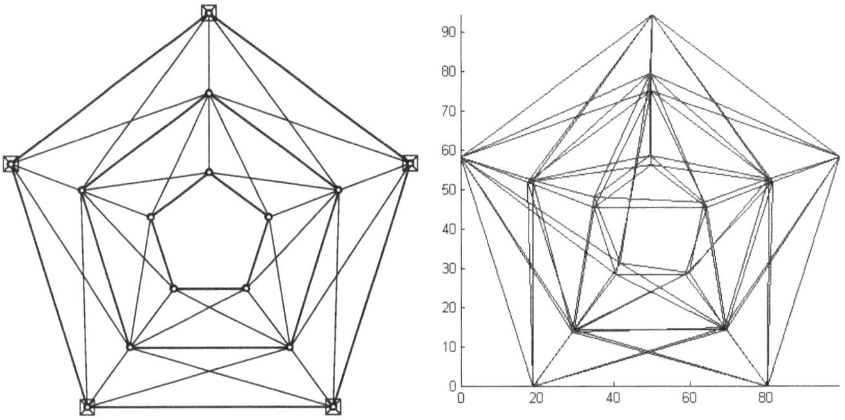

(a) (b)

Fig. 3.19 **a** The regular structure obtained by deleting the excessive members, **b** the deformed shape of the structure of Example 1

In this relation the matrices $\boldsymbol{A}_i$ and $\boldsymbol{P}_i$ are the submatrices constituting the matrix S.

Using the method presented in Kaveh and Rahami [8] the inverse of the matrix S is formed by using the eigenvalues and eigenvectors of five 6×6 matrices.

s is a diagonal matrix of dimension 45×45 containing the stiffness matrices of the members of the regular structure and F is a diagonal matrix of dimension 47×47 consisting of the flexibility matrices of the irregular structure. Therefore the matrices $\boldsymbol{B_0}$ and $\boldsymbol{B_1}$ of dimension 47×30 and 47×2 can be obtained using Eq. (3.24).

The matrices $\boldsymbol{D}_1$ and $\boldsymbol{D}_2$ of dimension 2×30 and 2×2 obtained by using Eq. (3.28) as follows:

$$\boldsymbol{D}_1 = \begin{bmatrix} \overset{1}{\vdots} & \overset{3}{5.6163} & \ldots & \overset{14}{-30.9438} & \overset{30}{\vdots} \\ \ldots & -62.3208 & \ldots & -29.6975 & \ldots \end{bmatrix}$$

$$\boldsymbol{D}_2 = \begin{bmatrix} 426.7122 & -156.126 \\ -156.126 & 267.2256 \end{bmatrix}$$

The vector of the internal forces of the excessive members $\boldsymbol{X}$ can be calculated from Eq. (3.29) as:

$$\boldsymbol{X} = -inv\left(\begin{bmatrix} 426.7122 & -156.126 \\ -156.126 & 267.2256 \end{bmatrix}\right) \cdot \begin{bmatrix} \overset{1}{\vdots} & \overset{3}{5.6163} & \ldots & \overset{14}{-30.9438} & \overset{30}{\vdots} \\ \ldots & -62.3208 & \ldots & -29.6975 & \ldots \end{bmatrix}$$
$$\times \begin{bmatrix} \overset{1}{0} \ldots \overset{3}{10} \ldots \overset{14}{20} \ldots \overset{30}{0} \end{bmatrix}^t = \{-1.9283, 0.7148\}^t$$

The matrix of the internal forces of the irregular structure can be obtained by substituting X in Eq. (3.23) as follows:

$$\boldsymbol{Q} = \begin{bmatrix} \overset{1}{8.7095} \; \overset{2}{0.1002} \; \overset{3}{-0.6626} \; \overset{4}{-2.1497} \ldots \overset{44}{-0.6706} \; \overset{45}{0.4741} \; \overset{46}{-1.9283} \; \overset{47}{0.7148} \end{bmatrix}^t$$

We substitute X in Eq. (3.30), and substitute the vector of the equivalent external forces of the regular structure in Eq. (3.31). The vector of the displacements for the irregular structure can then be obtained using the inverse of the stiffness matrix of the regular structure.

$$\boldsymbol{\Delta} = \begin{bmatrix} \overset{1}{-70.6421} \; \overset{2}{-47.0104} \; \overset{3}{399.9523} \; \overset{4}{117.5682} \ldots \overset{27}{-9.5185} \; \overset{28}{110.8598} \; \overset{29}{56.1357} \; \overset{30}{90.5555} \end{bmatrix}^t$$

The deformed shape of the structure is shown in Fig. 3.19b.

If we solve the structure by a conventional method we have to find the inverse of a matrix of dimension 30×30, while the present approach requires the inverse of 5 matrices of dimension 6×6 and the inverse of the matrix $\boldsymbol{D}_1$ of dimension 2×2.

Example 2 A communication space tower studied in [8], is considered in here. This model can be expressed as the product two graphs. In practice the towers have horizontal belts employed in different heights. These belts make their model irregular. In here we want to analyze those towers which have belts at the central core of the structure (Fig. 3.20). If we consider these members as additional ones, we will obtain a regular structure. Such a regular structure can be generated by rotation of one of its faces. Using the force method the internal forces of the excessive members will be calculated and together with other external loads will be applied to the regular structure.

This tower has 84 nodes and 330 members. The load applied to the structure is $P = 1\,\text{kN}$ applied in all DOFs of the 4 upper nodes of the tower. The value of $EA = 100\,\text{N}$ for all members are considered to be identical.

The structure has 80 free nodes and 330 members and 10 members belong to the belt of the structure.

Forming the equilibrium matrix of the irregular structure which is of dimension 240×330, and its partitioning by employing Eq. (3.20), the matrices $\boldsymbol{A}$ and $\boldsymbol{N}$ of dimensions 240×320 and 240×10 can be obtained.

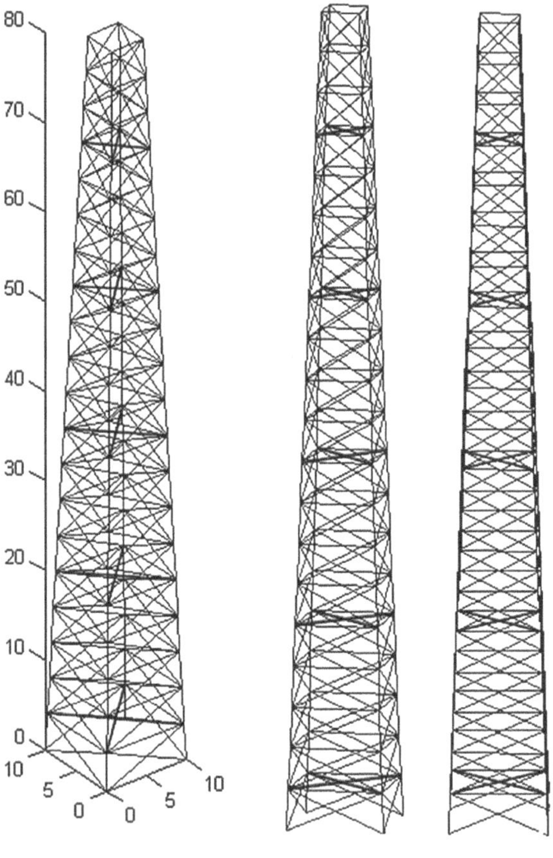

Fig. 3.20 A communication space tower with 5 pairs of excessive members being shown at the central core of the structure

The stiffness matrix $\boldsymbol{S}$ of the regular structure is of 240×240 and has the following form:

$$\boldsymbol{S} = \sum_{i=1}^{4} (\boldsymbol{P}_i \otimes \boldsymbol{A}_i)$$

Using the method presented in [8], the inverse of the stiffness matrix of the regular structure can be obtained using the eigenvalues and eigenvectors of four 60×60 matrices. In this example, since the structure is truss, s is a diagonal matrix of dimension 320×320 containing the stiffness matrices of the members of the regular structure. Having the above mentioned matrices and using Eq. (3.30) one can easily form the $\boldsymbol{B_0}$ and $\boldsymbol{B_1}$ matrices which are of dimension 330×240 and 330×10. For this structure, the flexibility matrix F of the irregular structure has dimension 330×330. Having the matrices $\boldsymbol{B_0}$, $\boldsymbol{B_1}$ and $\boldsymbol{F}$, the matrices $\boldsymbol{D_1}$ and $\boldsymbol{D_2}$ of dimensions 10×320 and 10×10 are obtained from Eq. (3.28). Here $\boldsymbol{D_2}$ is as follows:

$$\boldsymbol{D_2} = \begin{bmatrix} 0.2999 & -0.1033 & 0.2226 & -0.2225 & 0.3121 & -0.3121 & 0.4017 & -0.4017 & 0.4912 & -0.4912 \\ -0.1033 & 0.2999 & -0.2225 & 0.2226 & -0.3121 & 0.3121 & -0.4017 & 0.4017 & -04912 & 0.4912 \\ 0.2226 & -0.2225 & 2.9468 & -2.7704 & 5.5850 & -5.5850 & 8.3715 & -8.3715 & 11.1580 & -11.1581 \\ -0.2225 & 0.2226 & -2.7704 & 2.9468 & -5.5850 & 5.5850 & -8.3715 & 8.3715 & -11.158 & 11.1581 \\ 0.3121 & -0.3121 & 5.5850 & -5.5850 & 15.3371 & -15.1836 & 25.2968 & -25.296 & 35.3849 & -35.385 \\ -0.3121 & 0.3121 & -5.5850 & 5.5850 & -15.183 & 15.3371 & -25.296 & 25.2968 & -35.385 & 35.3849 \\ 0.4017 & -0.4017 & 8.3715 & -8.3715 & 25.2968 & -25.2968 & 48.2162 & -48.085 & 71.7608 & -71.760 \\ -0.4017 & 0.4017 & -8.3715 & 8.3715 & -25.296 & 25.2968 & -48.0854 & 48.2162 & -71.760 & 71.7608 \\ 0.4912 & -0.4912 & 11.1580 & -11.158 & 35.3849 & -35.385 & 71.7608 & -71.760 & 116.622 & -116.514 \\ -0.4912 & 0.4912 & -11.158 & 11.1580 & -35.385 & 35.3849 & -71.760 & 71.7608 & -116.514 & 116.622 \end{bmatrix}$$

In this way and using Eq. (3.29), the vector of internal forces of the excessive members can be obtained as

$$\boldsymbol{X} = \{-84.7178, -84.7178, -85.0153, -85.0153, -106.172, -106.172, \\ -133.071, -133.071, -163.915, -163.915\}^t$$

Adding $\boldsymbol{X}$ to the external load vector using Eq. (3.30) and applying the load to the regular structure in Eq. (3.31), the vector displacements for the irregular structure is obtained. The nodal displacements in some DOFs are as follows:

$$\boldsymbol{\Delta} = \left[.\overset{1}{.}.\ 951\overset{61}{3}.052 \ldots 314\overset{121}{6}8.4 \ldots 653\overset{181}{1}8.71 \overset{240}{\ldots} \right]^t$$

The deformed shape of the structure is shown in Fig. 3.21.

For solution of this structure using a conventional stiffness method we have to find the inverse of 240×240, while the present approach requires the inverse of 4 matrices of dimension 60×60 and the inverse of the matrix $\boldsymbol{D}_1$ of dimension 10×10 to complete the analysis of the irregular structure.

Example 3 Consider a 43-bar truss structure shown in Fig. 3.22a. This structure becomes a cyclically symmetric structure by addition of two members between the nodes 12 and 14, and nodes 11 and 13.

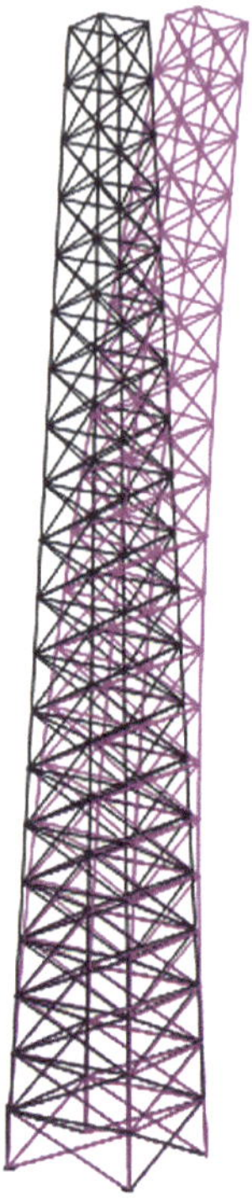

Fig. 3.21 A communication transmission tower together with its deformation

A pair of members with identical geometry and equal modulus of elasticity having different signs, are added where we have lack of members for regularity. In the next step the members with negative modulus of elasticity are considered as excessive members are separated from the structure. The external forces consist of $P_z^2 = P_y^8 = 10\,\mathrm{N}$. For all the member we consider $EA = 1\,\mathrm{N}$.

Forming the equilibrium matrix $\boldsymbol{A_T}$ of the irregular structure according to Eq. (3.25), which is of dimension 30 × 47, and its partitioning by employing Eq. (3.20), the matrices $\boldsymbol{A}$ and $\boldsymbol{N}$ of dimensions 30 × 45 and 30 × 2 are obtained. It should be noted that the matrix $\boldsymbol{A_T}$ corresponds to the irregular structure which has both members of positive and negative modulus of elasticity.

$$\boldsymbol{N} = \begin{bmatrix} \overset{1}{0} & \ldots & \overset{19}{0.0248} & 0.8650 & 0.5010 & 0 & 0 & 0 & 0 & 0 & \overset{27}{0} & \ldots & \overset{30}{0} \\ 0 & \ldots & 0 & 0 & 0 & -0.1375 & 0.8347 & 0.5332 & 0.1375 & -0.8347 & -0.5332 & \ldots & 0 \end{bmatrix}^t$$

The matrix $\boldsymbol{S}$ corresponding to Fig. 3.22b can be expressed as

$$\boldsymbol{S} = \sum_{i=1}^{5} (P_i \otimes A_i)$$

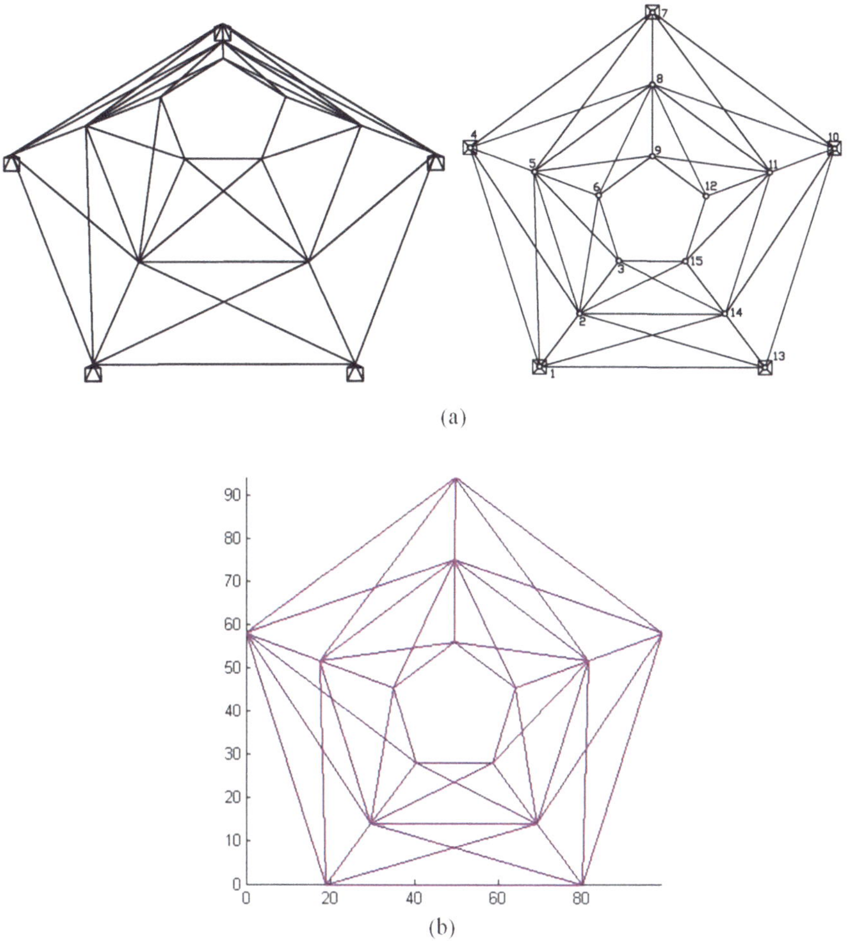

Fig. 3.22 **a** Two and three dimensional representations of the 43-bar structure, **b** the corresponding regular structure

Thus as we observed in Example 1, the inverse of the stiffness matrix which is of dimension 30 × 30 can be calculated by evaluating the eigenvalues and eigenvectors of 5 matrices of dimension 6 × 6. The diagonal matrix s is of dimension 45 × 45 containing the stiffness matrices of the members of the structure shown in Fig. 3.22b. The diagonal matrix F is of dimension 47 × 47 containing the flexibility matrices of the members of the regular structure together with the excessive members.

Using Eq. (3.24) one can easily form the $\boldsymbol{B_0}$ and $\boldsymbol{B_1}$ matrices which are of dimension 47 × 30 and 47 × 2. Having the matrices $\boldsymbol{B_0}$, $\boldsymbol{B_1}$ and $\boldsymbol{F}$, the matrices $\boldsymbol{D_1}$ and $\boldsymbol{D_2}$ of dimensions 2 × 30 and 2 × 2 are obtained from Eq. (3.28). Here $\boldsymbol{D_2}$ is as follows:

$$\boldsymbol{D}_2 = \begin{bmatrix} -19.4620 & 1.8186 \\ 1.8186 & -10.7629 \end{bmatrix}$$

Equation (3.29) can be employed to find the vector $\boldsymbol{X}$ as:

$$\boldsymbol{X} = -\boldsymbol{D}_2^{-1}.\boldsymbol{D}_1.\boldsymbol{P}$$

$$\boldsymbol{X} = -\begin{bmatrix} -19.4620 & 1.8186 \\ 1.8186 & -10.7629 \end{bmatrix}^{-1} . \begin{bmatrix} \ldots & \overset{3}{0.3207} & \ldots & \overset{14}{1.3854} & \ldots \\ \ldots & -0.0640 & \ldots & -0.3018 & \ldots \end{bmatrix}$$

$$\times \begin{bmatrix} \overset{1}{0} \ldots \overset{3}{10} \ldots \overset{14}{10} \ldots \overset{30}{0} \end{bmatrix}^t = \begin{bmatrix} 0.8584 \\ -0.1949 \end{bmatrix}$$

Using Eq. (3.30) the equivalent external forces of the regular structure are obtained, and using the inverse of $\boldsymbol{S}$ one can easily find the nodal displacements of the irregular structure by Eq. (3.31).

$$\boldsymbol{P}^* = \begin{bmatrix} \overset{1}{0} & \ldots & \overset{3}{10} & \ldots & \overset{14}{10} & \ldots & \overset{19}{0.0213} & 0.7426 & 0.4301 & 0.0268 & -0.1627 & -0.1039 & -0.0268 & 0.1627 & \overset{27}{0.1039} & \ldots & \overset{30}{0} \end{bmatrix}^t$$

$$\Delta = \begin{bmatrix} \overset{1}{-105.512} & \overset{2}{-134.943} & 380.134 & 51.383 & \ldots & -18.489 & 50.4185 & \overset{29}{-75.205} & \overset{30}{116.529} \end{bmatrix}^t$$

The internal forces of the regular structure and the excessive members can be found using Eq. (3.23) as

$$\boldsymbol{Q} = \begin{bmatrix} \overset{1}{6.8906} & \overset{2}{-0.9281} & \overset{3}{1.7904} & \overset{5}{-0.950} & \overset{6}{-5.60} & \ldots & \overset{43}{-0.0003} & \overset{44}{0.8584} & \overset{45}{-0.1949} & \overset{46}{0.8584} & \overset{47}{-0.1949} \end{bmatrix}^t$$

As it can be seen, the internal forces in members 44, 46 and 45, 47 which are the added pairs of members are the same as the entries of $\boldsymbol{X}$.

For solution of this irregular structure using a conventional stiffness method we have to find the inverse of 30×30, while the present approach requires the inverse of 5 matrices of dimension 6×6 and the inverse a matrix of dimension 2×2 to complete the analysis of the irregular structure.

Example 4 A 24-story three dimensional frame is shown in Fig. 3.23a, with 49 columns and 84 beam in each story. The dimensions of all the beams and columns are assumed to be identical in all the stories. In each face of the building eight bracing elements are added to increase the stiffness of the structure. Naturally these elements make the model irregular. Here using the presented method, the bracing elements are decomposed from the structure the analysis is performed for two separate parts, namely the regular bending frame and the excessive bracing elements. In fact this is a combined problem which utilizes the methods of Sects. 3.3.2 and 3.3.3.

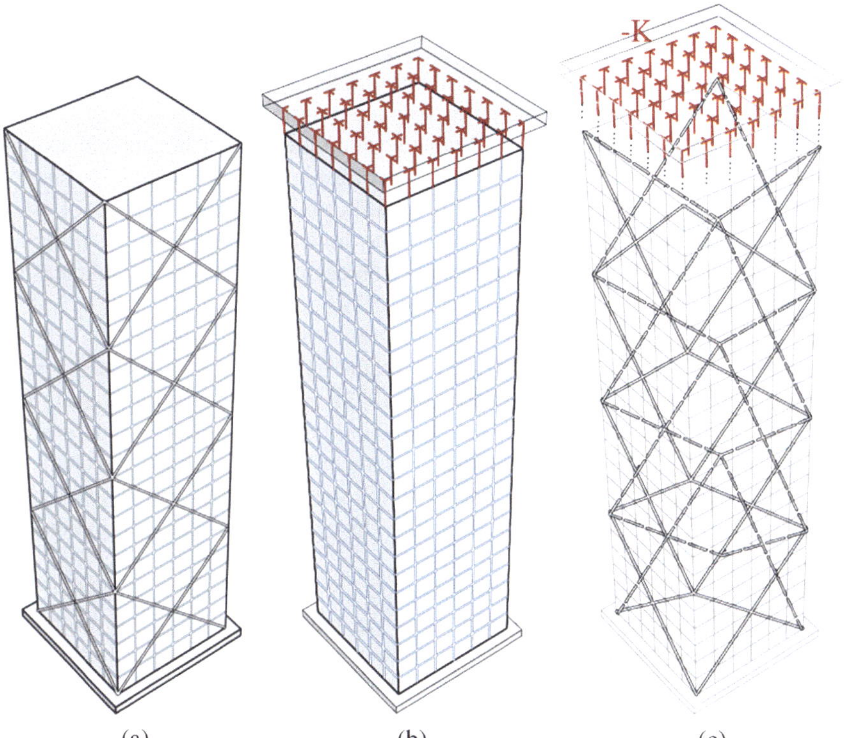

Fig. 3.23 **a** A 24-story irregular frame with bracing, **b** the regular part of the irregular frame with fictitious columns of positive modulus of elasticity being added, **c** the bracing part consisting of 32 bracing elements and 49 fictitious bending elements with negative modulus of elasticity being added as shown at the top of the structure

If the top part of the structure is fixed similar to the bottom part, then the bending frame structure can be easily analyzed using the method presented in [9]. Therefore, here we consider pairs of bending elements with positive and negative modulus of elasticity at end part of the structure, similar to the columns of the other stories, as shown in Fig. 3.23b. The columns are connected to the top part of the structure and fixed at the other ends. The elements with negative modulus of elasticity are considered as excessive members and are separated from the structure. Therefore, the excessive members consist of 32 bracing elements and 49 bending elements with negative modulus of elasticity. These elements are highlighted in Fig. 3.23c. In this way, the regular structure consists of a 24-story frame together with additional bending elements with positive modulus of elasticity as shown in Fig. 3.23b. The required parameters for the analysis are as $k = 7056$, $\mathrm{t} = 326$, and $e = 19446$. Using Eqs. (3.13) and (3.14) and employing the rotation and stiffness matrices of the elements in their local coordinate systems, the matrix A_T of dimension 7056×19772 can be constructed. For the formation of the stiffness matrices of the elements, the local coordinate systems should be selected such that the form given in Eq. (3.26) is formed. For each bending elements, six internal forces, and for bracing elements only one axial force are assumed. It should be noted that the fictitious elements with $\pm$ modulus of elasticity contribute in the formation of this matrix.

By partitioning the matrix $\boldsymbol{A}_T$ we obtain two matrices $\boldsymbol{A}$ and $\boldsymbol{N}$ having dimensions 7056×19446 and 7056×326, respectively. Since the regular part contains 3241 bending elements, thus the unassembled stiffness matrix $\boldsymbol{s}$ is of dimension 19446×19446. The assembled matrix S of the regular part has dimension 7056×7056. Utilizing the method of [9], the inverse of this matrix can easily be obtained calculating the eigenvalues of 24 matrices of dimension 294×294.

In this way forming the inverse of the stiffness matrix and using Eq. (3.17), the matrix $\boldsymbol{R}$ of dimension 19446×7056 can be obtained. Having this matrix the matrices $\boldsymbol{B}_0$ and $\boldsymbol{B}_1$ of dimensions 19772×7056 and 19772×326 will be formed using Eq. (3.24). The flexibility matrix $\boldsymbol{F}$ contains the flexibility of all the elements of the near-regular structure (bending elements, bracing elements, and pair of fictitious elements with $+$ and $-$ signs). This matrix is a block matrix such that for the bending members blocks are 6×6 and for the bracing members the blocks are 1×1. Thus, the dimension of F is 19772×19772.

With help of Eq. (3.28) the matrices $\boldsymbol{D}_1$ and $\boldsymbol{D}_2$ of dimensions 326×7056 and 326×326 are obtained, respectively. Using Eq. (3.29) and finding the inverse of $\boldsymbol{D}_2$ leads to the vector of unknown $\boldsymbol{X}$ of dimension 326×1. Substituting this in Eq. (3.30), the equivalent external force vector of the regular part of the structure is obtained. Multiplying the inverse of the stiffness matrix of the regular part, the displacement vector of the near-regular structure of dimension 7056×1 is obtained.

It can be observed that the analysis of the problem with the help of this method for frame structures is the same as that of the trusses which were discussed in the previous examples, with the only difference that the rotation and stiffness matrices for the bending elements in the local coordinate systems should be defined according to Eq. (3.26). In this problem instead of inverting the stiffness matrix of dimension 7056×7056 in direct analysis of the near-regular structure, one needs to find the inverse of the matrix $\boldsymbol{D}_2$ of dimension 326×326, and calculate the eigenvalues of 24 matrices of dimension 294×294. This shows the efficiency of the present method. Obviously increasing the number of stories this efficiency will become more apparent. In other words in this method a matrix of dimension 7056 is decomposed into 24 matrices of dimension 294.

3.3.5 Appendix: Calculation of the Flexibility Matrix B Without the Formation of the Equilibrium Matrix A

In this method the matrix $\boldsymbol{B}$ is formed directly. This means with the help of the rotation matrices of the members of the irregular structure, and without forming the equilibrium matrix, the matrix $\boldsymbol{B}$ can be obtained. In fact this method is the improved method for calculating the flexibility matrix which is described in Sect. 3.3.1.

For this purpose first the matrix L is defined as

$$L = \left[S^{-1} \; S^{-1}.N \right] \tag{3.33}$$

Here, S^{-1} is the inverse of the stiffness matrix of the regular structure. The matrix N is of order $k \times t$ and can be formed as follows:

If i is the DOFs of the two ends of the imaginary member j in the local coordinate system and r are the DOFs of the two ends of the assumed member j in the global coordinate system, the columns corresponding to i in the matrix N will be as follows:

$$N(r, i) = T_j^t \tag{3.34}$$

The remaining rows of these columns are equal to zero. This work is performed for all the excessive members of the structure. Defining $B(i, :)$ as those rows of B which correspond to the internal forces of the member j, the matrix B is formed as follows:

$$\begin{aligned} \boldsymbol{B}(i,:) &= \boldsymbol{s}_j.\boldsymbol{T}_j.[\boldsymbol{L}(r,:)], \quad j = 1:e \\ \boldsymbol{B}(i,:) &= \left[\boldsymbol{Z}_{t\times k} \; \boldsymbol{I}_{t\times t} \right], \quad j = e+1:e+t \\ \boldsymbol{B} &= \left[\boldsymbol{B}_0 \; \boldsymbol{B}_1 \right] \end{aligned} \tag{3.35}$$

The matrix $\boldsymbol{s}_j$ is the jth block of the matrix s and $\boldsymbol{T}_j$ is the modified rotation matrix of a typical excessive member. $\boldsymbol{B}$ is the flexibility matrix of the irregular structure which is of dimension $(e+t)\times(k+t)$. $\boldsymbol{I}$ and $\boldsymbol{Z}$ are unit and zero matrices, respectively.

Here, an example of calculating the matrix $\boldsymbol{B}$ using the above mentioned approach is presented. If we want to form the matrix $\boldsymbol{B}$ for the example presented in Sect. 3.3.1, we perform the calculation according to the Fig. 3.17c.

The following matrix $\boldsymbol{C}$ contains the direction cosines of the members of the irregular structure shown in Fig. 3.17b. Using this, the rotation matrices of the excessive members can be formed.

$$\boldsymbol{C} = \begin{bmatrix} \overset{1}{1} & \overset{2}{1} & \overset{3}{0} & \overset{4}{0.8944} & \overset{5}{0.8944} & \overset{6}{0.8944} \\ 0 & 0 & -1 & -0.4472 & 0.4472 & 0.4472 \\ 0 & 0 & 0 & 0 & 0 & 0 \end{bmatrix}$$

$$\boldsymbol{T}_j^t = \begin{bmatrix} 1 \\ -1 \end{bmatrix} \otimes \boldsymbol{C}(:, j)$$

Using Eq. (3.34) the matrix $\boldsymbol{N}$ can be obtained from the 6th column of the $\boldsymbol{T}^t$.

$$\boldsymbol{N} = \left[0 \; 0 \; 0.8944 \; 0.4472 \right]^t$$

Since the member 6 is only associated with the DOFs 3 and 4, therefore the entries of the rotation matrix $\boldsymbol{T}_6^t$ exist only in theses DOFs. The inverse of the stiffness matrix of the regular structure $\boldsymbol{S}^{-1}$ shown in Fig. 3.17c and the diagonal matrix $\boldsymbol{s}$ containing the stiffnesses of the members of this structure are formed as

$$\boldsymbol{S}^{-1} = \begin{bmatrix} 1.5935 & 2.0508 & -0.4065 & 1.9491 \\ 2.0508 & 9.8338 & -1.9491 & 9.3465 \\ -0.4065 & -1.9491 & 1.5935 & -2.0508 \\ 1.9491 & 9.3465 & -2.0508 & 9.8338 \end{bmatrix}$$

$$s = diag\{0.5\ 0.5\ 1\ 0.4472\ 0.4472\}$$

For the formation of $\boldsymbol{L}$ one should multiply the matrix $\boldsymbol{N}$ in $\boldsymbol{S}^{-1}$, and add the result to the $\boldsymbol{S}^{-1}$ as follows:

$$\boldsymbol{S}^{-1}\boldsymbol{N} = \begin{bmatrix} 0.5081 \\ 2.4364 \\ 0.5081 \\ 2.5635 \end{bmatrix}$$

$$\boldsymbol{L} = \begin{bmatrix} 1.5935 & 2.0508 & -0.4065 & 1.9491 & 0.5081 \\ 2.0508 & 9.8338 & -1.9491 & 9.3465 & 2.4364 \\ -0.4065 & -1.9491 & 1.5935 & -2.0508 & 0.5081 \\ 1.9491 & 9.3465 & -2.0508 & 9.8338 & 2.5635 \end{bmatrix}$$

Separating the rows of $\boldsymbol{L}$ according to the DOFs both ends of each member in the global coordinate system and substituting in Eq. (3.35) one can form the row of the matrix $\boldsymbol{B}$ row by row.

$$\boldsymbol{B}(1,:) = 0.5 \times [1\ 0\ 0\ 0]. \begin{bmatrix} 1.5935 & 2.0508 & -0.4065 & 1.9491 & 0.5081 \\ 2.0580 & 9.8338 & -1.9491 & 9.3465 & 2.4364 \\ 0 & 0 & 0 & 0 & 0 \\ 0 & 0 & 0 & 0 & 0 \end{bmatrix}$$

$$\Rightarrow \boldsymbol{B}(1,:) = [0.7967\ 1.0254\ -0.2032\ 0.9745\ 0.2540]$$

$$\boldsymbol{B}(2,:) = 0.5 \times [1\ 0\ 0\ 0]. \begin{bmatrix} -0.4065 & -1.9491 & 1.5935 & -2.0508 & 0.5081 \\ 1.9491 & 9.3465 & -2.0508 & 9.8338 & 2.5635 \\ 0 & 0 & 0 & 0 & 0 \\ 0 & 0 & 0 & 0 & 0 \end{bmatrix}$$

$$\Rightarrow \boldsymbol{B}(2,:) = [-0.2032\ -0.9745\ 0.7967\ -1.0254\ 0.2540]$$

$$\boldsymbol{B}(3,:) = 1 \times [0\ -1\ 0\ 1]. \begin{bmatrix} 1.5935 & 2.0508 & -0.4065 & 1.9491 & 0.5081 \\ 2.0508 & 9.8338 & -1.9491 & 9.3465 & 2.4364 \\ -0.4065 & -1.9491 & 1.5935 & -2.0508 & 0.5081 \\ 1.9491 & 9.3465 & -2.0508 & 9.8338 & 2.5635 \end{bmatrix}$$

$$\Rightarrow \boldsymbol{B}(3,:) = [-0.1016\ -0.4873\ -0.1016\ 0.4872\ 0.1270]$$

$$\boldsymbol{B}(4,:) = 0.4472 \times \begin{bmatrix} 0.8944 & -0.4472 & 0 & 0 \end{bmatrix}.$$

$$\begin{bmatrix} 1.5935 & 2.0508 & -0.4065 & 1.9491 & 0.5081 \\ 2.0508 & 9.8338 & -1.9491 & 9.3465 & 2.4364 \\ 0 & 0 & 0 & 0 & 0 \\ 0 & 0 & 0 & 0 & 0 \end{bmatrix}$$

$$\Rightarrow \boldsymbol{B}(4,:) = \begin{bmatrix} 0.2272 & -1.1464 & 0.2272 & -1.0896 & -0.2840 \end{bmatrix}$$

$$\boldsymbol{B}(5,:) = 0.4472 \times \begin{bmatrix} 0.8944 & 0.4472 & 0 & 0 \end{bmatrix}.$$

$$\begin{bmatrix} -0.4065 & -1.9491 & 1.5935 & -2.0508 & 0.5081 \\ 1.9491 & 9.3465 & -2.0508 & 9.8338 & 2.5635 \\ 0 & 0 & 0 & 0 & 0 \\ 0 & 0 & 0 & 0 & 0 \end{bmatrix}$$

$$\Rightarrow \boldsymbol{B}(5,:) = \begin{bmatrix} 0.2272 & 1.0896 & 0.2272 & 1.1464 & 0.7159 \end{bmatrix}$$

$$\boldsymbol{B}(6,:) = \begin{bmatrix} 0 & 0 & 0 & 0 & 1 \end{bmatrix}$$

Finally by assembling the rows of the matrix $\boldsymbol{B}$ according to the numbers assigned to the internal forces of the members of the irregular structure, the following matrix is obtained. This is identical to the matrix obtained by the method presented in Sect. 3.3.1.

$$\boldsymbol{B} = \begin{bmatrix} 0.7967 & 1.0254 & -0.2032 & 0.9745 & 0.2540 \\ -0.2032 & -0.9745 & 0.7967 & -1.0254 & 0.2540 \\ -0.1016 & -0.4873 & -0.1016 & 0.4872 & 0.1270 \\ 0.2272 & -1.1464 & 0.2272 & -1.0896 & -0.2840 \\ 0.2272 & 1.0896 & 0.2272 & 1.1464 & 0.7159 \\ 0 & 0 & 0 & 0 & 1 \end{bmatrix}$$

References

1. Shojaei, I., Kaveh, A., Rahami, H.: Analysis of structures convertible to repeated structures using graph products. Comput. Struct. **125**, 153–163 (2013)
2. Chan, H.C., Cai, C.W., Cheung, Y.K.: Exact Analysis of Structures with Periodicity Using U-Transformation. World Scientific, Singapore (1998)
3. Kaveh, A., Sayarinejad, M.A.: Eigensolutions for matrices of special patterns. Commun. Numer. Methods Eng. **19**, 25–136 (2003)
4. Kaveh, A., Rahami, H., Mirghaderi, S.R., Ardalan, Asl M.: Analysis of near-regular structures using the force method. Eng. Comput. **1**(30), 21–48 (2013)
5. Kaveh, A.: Optimal Analysis of Structures by Concepts of Symmetry and Regularity. Springer, New York (2013)
6. Kaveh, A.: Structural Mechanics: Graph and Matrix Methods, 3rd edn. Research Studies Press (Wiley), UK (2006)

7. Kaveh, A., Rahami, H.: An efficient analysis of repetitive structures generated by graph products. Int. J. Numer. Methods Eng. **84**, 108–126 (2010)
8. Kaveh, A., Rahami, H.: Block circulant matrices and applications in free vibration analysis of cyclically repetitive structures. Acta Mech. **217**, 51–62 (2011)
9. Kaveh, A., Rahami, H.: Eigenvalues of the adjacency and Laplacian matrices for modified regular structural models. Int. J. Numer. Methods Bio Eng. **26**, 1836–1855 (2010)

Chapter 4
Static Analysis of Near-Regular Skeletal Structures: Additional Nodes

4.1 Introduction

In Chap. 3, methods for calculation of eigenpairs and inverse of matrices of near-regular structures with additional members (Type I) were developed. These near-regular structures were convertible to regular structures by adding or removing some members where the number of degrees of freedom was the same for the near-regular and the regular structure. There is another group of near-regular structures (hereafter called near-regular structures Type II) wherein the number of degrees of freedom for the near-regular structure is larger than that of the regular structure. Therefore, a near-regular structure Type II can be decomposed to regular and irregular substructures to get advantages of swift solution of regular structures and parallel computations. This chapter is devoted to the analysis of near-regular Type II structures using displacement and force methods as well as eigensolution of these structures.

4.2 Analysis of Near-Regular Structures with Additional Nodes Using Displacement Methods

A near-regular structure Type II is composed of a regular structure and an irregular part where the number of degrees of freedom in the near-regular structure is larger than that of the regular structure [1]. In other words, the irregular part in a near-regular structure includes some degrees of freedom that are not shared by the regular structure. To solve these near-regular structures, an algorithm was developed wherein the near-regular structure is split into the regular and irregular parts and each part is solved separately. Specifically, the regular part is solved using the developed relations in Chap. 2. The solutions are then combined using matrix operations to obtain the solution of the near-regular structure. The proposed algorithm, developed bellow, is computationally efficient only if the irregular part, compared to the regular part, is small enough.

A. Kaveh et al., *Swift Analysis of Civil Engineering Structures Using Graph Theory Methods*, Studies in Systems, Decision and Control 290,
https://doi.org/10.1007/978-3-030-45549-1_4

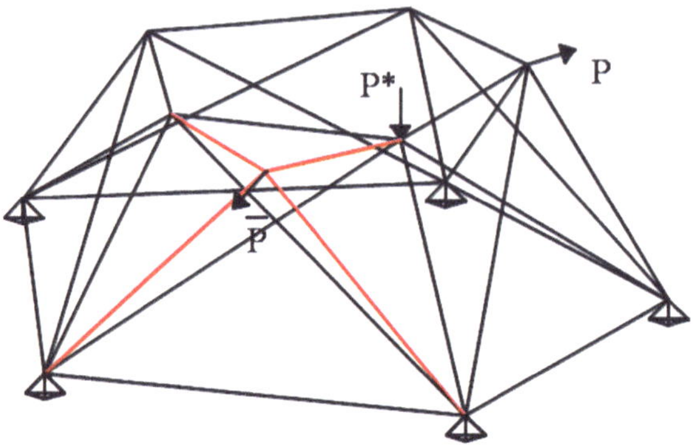

Fig. 4.1 A near-regular structure composed of regular and irregular parts

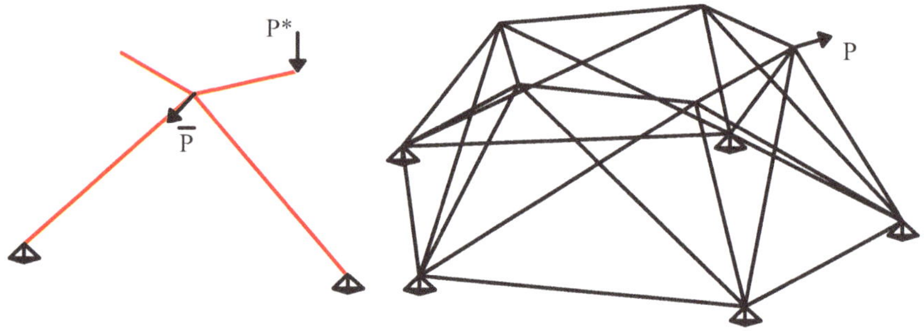

Fig. 4.2 The regular and irregular parts under the applied loads

Let us show the algorithm using a simple example. Consider the near-regular structure in Fig. 4.1.

The regular and irregular parts are separated and the load vector $\boldsymbol{P}^*$ is considered for the irregular part (Fig. 4.2).

Now, let's construct the relationship $\boldsymbol{F} = \boldsymbol{K}\boldsymbol{\Delta}$ for the irregular part:

$\boldsymbol{K}$: The stiffness matrix of the irregular structure.

$$\begin{matrix} l\{ \\ j\{ \end{matrix}\begin{bmatrix} \boldsymbol{P}^* - \boldsymbol{F}^* \\ \cdots \\ \bar{\boldsymbol{P}} \end{bmatrix}_{(j+l)\times 1} = \begin{bmatrix} \boldsymbol{K}_1 & \vdots & \boldsymbol{K}_2 \\ \cdots & \vdots & \cdots \\ \boldsymbol{K}_3 & \vdots & \boldsymbol{K}_4 \end{bmatrix}_{(j+l)\times(j+l)} \begin{bmatrix} \boldsymbol{\Delta}^* \\ \cdots \\ \bar{\boldsymbol{\Delta}} \end{bmatrix}_{(j+l)\times 1} \tag{4.1}$$

l The degrees of freedom that two parts (regular and irregular) have in common.

j The degrees of freedom belonging to the irregular part.

$\boldsymbol{F}^*$ The internal force vector corresponding to the degrees of freedom l.

$\boldsymbol{P}^*$, $\boldsymbol{\Delta}^*$ The load and displacement vectors corresponding to the degrees of freedom l.

Constructing the relationship $\boldsymbol{F} = \boldsymbol{S\Delta}$ for the regular part:

$$\begin{matrix} l\{ \\ i\{ \end{matrix} \begin{bmatrix} \boldsymbol{F}^* \\ \cdots \\ \boldsymbol{P} \end{bmatrix}_{(l+i)\times 1} = \begin{bmatrix} \boldsymbol{S}_1 \\ \cdots \\ \boldsymbol{S}_2 \end{bmatrix}_{(l+i)\times(l+i)} \begin{bmatrix} \boldsymbol{\Delta}^* \\ \cdots \\ \boldsymbol{\Delta} \end{bmatrix}_{(l+i)\times 1} \tag{4.2}$$

i The degrees of freedom belonging to the regular part.
$\boldsymbol{P}$, $\boldsymbol{\Delta}$ The load and displacement vectors corresponding to the degrees of freedom i.
$\boldsymbol{S}$ The stiffness matrix of the regular structure.

Now, we calculate the relationship between $\boldsymbol{F}^*$ and $\boldsymbol{\Delta}^*$ in the regular structure. For the regular structure we have

$$\boldsymbol{F} = \boldsymbol{S\Delta} \rightarrow \boldsymbol{\Delta} = \boldsymbol{DF} \tag{4.3}$$

$\boldsymbol{D}$ The inverse of the matrix $\boldsymbol{S}$.

where matrix $\boldsymbol{D}$ is calculated using the relationships in Chap. 2. We can write

$$\begin{bmatrix} \boldsymbol{\Delta}^* \\ \cdots \\ \boldsymbol{\Delta} \end{bmatrix} = \begin{bmatrix} \boldsymbol{D}_1 \\ \cdots \\ \boldsymbol{D}_2 \end{bmatrix} \begin{bmatrix} \boldsymbol{F}^* \\ \cdots \\ \boldsymbol{P} \end{bmatrix} \Rightarrow \boldsymbol{\Delta}^* = \boldsymbol{D}_1 \begin{bmatrix} \boldsymbol{F}^* \\ \cdots \\ \boldsymbol{P} \end{bmatrix}, \boldsymbol{\Delta} = \boldsymbol{D}_2 \begin{bmatrix} \boldsymbol{F}^* \\ \cdots \\ \boldsymbol{P} \end{bmatrix} \tag{4.4}$$

$$\boldsymbol{\Delta}^* = \begin{bmatrix} \boldsymbol{D}_{11} \vdots \boldsymbol{D}_{12} \end{bmatrix} \begin{bmatrix} \boldsymbol{F}^* \\ \cdots \\ \boldsymbol{P} \end{bmatrix} \text{ and } \boldsymbol{\Delta} = \begin{bmatrix} \boldsymbol{D}_{21} \vdots \boldsymbol{D}_{22} \end{bmatrix} \begin{bmatrix} \boldsymbol{F}^* \\ \cdots \\ \boldsymbol{P} \end{bmatrix} \tag{4.5}$$

$$\boldsymbol{\Delta}^* = \boldsymbol{D}_{11}\boldsymbol{F}^* + \boldsymbol{D}_{12}\boldsymbol{P} \tag{4.6}$$

$$\boldsymbol{A} = \boldsymbol{D}_{11}^{-1}, \boldsymbol{A}^* = \boldsymbol{D}_{12} \Rightarrow \boldsymbol{F}^*_{l\times 1} = \boldsymbol{A}_{l\times l}\boldsymbol{\Delta}^*_{l\times 1} - \underbrace{\boldsymbol{A}_{l\times l}\boldsymbol{A}^*_{l\times i}\boldsymbol{P}_{i\times 1}}_{\boldsymbol{B}_{l\times 1}} \tag{4.7}$$

Substituting $\boldsymbol{F}^*$ into Eq. (4.1)

$$\begin{bmatrix} \boldsymbol{P}^* - (\boldsymbol{A\Delta}^* - \boldsymbol{B}) \\ \cdots \\ \bar{\boldsymbol{P}} \end{bmatrix} = \begin{bmatrix} \boldsymbol{K}_1 \vdots \boldsymbol{K}_2 \\ \cdots \vdots \cdots \\ \boldsymbol{K}_3 \vdots \boldsymbol{K}_4 \end{bmatrix} \begin{bmatrix} \boldsymbol{\Delta}^* \\ \cdots \\ \bar{\boldsymbol{\Delta}} \end{bmatrix}$$

$$\begin{bmatrix} \boldsymbol{P}^* + \boldsymbol{B} \\ \cdots \\ \bar{\boldsymbol{P}} \end{bmatrix} = \begin{bmatrix} \boldsymbol{K}_1 + \boldsymbol{A} & \vdots & \boldsymbol{K}_2 \\ \cdots & \vdots & \cdots \\ \boldsymbol{K}_3 & \vdots & \boldsymbol{K}_4 \end{bmatrix} \begin{bmatrix} \boldsymbol{\Delta}^* \\ \cdots \\ \bar{\boldsymbol{\Delta}} \end{bmatrix}$$

And

$$\begin{bmatrix} \boldsymbol{\Delta}^* \\ \cdots \\ \bar{\boldsymbol{\Delta}} \end{bmatrix} = \boldsymbol{C} \begin{bmatrix} \boldsymbol{P}^* + \boldsymbol{B} \\ \cdots \\ \bar{\boldsymbol{P}} \end{bmatrix}, \boldsymbol{C} = \begin{bmatrix} \boldsymbol{K}_1 + \boldsymbol{A} & \vdots & \boldsymbol{K}_2 \\ \cdots & \vdots & \cdots \\ \boldsymbol{K}_3 & \vdots & \boldsymbol{K}_4 \end{bmatrix}^{-1} \tag{4.8}$$

Calculating inverse of the stiffness matrix for the near-regular structure

$$\begin{bmatrix} \boldsymbol{\Delta}^* \\ \cdots \\ \bar{\boldsymbol{\Delta}} \end{bmatrix} = \begin{bmatrix} \boldsymbol{C}_1 & \vdots & \boldsymbol{C}_2 \\ \cdots & \vdots & \cdots \\ \boldsymbol{C}_3 & \vdots & \boldsymbol{C}_4 \end{bmatrix} \begin{bmatrix} \boldsymbol{P}^* + \boldsymbol{B} \\ \cdots \\ \bar{\boldsymbol{P}} \end{bmatrix}$$

$$\boldsymbol{B} = \boldsymbol{A}\boldsymbol{A}^*\boldsymbol{P} = \boldsymbol{B}^*\boldsymbol{P}$$

$$\begin{bmatrix} \boldsymbol{\Delta}^* \\ \cdots \\ \bar{\boldsymbol{\Delta}} \end{bmatrix} = \begin{bmatrix} \boldsymbol{C}_1 & \vdots & \boldsymbol{C}_2 \\ \cdots & \vdots & \cdots \\ \boldsymbol{C}_3 & \vdots & \boldsymbol{C}_4 \end{bmatrix} \begin{bmatrix} \boldsymbol{P}^* + \boldsymbol{B}^*\boldsymbol{P} \\ \cdots \\ \bar{\boldsymbol{P}} \end{bmatrix} \tag{4.9}$$

We can write

$$\begin{bmatrix} \boldsymbol{\Delta}^* \\ \cdots \\ \bar{\boldsymbol{\Delta}} \\ \cdots \\ \boldsymbol{\Delta} \end{bmatrix} = \begin{bmatrix} \boldsymbol{C}_1 & \vdots & \boldsymbol{C}_2 & \vdots & \boldsymbol{C}_1\boldsymbol{B}^* \\ \cdots & \vdots & \cdots & \vdots & \cdots \\ \boldsymbol{C}_3 & \vdots & \boldsymbol{C}_4 & \vdots & \boldsymbol{C}_3\boldsymbol{B}^* \\ \cdots & \vdots & \cdots & \vdots & \cdots \\ \boldsymbol{C}_1^* & \vdots & \boldsymbol{C}_2^* & \vdots & \boldsymbol{C}_3^* \end{bmatrix} \begin{bmatrix} \boldsymbol{P}^* \\ \cdots \\ \bar{\boldsymbol{P}} \\ \cdots \\ \boldsymbol{P} \end{bmatrix} \tag{4.10}$$

where the matrix in Eq. (4.10) is inverse of the stiffness matrix of near-regular structure. $\boldsymbol{C}_1$, $\boldsymbol{C}_2$, $\boldsymbol{C}_3$, $\boldsymbol{C}_4$ and $\boldsymbol{B}^*$ were already obtained using Eqs. (4.8) and (4.9), and $\boldsymbol{C}_1^*$, $\boldsymbol{C}_2^*$, and $\boldsymbol{C}_3^*$ are calculated as follows:

$$\boldsymbol{\Delta} = \boldsymbol{D}_2 \begin{bmatrix} \boldsymbol{F}^* \\ \cdots \\ \boldsymbol{P} \end{bmatrix} = \boldsymbol{D}_{21}\boldsymbol{F}^* + \boldsymbol{D}_{22}\boldsymbol{P}$$

$$\boldsymbol{\Delta} = \boldsymbol{D}_{21}\left[\boldsymbol{A}\boldsymbol{\Delta}^* - \boldsymbol{B}^*\boldsymbol{P}\right] + \boldsymbol{D}_{22}\boldsymbol{P}$$

$$\boldsymbol{\Delta} = \boldsymbol{D}_{21}\left[\boldsymbol{A}\left[\boldsymbol{C}_1\boldsymbol{P}^* + \boldsymbol{C}_2\bar{\boldsymbol{P}} + \boldsymbol{C}_1\boldsymbol{B}^*\boldsymbol{P}\right] - \boldsymbol{B}^*\boldsymbol{P}\right] + \boldsymbol{D}_{22}\boldsymbol{P}$$

And

$$\boldsymbol{\Delta} = \underbrace{\left[\boldsymbol{D}_{21}\left[\boldsymbol{A}\boldsymbol{C}_1 - \boldsymbol{I}\right]\boldsymbol{B}^* + \boldsymbol{D}_{22}\right]}_{\boldsymbol{C}_3^*}\boldsymbol{P} + \underbrace{\left[\boldsymbol{D}_{21}\boldsymbol{A}\boldsymbol{C}_1\right]}_{\boldsymbol{C}_1^*}\boldsymbol{P}^* + \underbrace{\left[\boldsymbol{D}_{21}\boldsymbol{A}\boldsymbol{C}_2\right]}_{\boldsymbol{C}_2^*}\bar{\boldsymbol{P}} \tag{4.11}$$

Therefore, all blocks of the inverse of the stiffness matrix for the main structure are obtained.

The summary of the algorithm would be as follows:

- Separating the regular and the irregular parts and considering the load vector $\boldsymbol{P}^*$ for the irregular part
- Constructing the relationship $\boldsymbol{F} = \boldsymbol{K}\boldsymbol{\Delta}$ for the irregular part using Eq. (4.1)
- Constructing the relationship $\boldsymbol{F} = \boldsymbol{S}\boldsymbol{\Delta}$ for the regular part using Eq. (4.2)
- Calculating the relationship between $\boldsymbol{F}^*$ and $\boldsymbol{\Delta}^*$ in the regular structure through Eqs. (4.3) to (4.7)
- Substituting $\boldsymbol{F}^*$ into Eq. (4.1) using Eq. (4.8)
- Calculating inverse of the stiffness matrix for the near-regular structure using Eqs. (4.9) to (4.11).

It should be noted that in the present algorithm every structure with an existing inverted matrix can be used. Therefore, we can further generalize our method by using inverted matrix of near-regular structures with additional members that previously obtained in Chap. 3 by Eq. (3.12)

$$\begin{bmatrix} \boldsymbol{\Delta}'_I \\ \cdots \\ \boldsymbol{\Delta}'_{II} \end{bmatrix} = \begin{bmatrix} \bar{\boldsymbol{D}}_{I,I}\boldsymbol{D}_{I,I} & \vdots & \bar{\boldsymbol{D}}_{I,I}\boldsymbol{D}_{I,II} \\ \cdots & \vdots & \cdots \\ -\boldsymbol{D}_{II,I}\boldsymbol{K}'_{I,I}\bar{\boldsymbol{D}}_{I,I}\boldsymbol{D}_{I,I} + \boldsymbol{D}_{II,I} & \vdots & -\boldsymbol{D}_{II,I}\boldsymbol{K}'_{I,I}\bar{\boldsymbol{D}}_{I,I}\boldsymbol{D}_{I,II} + \boldsymbol{D}_{II,II} \end{bmatrix} \begin{bmatrix} \boldsymbol{F}_I \\ \cdots \\ \boldsymbol{F}_{II} \end{bmatrix}$$

Or in a compact form

$$\boldsymbol{\Delta}' = \boldsymbol{D}'\boldsymbol{F} \tag{4.12}$$

Now, the blocks of matrix $\boldsymbol{D}'$ should replace the blocks of matrix $\boldsymbol{D}$ in Eqs. (4.3) through (4.7) of the present algorithm. The dimension of the blocks of matrix $\boldsymbol{D}$ in the Eq. (4.10) were already defined in Sect. 3.2 and [2]. However, the dimension of the blocks should be updated based on the dimension of the blocks $\boldsymbol{D}_{11}$, $\boldsymbol{D}_{12}$, $\boldsymbol{D}_{21}$ and $\boldsymbol{D}_{22}$ to be compatible with the current algorithm. Therefore, the matrix $\boldsymbol{D}'$ is rewritten as follows:

$$D' = \begin{bmatrix} D'_{I,I} & \vdots & D'_{I,II} \\ \dots & \vdots & \dots \\ D'_{II,I} & \vdots & D'_{II,II} \end{bmatrix} \quad (4.13)$$

where the block $D'_{\mathrm{I,I}}$ is of the dimension $m \times m$, $D'_{\mathrm{I,II}}$ is of the dimension $m \times i$, $D'_{\mathrm{II,I}}$ is of the dimension $i \times m$, and $D'_{\mathrm{II,II}}$ is of the dimension $i \times i$.

The rest of the present algorithm is repeated similarly and the solution is obtained.

4.2.1 Computational Complexity of the Method

Since the regular part of the structure can efficiently be solved and the irregular part has a small number of degrees of freedom, it is expected that the present algorithm have lower computational complexity than conventional algorithms. By small number of degrees of freedom in the irregular part we mean that in calculating computational complexity of the algorithm when n (number of degrees of freedom in the main structure) approaches infinity, the number of degrees of freedom in the irregular part does not change. In the present algorithm, we solve the equation $\boldsymbol{F} = \boldsymbol{K\Delta}$ and then find $\boldsymbol{K}^{-1}$. The matrix $\boldsymbol{K}^{-1}$ is not obtained through a direct time-consuming approach, but instead it is calculated indirectly in the process of solving the equation $\boldsymbol{F} = \boldsymbol{K\Delta}$. The blocks of $\boldsymbol{K}^{-1}$ are efficiently found because a divide and conquer algorithm is used. The main structure is divided into regular and irregular parts (divide). The regular part is efficiently solved using the relationships developed in Chap. 2 and the irregular part is swiftly solved because of its small dimension (conquer). Our interest in finding $\boldsymbol{K}^{-1}$, in addition to solving $\boldsymbol{F} = \boldsymbol{K\Delta}$ (finding $\boldsymbol{\Delta}$), is because of application of $\boldsymbol{K}^{-1}$ in problems involving reanalysis, design, rehabilitation, and constructional imperfection.

To define computational complexity of inverting the stiffness matrix, it is enough to calculate computational complexity of blocks of the inverted matrix. In Eq. (4.10), the blocks $\boldsymbol{C}_1$, $\boldsymbol{C}_2$, $\boldsymbol{C}_3$, and $\boldsymbol{C}_4$ purely belong to the irregular part, which is of a small dimension, and are not considered in the complexity computation. For the remaining blocks, components of the regular structure are present in the blocks. Therefore, let's first calculate the computational complexity of the regular part. As mentioned earlier, the relationship $\boldsymbol{A}_1\boldsymbol{A}_2 = \boldsymbol{A}_2\boldsymbol{A}_1$ holds for regular structures leading to stiffness matrices of the forms $\boldsymbol{G}_{\mathrm{n}}(\boldsymbol{A}_{\mathrm{m}}, \boldsymbol{B}_{\mathrm{m}}, \boldsymbol{A}_{\mathrm{m}})$ or $\boldsymbol{F}_{\mathrm{n}}(\boldsymbol{A}_{\mathrm{m}}, \boldsymbol{B}_{\mathrm{m}}, \boldsymbol{C}_{\mathrm{m}})$. These matrices are decomposable to n blocks of dimension m. Consequently, eigenvalues of matrices of dimension m should be calculated n times in order to find inverse of the stiffness matrix using the eigenvalues. When the dimension of the matrix approaches infinity, m does not change while n grows. In fact m is the number of nodes in a sector and is a constant value, whereas n is the number of the repetition of a sector that enlarges in our problems. As a result, computational complexity of calculating eigenvalues for a regular structure is $O(n)$.

Now, using the computational complexity of solving a regular structure, computational complexity of calculating blocks of the matrix in Eq. (4.10) is obtained as

$$\boldsymbol{B}^* = \boldsymbol{A}\boldsymbol{A}^* = \boldsymbol{A}\boldsymbol{D}_{12} \rightarrow O(n) \tag{4.14}$$

$$\boldsymbol{C}_3^* = \boldsymbol{D}_{21}\left[\boldsymbol{A}\boldsymbol{C}_1 - \boldsymbol{I}\right]\boldsymbol{B}^* + \boldsymbol{D}_{22} \rightarrow O\left(n^2\right) + O(n) \tag{4.15}$$

$$\boldsymbol{C}_1^* = \boldsymbol{D}_{21}\boldsymbol{A}\boldsymbol{C}_1 \rightarrow O(n) \tag{4.16}$$

$$\boldsymbol{C}_2^* = \boldsymbol{D}_{21}\boldsymbol{A}\boldsymbol{C}_2 \rightarrow O(n) \tag{4.17}$$

Consequently, the computational complexity of the algorithm is the dominant complexity $O(n^2)$ obtained from Eq. (4.15). In literature, computational complexity of inverting a matrix of dimension n has been reported to be $O(n^3)$ in the Gauss–Jordan elimination method, $O(n^{2.807})$ in the Strassen algorithm, and $O(n^{2.376})$ in the Coppersmith–Winograd algorithm. Although the present method has the efficient computational complexity $O(n^2)$, the mathematically strong condition $\boldsymbol{A}_1\boldsymbol{A}_2 = \boldsymbol{A}_2\boldsymbol{A}_1$ is required. Fortunately, for many real life structures, due to symmetry and repetition, the above condition holds.

4.2.2 Complementary Examples

Example 1 Consider the spatial truss under the loading condition in Fig. 4.3. The side view of the truss is indicated in x-z plane (Fig. 4.4). The degrees of freedom of the truss at nodes 1–4 are constrained. Due to constructional imperfection the central top node (i.e., node 9) is not exactly located in the center (Figs. 4.5 and 4.6). The cross sectional area for all members is equal to 5 cm^2, the elastic modulus is taken as 200 kN/mm^2, and the length of truss members is 50 cm.

The stiffness matrix for a perfect truss holds the following pattern that was studied in Chap. 2 or [3].

$$\boldsymbol{K} = \begin{bmatrix} \boldsymbol{A}_1 & \boldsymbol{A}_2 & \cdots & \boldsymbol{A}_{n-1} & \boldsymbol{A}_n \\ \boldsymbol{A}_n & \boldsymbol{A}_1 & \boldsymbol{A}_2 & \cdots & \boldsymbol{A}_{n-1} \\ \vdots & \vdots & \vdots & \vdots & \vdots \\ \boldsymbol{A}_3 & \boldsymbol{A}_4 & \cdots & \boldsymbol{A}_1 & \boldsymbol{A}_2 \\ \boldsymbol{A}_2 & \boldsymbol{A}_3 & \cdots & \boldsymbol{A}_n & \boldsymbol{A}_1 \end{bmatrix}$$

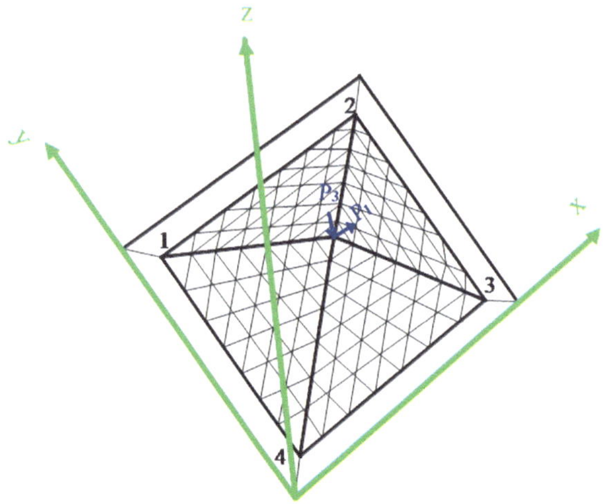

Fig. 4.3 The spatial truss under the applied loads in the Cartesian coordinate system

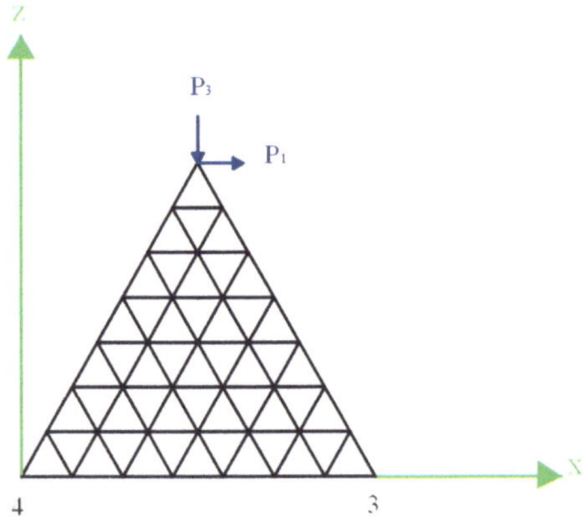

Fig. 4.4 The side view of the truss

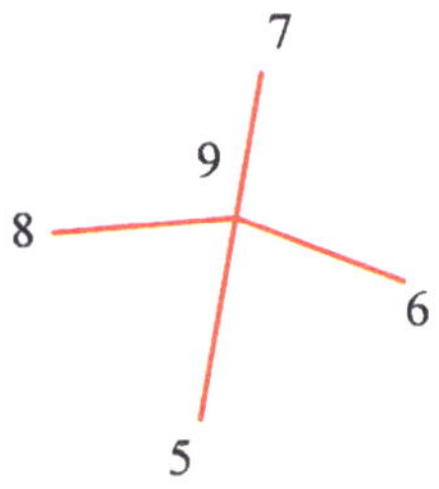

Fig. 4.5 Constructional imperfection of node 9

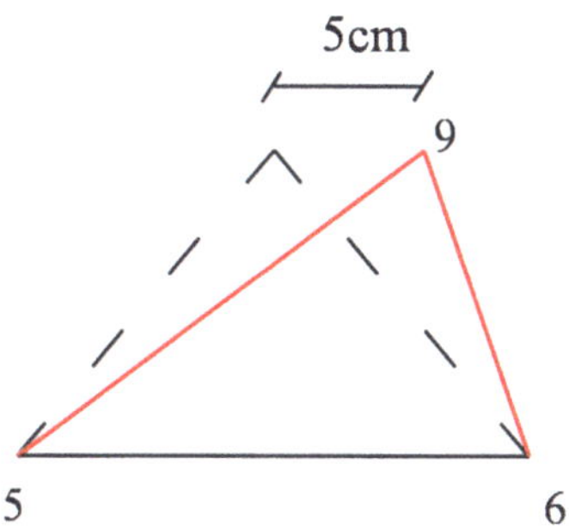

Fig. 4.6 A 5 cm shift of node 9 toward right side in the x direction

Due to constructional imperfection, however, the symmetry of the structure does not hold anymore. Alternatively, the truss can be regarded as a near-regular structure composed of regular and irregular parts (Fig. 4.7).

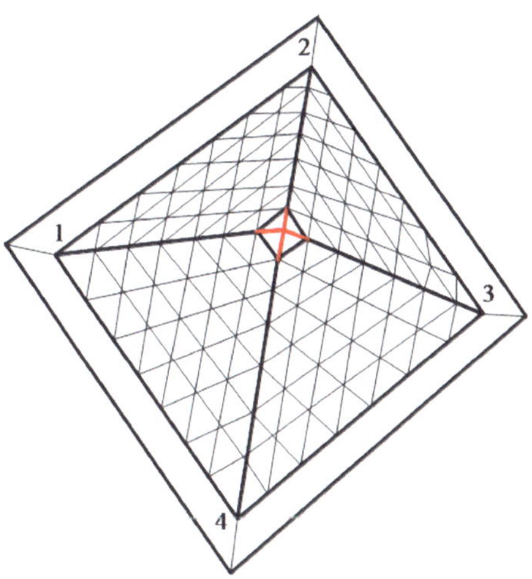

Fig. 4.7 The near-regular structure composed of regular (black) and irregular (red) parts

By forming the stiffness matrix of the regular structure in the cylindrical coordinate system, we will have

$$\boldsymbol{M} = \boldsymbol{A}_1 \otimes \boldsymbol{B}_1 + \boldsymbol{A}_2 \otimes \boldsymbol{B}_2 + \boldsymbol{A}_3 \otimes \boldsymbol{B}_3;$$
$$\boldsymbol{A}_1 = \boldsymbol{I}; \quad \boldsymbol{A}_3 = \boldsymbol{A}_2^t; \quad \boldsymbol{B}_3 = \boldsymbol{B}_2^t$$

where matrices $\boldsymbol{B}_1$ and $\boldsymbol{B}_2$ are of dimension 27 (the number of degrees of freedom at each side) and matrices $\boldsymbol{A}_2$ and $\boldsymbol{A}_3$, of dimension 4, were introduced in Chap. 2 or [2, 3].

The problem is solved using the steps below:

(1) Solving the regular structure and finding matrix $\boldsymbol{D}$ using the methods developed in Chap. 2.
(2) Inverting the block $\boldsymbol{D}_{11}$ of dimension 12 to obtain matrix $\boldsymbol{A}$, multiplying $\boldsymbol{D}_{12}$ by $\boldsymbol{A}$ to find matrix $\boldsymbol{B}^*$.
(3) Adding matrix $\boldsymbol{A}$ to block $\boldsymbol{K}_1$ and inverting the resultant matrix, of dimension 15, to find $\boldsymbol{C}_1$, $\boldsymbol{C}_2$, $\boldsymbol{C}_3$ and $\boldsymbol{C}_4$.
(4) Using Eq. (4.11) to find $\boldsymbol{C}_1^*$, $\boldsymbol{C}_2^*$, and $\boldsymbol{C}_3^*$.

Displacements of nodes 5–9 in the Cartesian coordinate system are defined as:

$$\begin{bmatrix} \boldsymbol{\Delta}_x \\ \boldsymbol{\Delta}_y \\ \boldsymbol{\Delta}_z \end{bmatrix} = \begin{bmatrix} 0.48 & 0.53 & 0.53 & 0.48 & 0.65 \\ 0 & 0 & 0 & 0 & 0 \\ -0.30 & -0.42 & -0.42 & -0.30 & -0.74 \end{bmatrix}$$

Example 2 Consider the two-dimensional truss and its loading in Fig. 4.8. The cross sectional area for all members is taken as 6 cm^2, the elastic modulus is 200 kN/mm^2, and the loads $\boldsymbol{P}_1$, $\boldsymbol{P}_2$ and $\boldsymbol{P}_3$ are equal to 40, 30 and 45 kN, respectively.

According to Fig. 4.9, the substructures in black and green are regular or near-regular Type I structures. Therefore, one of them can be selected as the regular part and the rest of the structure as the irregular part. Since the green substructure has more degrees of freedom than the black substructure, the green one is regarded as the regular part. However, it is possible to consider both black and green parts as a single regular structure (Figs. 4.10 and 4.11).

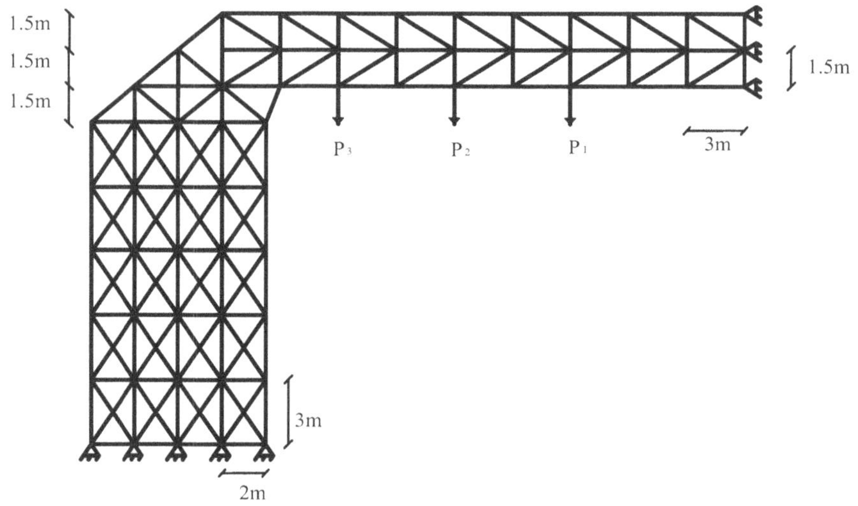

Fig. 4.8 Representation of dimensions and the loading condition for the two-dimensional truss

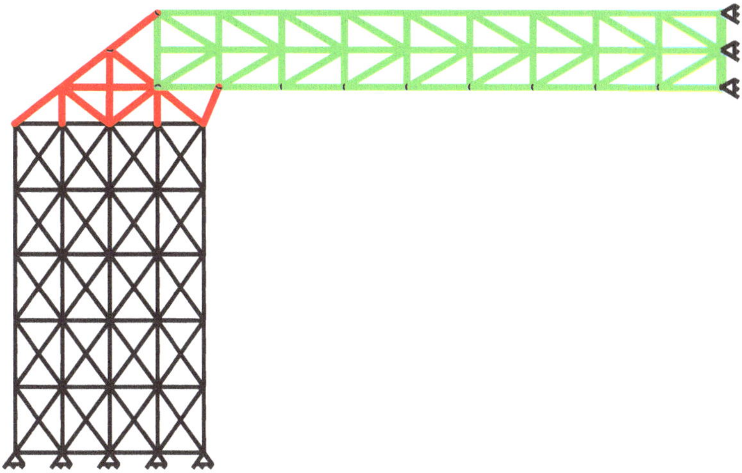

Fig. 4.9 Different parts of the structure

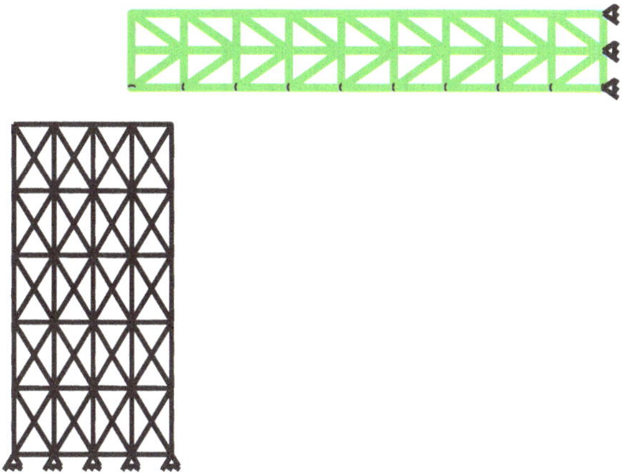

Fig. 4.10 The regular structure composed of two regular parts

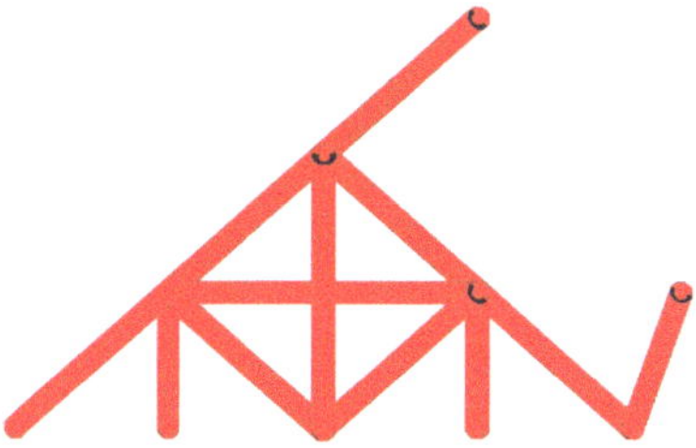

Fig. 4.11 The irregular part

Since each part of the regular structure is stable separately, the stiffness matrix of the regular structure is invertible.

$$\boldsymbol{K}_{reg} = \begin{bmatrix} \boldsymbol{K}_{reg1} & \boldsymbol{0} \\ \boldsymbol{0} & \boldsymbol{K}_{reg2} \end{bmatrix}$$

$$\boldsymbol{K}_{reg}^{-1} = \begin{bmatrix} \boldsymbol{K}_{reg1}^{-1} & \boldsymbol{0} \\ \boldsymbol{0} & \boldsymbol{K}_{reg2}^{-1} \end{bmatrix}$$

where $\boldsymbol{K}_{\text{reg1}}$ and $\boldsymbol{K}_{\text{reg2}}$ are the stiffness matrices of the black and green parts, respectively. It is notable both parts of the regular structure are, in fact, near-regular Type I structures and are convertible to the regular structures (Figs. 4.12 and 4.13).

Fig. 4.12 The near-regular structure convertible to a regular one

The inverse of the stiffness matrix (i.e., $\boldsymbol{K}_{reg1}^{-1}$ and $\boldsymbol{K}_{reg2}^{-1}$) for these near-regular Type I structures is calculated using Eqs. (3.12) or (4.10). Finally, using the proposed algorithm, the blocks of the inverted stiffness matrix of the initial near-regular Type II structure are obtained. Displacements of the three nodes at which the loads are applied are calculated as

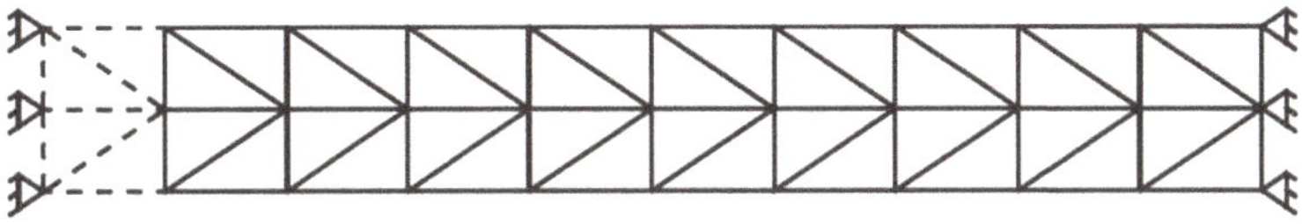

Fig. 4.13 The near-regular structure convertible to a regular one

$$\begin{bmatrix} \mathbf{\Delta}_x \\ \mathbf{\Delta}_y \end{bmatrix} = \begin{bmatrix} 0.08 & 0.11 & 0.09 \\ -0.47 & -0.64 & -0.73 \end{bmatrix}$$

Example 3 Consider the 40-story high-rise building shown in Fig. 4.14. The structure has 18,480 degrees of freedom and, therefore, a stiffness matrix of dimension 18,480 should be solved. The structure is, however, composed of regular and irregular parts (Fig. 4.14) and can be solved using the developed algorithm for near-regular Type II structures. The regular part, which is a near-regular Type I structure, is convertible to a regular structure via adding supports to the degrees of freedom at the top of the structure (4.17).

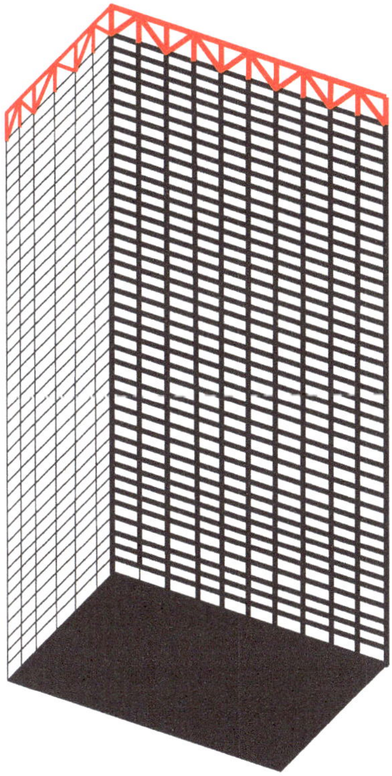

Fig. 4.14 The structure composed of regular and irregular parts

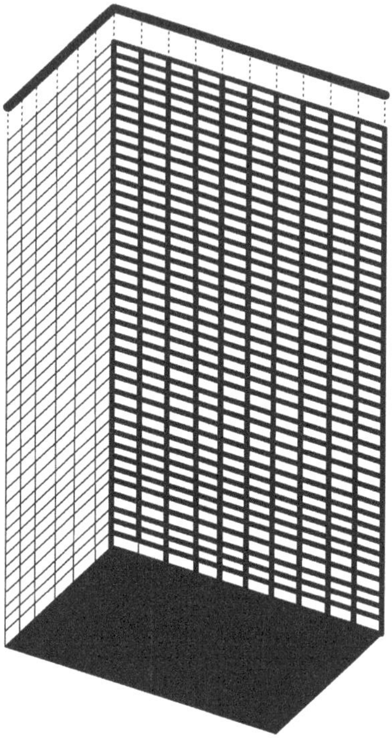

Fig. 4.15 The near-regular structure convertible to a regular one

The stiffness matrix of the regular structure (Fig. 4.15) holds the form $\boldsymbol{G}_{\mathrm{n}}(\boldsymbol{A}_{\mathrm{m}}, \boldsymbol{B}_{\mathrm{m}}, \boldsymbol{A}_{\mathrm{m}})$ with $m = 462$ and $n = 39$. Therefore, instead of solving a matrix of dimension 18,018 (462 $\times$ 39), matrices of dimension 462 should be solved. Furthermore, the stiffness matrix holds the condition for full decomposition and finally the matrices are converted to scalars. To compensate for the applied changes, the matrix $\boldsymbol{I} + \boldsymbol{D}_{1,1}\boldsymbol{K}'_{1,1}$ of dimension 462 should be inverted.

For the irregular part after adding matrix $\boldsymbol{A}$ to the block $\boldsymbol{K}_1$ of stiffness matrix $\boldsymbol{K}_{\mathrm{irreg}}$, a matrix of dimension 924 should be inverted. Finally, using Eq. (4.10), the inverted matrix of the initial structure is obtained. In summary, while in a conventional method a matrix of dimension 18,480 should be solved, the proposed algorithm only need to solve matrices of dimension 462 and 924.

4.3 Analysis of Near-Regular Structures with Additional Nodes Using Force Methods

In this section, a new force method for the analysis of near-regular structures Type II is presented. Unlike conventional force methods where a structure is converted to a determinate structure plus additional unknowns, here we convert a near-regular structure to an indeterminate structure (i.e. a regular structure and the determinate part of irregular structure) plus unknown (i.e., from the irregular structure) [4]. Such a conversion is possible because solution of regular structures is available from the methods developed in Chap. 2 or [3]. Assume the initial near-regular structure is under the load $\boldsymbol{P}$ as follow: [5]

$$\boldsymbol{P} = \left[\, P_1 \; P_2 \; \cdots \; P_n \,\right] \tag{4.18}$$

Suppose the structure includes m unknowns that should be calculated

$$\boldsymbol{X} = \left[\, X_1 \; X_2 \; \cdots \; X_m \,\right] \tag{4.19}$$

Consider the internal force of members in the initial near-regular structure as

$$\boldsymbol{S} = \left[\, S_1 \; S_2 \; \cdots \; S_j \,\right] \tag{4.20}$$

where j is the number of members in the initial structure. The internal forces vector $\boldsymbol{S}$ can be represented as a function of vectors $\boldsymbol{X}$ and $\boldsymbol{P}$

$$\boldsymbol{S} = \boldsymbol{B}_0\boldsymbol{P} + \boldsymbol{B}_1\boldsymbol{X} \tag{4.21}$$

Or

$$\boldsymbol{S} = \begin{bmatrix} \boldsymbol{B}_0 & \vdots & \boldsymbol{B}_1 \end{bmatrix} \begin{bmatrix} \boldsymbol{P} \\ \cdots \\ \boldsymbol{X} \end{bmatrix} \tag{4.22}$$

where ith column of matrix $\boldsymbol{B}_0$ includes internal forces of the structure under a unit load applied to ith degree of freedom ($\boldsymbol{P}_i = 1$). Similarly, ith column of matrix $\boldsymbol{B}_1$ includes internal forces of the structure under a unit load applied to ith redundant of the structure ($\boldsymbol{X}_i = 1$). Matrix $\boldsymbol{B} = \begin{bmatrix} \boldsymbol{B}_0 & \vdots & \boldsymbol{B}_1 \end{bmatrix}$ is known as flexibility matrix. In order to obtain the unknowns, deformation energy is minimized:

$$\boldsymbol{U} = \frac{1}{2}\left[\, S_1 \; S_2 \; \cdots \; S_j \,\right] \begin{bmatrix} F_1 & & & \\ & F_2 & & \\ & & \ddots & \\ & & & F_j \end{bmatrix} \begin{bmatrix} S_1 \\ S_2 \\ \vdots \\ S_j \end{bmatrix} \tag{4.23}$$

Or in a compact form

$$\boldsymbol{U} = \frac{1}{2}\boldsymbol{S}^T\boldsymbol{F}\boldsymbol{S} \tag{4.24}$$

where components of matrix $\boldsymbol{F}$ are the flexibility of members. To calculate the deformation energy as a function of the unknown vector $\boldsymbol{X}$, we can write

$$\boldsymbol{S}^T = \begin{bmatrix} \boldsymbol{P} \\ \cdots \\ \boldsymbol{X} \end{bmatrix}^T \begin{bmatrix} \boldsymbol{B}_0 & \vdots & \boldsymbol{B}_1 \end{bmatrix}^T = \begin{bmatrix} \boldsymbol{P} & \vdots & \boldsymbol{X} \end{bmatrix} \begin{bmatrix} \boldsymbol{B}_0 & \vdots & \boldsymbol{B}_1 \end{bmatrix}^T \tag{4.25}$$

Substituting in Eq. (4.24) results in

$$U = \frac{1}{2} \begin{bmatrix} \boldsymbol{P} & \vdots & \boldsymbol{X} \end{bmatrix} [\boldsymbol{H}] \begin{bmatrix} \boldsymbol{P} \\ \cdots \\ \boldsymbol{X} \end{bmatrix} \tag{4.26}$$

where

$$\boldsymbol{H} = \begin{bmatrix} \boldsymbol{B}_0 & \vdots & \boldsymbol{B}_1 \end{bmatrix}^T [\boldsymbol{F}] \begin{bmatrix} \boldsymbol{B}_0 & \vdots & \boldsymbol{B}_1 \end{bmatrix} \tag{4.27}$$

We can rewrite Eq. (4.26) as

$$U = \frac{1}{2} \begin{bmatrix} \boldsymbol{P} & \vdots & \boldsymbol{X} \end{bmatrix} \begin{bmatrix} \boldsymbol{H}_{pp} & \vdots & \boldsymbol{H}_{px} \\ \cdots & \vdots & \cdots \\ \boldsymbol{H}_{xp} & \vdots & \boldsymbol{H}_{xx} \end{bmatrix} \begin{bmatrix} \boldsymbol{P} \\ \cdots \\ \boldsymbol{X} \end{bmatrix} \tag{4.28}$$

Therefore,

$$U = \frac{1}{2} \boldsymbol{P}\boldsymbol{H}_{pp}\boldsymbol{P} + \boldsymbol{P}\boldsymbol{H}_{px}\boldsymbol{X} + \boldsymbol{X}\boldsymbol{H}_{xp}\boldsymbol{P} + \boldsymbol{X}\boldsymbol{H}_{xx}\mathbf{X} \tag{4.29}$$

By minimizing the energy we will have

$$\frac{\partial \boldsymbol{U}}{\partial \boldsymbol{X}} = \boldsymbol{H}_{xp}\boldsymbol{P} + \boldsymbol{H}_{xx}\boldsymbol{X} = 0 \tag{4.30}$$

And

$$\boldsymbol{X} = -\boldsymbol{H}_{xx}^{-1}\boldsymbol{H}_{xp}\boldsymbol{P} \tag{4.31}$$

Consider the near-regular structure shown in Fig. 4.16 and its regular and irregular parts in Fig. 4.17. Here, we have 10 statical indeterminacy and therefore in a traditional force method 10 redundants must be selected. However, in the present method the basic structure is chosen as a regular structure plus a determinate structure. It means one should only choose the redundants from the irregular part to convert it to

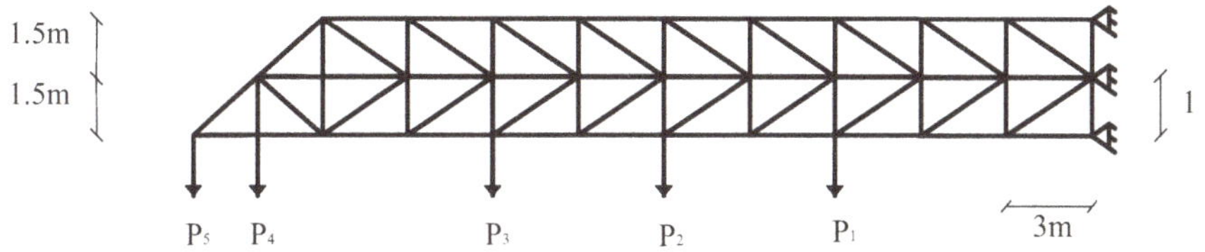

Fig. 4.16 A near-regular structure

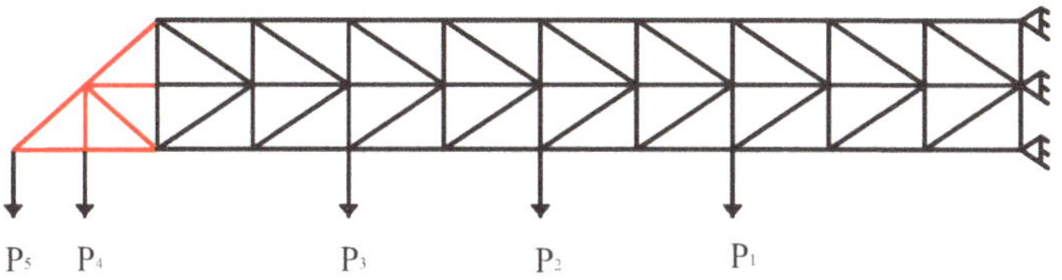

Fig. 4.17 Regular and irregular parts

a determinate structure. The irregular red part of the structure in Fig. 4.17 includes one statical indeterminacy. Now, the redundant element is considered as an external force and the basic structure is constructed as shown in Fig. 4.18.

The inverse of the stiffness matrix of the basic structure, similar to Eq. (4.10), is quickly obtained through

$$\boldsymbol{D} = \boldsymbol{K}^{-1} = \begin{bmatrix} \boldsymbol{C}_1 & \boldsymbol{C}_2 & \boldsymbol{C}_{11}\boldsymbol{B}^* \\ \boldsymbol{C}_3 & \boldsymbol{C}_4 & \boldsymbol{C}_3\boldsymbol{B}^* \\ \boldsymbol{C}_1^* & \boldsymbol{C}_2^* & \boldsymbol{C}_3^* \end{bmatrix} \tag{4.32}$$

The components of matrix $\boldsymbol{D}$ are detailed in Sect. 4.2. Now, we start forming the matrix $\boldsymbol{B}_0$ and $\boldsymbol{B}_1$ to find the matrix $\boldsymbol{H}$. The global displacements of the basic structure are obtained via Eq. (4.33).

$$\Delta = \boldsymbol{D}\bar{\boldsymbol{P}} \tag{4.33}$$

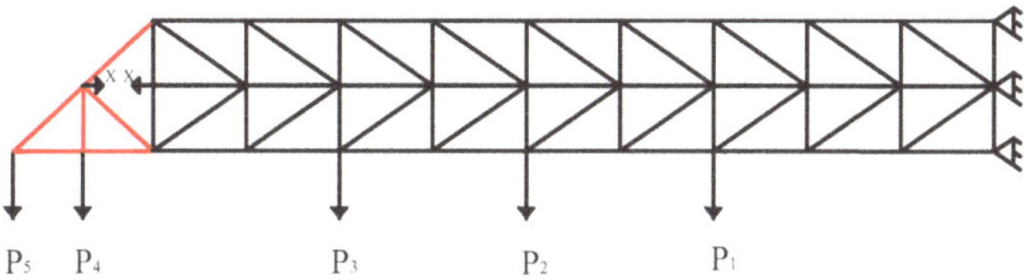

Fig. 4.18 The near-regular structure with a redundant element

where $\bar{\boldsymbol{P}}$ is the vector of global external forces that includes the vectors $\boldsymbol{P}$, $\boldsymbol{X}$ and zero components corresponding to the degrees of freedom with no load. After omitting the zero components of the vector $\bar{\boldsymbol{P}}$ and the corresponding columns of the matrix $\boldsymbol{D}$, the vector $\begin{bmatrix} \boldsymbol{P} \\ \cdots \\ \boldsymbol{X} \end{bmatrix}$ and matrix $\bar{\boldsymbol{D}}$ are obtained respectively. Now, to convert the global displacements to local (members) displacements, the equilibrium matrix is used [6].

$$\delta = \boldsymbol{A}^T \Delta = \boldsymbol{A}^T \bar{\boldsymbol{D}} \begin{bmatrix} \boldsymbol{P} \\ \cdots \\ \boldsymbol{X} \end{bmatrix} \tag{4.34}$$

where δ and $\boldsymbol{A}$ are local displacements and the equilibrium matrix, respectively [6]. Using the following equation the internal force of the members is calculated.

$$\boldsymbol{S} = \boldsymbol{k}\delta = \boldsymbol{k}\boldsymbol{A}^T \bar{\boldsymbol{D}} \begin{bmatrix} \boldsymbol{P} \\ \cdots \\ \boldsymbol{X} \end{bmatrix} \tag{4.35}$$

where, $\boldsymbol{k}$ is a block diagonal matrix in which the block diagonal components include the stiffness matrix. Considering Eq. (4.22)

$$\begin{bmatrix} \boldsymbol{B}_0 & \vdots & \boldsymbol{B}_1 \end{bmatrix} = \boldsymbol{k}\boldsymbol{A}^t \bar{\boldsymbol{D}} \tag{4.36}$$

Rewriting Eq. (4.27) results in

$$\boldsymbol{H} = \bar{\boldsymbol{D}}^T \boldsymbol{A}\boldsymbol{k}^T \boldsymbol{F}\boldsymbol{k}\boldsymbol{A}^T \bar{\boldsymbol{D}} \tag{4.37}$$

Since $\boldsymbol{k}^T \boldsymbol{F} = \boldsymbol{I}$, Therefore

$$\boldsymbol{H} = \boldsymbol{R}\boldsymbol{R}^T \tag{4.38}$$

where $\boldsymbol{R} = \bar{\boldsymbol{D}}^T \boldsymbol{A}$

Now, recalling Eq. (4.31) leads to finding the unknowns

$$\boldsymbol{X} = -\boldsymbol{H}_{xx}^{-1} \boldsymbol{H}_{xp} \boldsymbol{P} \tag{4.39}$$

The computational complexity of the present method is much less than that of the traditional method. Solving the problem using a traditional method leads to inverting the matrix $\boldsymbol{H}_{xx}$ of the dimension 10, while in the present method $\boldsymbol{H}_{xx}$ is a scalar (i.e. of the dimension 1). Moreover, in the traditional method for finding the flexibility matrix $\boldsymbol{B}$, the stiffness matrix of the basic structure of the dimension 60 should be inverted, while the basic structure in the present method is quickly inverted using Eq. (4.32).

Different steps of the method are summarized as follows

Step 1. Choosing the redundants from the irregular part of structure so that the basic structure is composed of a regular structure plus a determinate part.

Step 2. Forming the inverse of the stiffness matrix of the basic structure using Eq. (4.32) and omitting the columns corresponding to the zero forces to obtain matrix $\bar{\boldsymbol{D}}$

Step 3. Forming the equilibrium matrix $\boldsymbol{A}$ using the quick solution suggested in [7].

Step 4. Forming the matrix $\boldsymbol{H}$ using Eq. (4.38) and finding the unknowns using Eq. (4.39).

4.4 Eigensolution of Near-Regular Structures

In this section, a method is presented to find the eigenvalues of near-regular structures. To find eigenvalues of a near-regular structure, unlike the solution of $\boldsymbol{F} = \boldsymbol{K\Delta}$ for near-regular structures where manipulating the physics of the problem is helpful, rather pure mathematical calculation is required. A numerical method is proposed where using an appropriate ordering of the nodes of the structure and partitioning the stiffness matrix, the matrix becomes decomposable. The decomposed parts are solved using a bisection method with a proper initial start point until the solution is converged and eigenvalues are obtained [8].

4.4.1 Decomposition of Partitioned Block Matrices

This section is devoted to some techniques of partitioned block matrices. Manipulating a partitioned matrix is a basic and helpful approach in matrix analysis. The applied methods in partitioned matrices are similar to those of ordinary numerical matrices in some ways. Consider a 2×2 matrix as

$$\begin{bmatrix} a & b \\ c & d \end{bmatrix}, \quad a, b, c, d \in \boldsymbol{C} \tag{4.40}$$

Using a basic row operation, the second row of the matrix multiplied by 2 can be added to the first row so that the determinant of the matrix does not change:

$$\begin{bmatrix} 1 & 2 \\ 0 & 1 \end{bmatrix}\begin{bmatrix} a & b \\ c & d \end{bmatrix} = \begin{bmatrix} a+2c & b+2d \\ c & d \end{bmatrix} \tag{4.41}$$

This row or column operation can be generalized to partitioned block matrices as

1. Interchanging two block rows or columns,
2. Multiplying a block row or column from the left or right by a matrix of proper size,
3. Multiplying a block row or column by a matrix and then adding it to another row or column.

It should be noted that for maintaining the symmetry of the matrix, these operations are performed simultaneously on rows and columns and therefore the determinant does not change.

Now using the above three basic rules, one can convert a two-block matrix into a two-block diagonal matrix so that the determinant does not change:

$$\begin{bmatrix} \boldsymbol{I}_n & \boldsymbol{0} \\ -\boldsymbol{M}_{21}\boldsymbol{M}_{11}^{-1} & \boldsymbol{I}_m \end{bmatrix}\begin{bmatrix} \boldsymbol{M}_{11} & \boldsymbol{M}_{12} \\ \boldsymbol{M}_{21} & \boldsymbol{M}_{22} \end{bmatrix}\begin{bmatrix} \boldsymbol{I}_n & -\boldsymbol{M}_{11}^{-1}\boldsymbol{M}_{12} \\ \boldsymbol{0} & \boldsymbol{I}_m \end{bmatrix} = \begin{bmatrix} \boldsymbol{M}_{11} & \boldsymbol{0} \\ \boldsymbol{0} & \boldsymbol{M}_{22} - \boldsymbol{M}_{21}\boldsymbol{M}_{11}^{-1}\boldsymbol{M}_{12} \end{bmatrix} \tag{4.42}$$

where $\boldsymbol{M}_{11}$ is an invertible matrix of dimension n and $\boldsymbol{M}_{22}$ is a matrix of dimension m.

Consequently, we will have

$$\begin{vmatrix} \boldsymbol{M}_{11} & \boldsymbol{M}_{12} \\ \boldsymbol{M}_{21} & \boldsymbol{M}_{22} \end{vmatrix} = det\boldsymbol{M}_{11}\det\left(\boldsymbol{M}_{22} - \boldsymbol{M}_{21}\boldsymbol{M}_{11}^{-1}\boldsymbol{M}_{12}\right) \tag{4.43}$$

For finding the eigenvalues of the partitioned matrices, the matrix $\boldsymbol{M}_{11}$ is changed to the matrix $\boldsymbol{M}_{11} - \lambda\boldsymbol{I}_n$ and the matrix $\boldsymbol{M}_{22}$ is changed to the matrix $\boldsymbol{M}_{22} - \lambda\boldsymbol{I}_m$,

$$\begin{vmatrix} \boldsymbol{M}_{11} - \lambda\boldsymbol{I}_n & \boldsymbol{M}_{12} \\ \boldsymbol{M}_{21} & \boldsymbol{M}_{22} - \lambda\boldsymbol{I}_m \end{vmatrix} = det(\boldsymbol{M}_{11} - \lambda\boldsymbol{I}_n)\det\left(\boldsymbol{M}_{22} - \lambda\boldsymbol{I}_m - \boldsymbol{M}_{21}(\boldsymbol{M}_{11} - \lambda\boldsymbol{I}_n)^{-1}\boldsymbol{M}_{12}\right) \tag{4.44}$$

Equating this determinant to zero, results in:

$$det(\boldsymbol{M}_{11} - \lambda\boldsymbol{I}_n) = 0 \tag{4.45}$$

and

$$\det\left(\boldsymbol{M}_{22} - \lambda\boldsymbol{I}_m - \boldsymbol{M}_{21}(\boldsymbol{M}_{11} - \lambda\boldsymbol{I}_n)^{-1}\boldsymbol{M}_{12}\right) = 0 \tag{4.46}$$

Considering the condition of the decomposition, where $\boldsymbol{M}_{11}$ in Eq. (4.43) or $\boldsymbol{M}_{11} - \lambda \boldsymbol{I}_n$ in Eq. (4.44) is invertible, Eq. (4.45) does not hold. Consequently, for finding the eigenvalues of the two-block matrix, only Eq. (4.46) should be solved.

As a simple example consider the following matrix

$$\boldsymbol{M} = \begin{bmatrix} 0.8147 & 0.6324 & 0.9575 & 0.9572 \\ 0.9058 & 0.0975 & 0.9649 & 0.4854 \\ 0.1270 & 0.2785 & 0.1576 & 0.8003 \\ 0.9134 & 0.5469 & 0.9706 & 0.1419 \end{bmatrix}$$

Using Eq. (4.45), we will have

$$\det\left(\begin{bmatrix} 0.1576 & 0.8003 \\ 0.9706 & 0.1419 \end{bmatrix} - \lambda \boldsymbol{I} - \begin{bmatrix} 0.1270 & 0.2785 \\ 0.9134 & 0.5469 \end{bmatrix}\right.$$
$$\left.\left(\begin{bmatrix} 0.8147 & 0.6324 \\ 0.9058 & 0.0975 \end{bmatrix} - \lambda \boldsymbol{I}\right)^{-1} \begin{bmatrix} 0.9575 & 0.9572 \\ 0.9649 & 0.4854 \end{bmatrix}\right) = 0$$

The roots of this equation will be the eigenvalues of Matrix $\boldsymbol{M}$.

$$\lambda_M = \{2.4021, -0.0346, -0.7158, -0.4400\}$$

Solving Eq. (4.46) results in finding the characteristic equation of the two-block matrix. In general, finding the eigenvalues of matrices using their characteristic equation is not considered as a time-saving approach because both forming and solving the characteristic equation are difficult.

4.4.2 Computational Complexity of Eigensolution of Near-Regular Graphs via Solution of Characteristic Equation

Eigensolution of a matrix by solving the equation $det(\boldsymbol{M} - \lambda \boldsymbol{I}) = 0$ includes forming the characteristic equation and then solving it by an iterative method. In addition to the difficulties inherent to the formation of characteristic equation, using an iterative method for calculating eigenvalues makes the method very complicated. However, when it comes to near-regular forms, the method changes to an efficient approach via the application of the features of the regular patterns.

Complexity theory has been well developed in numerical linear algebra. However, for eigenvalue problems where iterative methods are required, the complexity of algorithms cannot be easy obtained since the number of steps for achieving a desirable accuracy is not definite. Another problem is concerned with the initial starting point

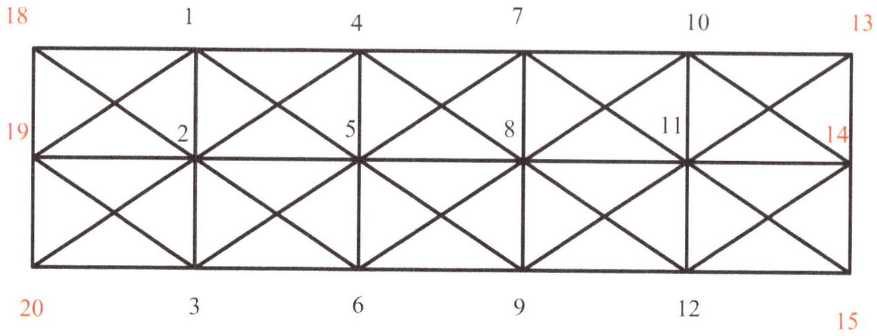

Fig. 4.19 A near-regular graph with a suitable ordering

that is very important in many iterative algorithms which affects the number of iterations and consequently the computational complexity.

In this study, the difficulties of solving the equation $det(\boldsymbol{M} - \lambda \boldsymbol{I}) = 0$, shown in Eq. (4.46), are smoothened for near-regular graphs. The method is carried out through

1. Removing the difficulty of inverting the matrix $\boldsymbol{M}_{11} - \lambda \boldsymbol{I}_n$, shown in the Eq. (4.46).
2. Decreasing the complexity of the applied iterative method.

As it was mentioned, the matrices in this study are near-regular matrices composed of regular and irregular parts. Numbering the nodes of the graph (structure) is performed in such a way that the regular and irregular parts are separated. Therefore, the Laplacian (stiffness) matrix will have a two-block matrix form $\boldsymbol{M}$ as shown in Eq. (4.42), where the block $\boldsymbol{M}_{11}$ represents the regular part and the block $\boldsymbol{M}_{22}$ represents the irregular part. Such an ordering can be seen in Fig. 4.19 for a near-regular graph.

Where the black numbers form the regular block $\boldsymbol{M}_{11}$ and the red ones form the irregular block $\boldsymbol{M}_{22}$. Previously, the inverse of the Laplacian matrix of regular graphs was obtained using its eigenpairs [9]:

$$\begin{gathered}
\boldsymbol{Ax} = \boldsymbol{b} \Rightarrow \{\varphi\}_j^T \boldsymbol{A}\{\varphi\}_j y_j = \lambda_j y_j = \{\varphi\}_j^T \boldsymbol{b} \\
y_j = \frac{b_j}{\lambda_j} \Rightarrow \{x\}_n = \sum_{i=1}^{n} \{\varphi\}_i \\
y_i = \sum_{i=1}^{n} \{\varphi\}_i \frac{b_i}{\lambda_i} = \sum_{i=1}^{n} \frac{\{\varphi\}_i \{\varphi\}_i^T}{\lambda_i} \boldsymbol{b}
\end{gathered} \tag{4.47}$$

where λ_i and $\{\varphi\}_i$ are the eigenpairs of the matrix $\boldsymbol{A}$.

Since the block $\boldsymbol{M}_{11}$ is regular, $\boldsymbol{M}_{11} - \lambda \boldsymbol{I}_n$ is also regular. But in references [9–11], the eigenpairs and the inverted matrix were found for a regular matrix filled with numbers. Inverting the parametrical matrix $\boldsymbol{M}_{11} - \lambda \boldsymbol{I}_n$, solving the parametrical determinant and finding the characteristic equation is rather difficult. However, since the characteristic equation should be ultimately solved iteratively to find the eigenvalues, in the present study an approach is utilized in which the iterative method is applied at earlier stage. In other words, the iterative solution is directly applied to the Eq. (4.46) without forming the characteristic equation. The advantage of applying the iterative solution in advance is that after making an initial guess for λ in the Eq. (4.46), the numerical matrix $\boldsymbol{M}_{11} - \lambda \boldsymbol{I}_n$ is quickly inverted using the Eq. (4.47) and subsequently the determinant is solved. The determinant is of the dimension of the irregular part ($\boldsymbol{M}_{22}$). Since in a near-regular graph we assume the irregular part to be small, the determinant is easily solved iteratively, until stopping criterion is satisfied. The iteration is performed using a bisection method to obtain all eigenvalues. Using strong initial guesses for the λ_is ensures the convergence of the solution. The Laplacian matrix for the graph shown in Fig. 4.19 can be written as

$$\boldsymbol{M} = \begin{bmatrix} \boldsymbol{M}_{11} & \boldsymbol{M}_{12} \\ \boldsymbol{M}_{21} & \boldsymbol{M}_{22} \end{bmatrix}; (\boldsymbol{M}_{11})_{nm} = \boldsymbol{F}_n(\boldsymbol{C}_m, \boldsymbol{B}_m, \boldsymbol{C}_m) = \boldsymbol{I}_n \otimes \boldsymbol{C}_m + \boldsymbol{T}_n \otimes \boldsymbol{B}_m \tag{4.48}$$

From Eq. (4.48) it can be seen that $\boldsymbol{T}_n = \boldsymbol{F}_n(0, 1, 0)$, and obviously $\boldsymbol{I}_n \boldsymbol{T}_n = \boldsymbol{T}_n \boldsymbol{I}_n$, i.e. the condition of Eq. (2.12) is satisfied and $\boldsymbol{M}_{11}$ has the property of being block disgonalized.

According to the definition of a near-regular structure, the regular part is much larger than the irregular part. In other words, in the computational complexity of a near-regular structure it is assumed that only the dimension of $\boldsymbol{M}_{11}$ (the regular part) approaches to infinity. Consequently, the initial guess is made using a matrix of the dimension of the matrix $\boldsymbol{M}$ but a pattern similar to the pattern of the regular matrix $\boldsymbol{M}_{11}$. Therefore, the following matrix is used for the initial guess

$$(\boldsymbol{M}^*)_{lm} = \boldsymbol{F}_l(\boldsymbol{C}_m, \boldsymbol{B}_m, \boldsymbol{C}_m) \tag{4.49}$$

Since the matrix $\boldsymbol{M}^*$ has a regular form, its eigenvalues are easily obtained. The eigenvalues of the matrix $\boldsymbol{M}^*$ are close to the eigenvalues of the matrix $\boldsymbol{M}$. The similarity becomes more obvious, when the regular part becomes larger.

Another way of finding the matrix $\boldsymbol{M}^*$ is to obtain the matrix $\boldsymbol{M}$ through a usual ordering as shown in Fig. 4.20. We will have

$$\boldsymbol{M}_{nm} = \boldsymbol{F}_n(\boldsymbol{A}_m, \boldsymbol{B}_m, \boldsymbol{C}_m) \tag{4.50}$$

Since the graph follows the pattern of the strong Cartesian product, $\boldsymbol{A}_m - \boldsymbol{B}_m \neq \boldsymbol{C}_m$ and it is classified as a near-regular form. For converting the near-regular matrix $\boldsymbol{M}$ to a regular one, the block $\boldsymbol{A}_m$ is changed to the block $\boldsymbol{C}_m$. This conversion causes the regular matrix $\boldsymbol{M}^*$ to be obtained again. In Ref. [12] it is shown that addition of members around a graph, the condition $A_m - B_m = C_m$ holds and the eigenvaluses

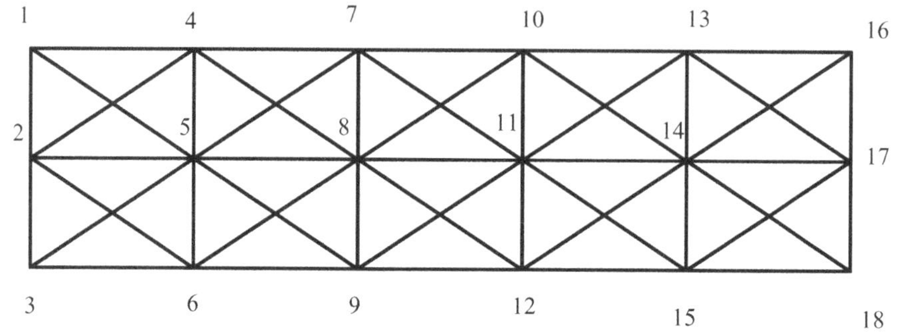

Fig. 4.20 A near-regular graph with a usual ordering

of the expanded graph are not much different with those of the graph before addition. Thus the eigenvalues of $\boldsymbol{M}$ and $\boldsymbol{M}^*$ are close to each other.

Up to now, a proper ordering was considered, the iterative method was applied in advance, the appropriate initial guess was made and the inverse of the matrix $\boldsymbol{M}_{11} - \lambda \boldsymbol{I}_n$ was found. Now, the complexity of the method is evaluated to show it is both measureable and less complicated. As it was mentioned, the computational complexity of many iterative methods is not measureable. The computational complexity of an iterative bisection method for finding the roots of the relationship $f(x) = 0$ is measureable only in the case of finding a certain root in the interval [*a*, *b*]. However, while finding the multiple roots of an equation without having a sense of the domain of the roots, obtaining the complexity of the method is impossible. The problem gets worse when the order of the equation (the characteristic equation in our problems) grows. Growing the order of the equation is certain because in calculating the complexity of the method the order of the equation (N) should approach to infinity.

Now, it is shown why calculating the complexity of the method in the near-regular structures is possible. In a bisection method for finding roots in the interval [*a*, *b*], the number of iterations (n) can be obtained as follows:

$$\frac{b-a}{2^{n+1}} < \varepsilon \tag{4.51}$$

where ε is the upper bound for the error of the answer. The Eq. (4.51) can also be written as

$$n > \frac{\ln(b-a) - \ln(\varepsilon)}{\ln(2)} - 1 \tag{4.52}$$

Which means for a specific ε, the number of iterations depends on the interval [*a*, *b*]. Consequently, the computational complexity of the method depends on the interval [*a*, *b*]. In obtaining the computational complexity of an arbitrary characteristic equation finding the proper *a* and *b* is impossible because one cannot predict

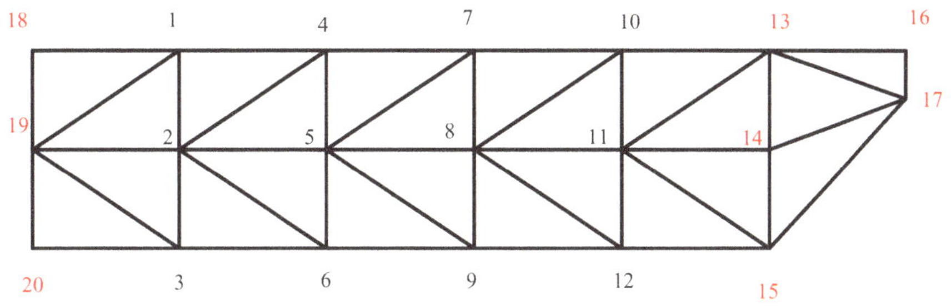

Fig. 4.21 A near-regular structure with a suitable ordering

the changes of the interval [a, b] when the order of the equation grows. In many cases, by increasing the order of the equation, the interval [a, b] grows too with an indefinite rate. This way, not only the number of iterations grows, but also finding the rate of the growth becomes impossible.

However, when we consider near-regular graphs, the interval [a, b] is limited and measurable. In a near-regular graph the regular part is the dominant part regarding the definition of near-regular graphs. Moreover, the eigenvalues of the main structure are close to those of the corresponding regular part. Consequently, for finding the domain of the problem (the interval [a, b]), one should obtain the domain of the regular structure, showing the lower and upper bounds are independent of the order of the equation. In graph product theory, the regular forms are presented in Kronecker forms, [Eq. (2.9)].

Since the matrices $\boldsymbol{A}_1$ and $\boldsymbol{A}_2$ represent the number of sectors in a graph, in comparison with the unchanged matrices $\boldsymbol{B}_1$ and $\boldsymbol{B}_2$ which represent the dimension of each sector, the matrices $\boldsymbol{A}_1$ and $\boldsymbol{A}_2$ are the basic matrices and their dimension will increase when the graph $\boldsymbol{M}$ expands.

Therefore, the matrices $\boldsymbol{A}_1$ and $\boldsymbol{A}_2$ control the bounds of the eigenvalues. Consider the Fig. 4.21.

In such a near-regular structure, the black nodes form the regular matrix $\boldsymbol{M}_{11}$ as

$$\boldsymbol{M}_{11} = \boldsymbol{F}(\boldsymbol{C}, \boldsymbol{B}, \boldsymbol{C}) = \boldsymbol{I} \otimes \boldsymbol{C} + \boldsymbol{T} \otimes \boldsymbol{B}; \boldsymbol{T} = \boldsymbol{F}(0, 1, 0). \tag{4.53}$$

The dimension of the matrices $\boldsymbol{I}$ and $\boldsymbol{T}$ represent the number of regular sectors that is equal to 4 in this example and the matrices $\boldsymbol{B}$ and $\boldsymbol{C}$ represent the number of nodes in each sector that is equal to 3 here. When the graph shown in the Fig. 4.21 grows, the number of sectors grows ($\boldsymbol{I}$ and $\boldsymbol{T}$), while the number of nodes in each sector does not change ($\boldsymbol{B}$ and $\boldsymbol{C}$). Thus, the basic matrices $\boldsymbol{A}_1$ and $\boldsymbol{A}_2$ control the bounds and computational complexity.

The known basic matrices in graph products include two groups: (1) tri-diagonal matrices (2) Circulant matrices

Consider the following tri-diagonal matrix for the first group:

$$Z_n = \begin{bmatrix} -\alpha + b & c & 0 & 0 & \dots & 0 & 0 \\ a & b & c & 0 & \dots & 0 & 0 \\ 0 & a & b & c & \dots & 0 & 0 \\ \dots & \dots & \dots & \dots & \dots & \dots & \dots \\ 0 & 0 & 0 & 0 & \dots & a & -\beta + b \end{bmatrix}_{n \times n} \tag{4.54}$$

The eigenvalues of such a matrix is presented in [13] using the following equation:

$$\lambda = b + 2\sqrt{ac}\cos\theta \; \theta \neq m\pi, m \in Z \tag{4.55}$$

where θ is calculated as follows:

$$ac\sin(n+1)\theta + (\alpha + \beta)\sqrt{ac}\sin n\theta + \alpha\beta\sin(n-1)\theta = 0 \tag{4.56}$$

In a regular graph α and β are equal to zero and $a = c$, then Eq. (4.56) is converted to

$$a^2\sin(n+1)\theta = 0 \tag{4.57}$$

And the Eq. (4.55) changes to the following:

$$\lambda = b + 2a\cos\theta \; \theta \neq m\pi, m \in Z \tag{4.58}$$

Using Eq. (4.58) we can define lower and upper bounds of the eigenvalues of matrices A_1 and A_2. Therefore, in a regular graph/structure, according to Eq. (4.58) and Eq. (2.16), as the size of the structure (reflected in n) approaches infinity, the bounds of eigenvalues stay limited.

Consequently, in a regular graph/structure when the number of sectors increases, according to Eq. (4.58) the lower and upper bounds of the eigenvalues does not change. This is because n in Eq. (4.57) influences only the cos θ in Eq. (4.58) that is always in the interval $[-1, 1]$. Therefore, the lower and upper bounds of the eigenvalues in Eq. (4.58) can be found. Using the Eq. (2.16), the eigenvalues for a block tri-diagonal matrix of form $\boldsymbol{F}_n(\boldsymbol{A}_m, \boldsymbol{B}_m, \boldsymbol{C}_m)$ are obtained.

Now, consider the following circulant matrices:

$$P = \begin{bmatrix} 0 & 1 & 0 & \dots & 0 \\ 0 & 0 & 1 & \dots & 0 \\ \vdots & \vdots & \vdots & \vdots & \vdots \\ 0 & 0 & 0 & \dots & 1 \\ 1 & 0 & 0 & \dots & 0 \end{bmatrix}_{n \times n} \tag{4.59}$$

It was already shown that the eigenvaluse of this matrix are obtained via solving

$$\det(P - \lambda I) = \lambda^n - 1 = 0 \tag{4.60}$$

The lower and upper bounds of λ in Eq. (4.60) are in the interval [−1, 1] for real values. These bounds are independent from the magnitude of n. Similarly, using Eq. (2.16) the eigenvalues for a block matrix of form $\boldsymbol{G}_{\mathrm{n}}(\boldsymbol{A}_{\mathrm{m}}, \boldsymbol{B}_{\mathrm{m}}, \boldsymbol{A}_{\mathrm{m}})$ are found.

Now, according to Eq. (4.52), since the value of $\ln(b - a)$ is quantifiable and limited, the number of iterations can be determined for a regular graph/structure with any arbitrary dimension. Therefore, the computational complexity of the proposed iterative method depends on the number of roots, which is n roots in a graph/structure of dimension n, and consequently the complexity of each iteration is $\mathbf{O}(n)$. It was mentioned that matrices of regular graphs/structures holding patterns of $\boldsymbol{G}_n(\boldsymbol{A}_m, \boldsymbol{B}_m, \boldsymbol{A}_m)$ and $\boldsymbol{F}_n(\boldsymbol{A}_m, \boldsymbol{B}_m, \boldsymbol{C}_m)$ can be decomposed to n blocks of dimension m. Therefore, eigenvalues of n matrices of dimension m should be calculated that is of order $\mathbf{O}(m^3 n) \sim \mathbf{O}(n)$. Then, the inverse of the matrix $\boldsymbol{M}_{11} - \lambda \boldsymbol{I}_n$ is obtained using the eigenvalues. It was mentioned as the dimension of the matrix $\boldsymbol{M}_{11} - \lambda \boldsymbol{I}_n$ approaches infinity, m (the number of nodes in a sector) does not change and only n (as the number of sectors) grows. As a result, the complexity of inverting the matrix $\boldsymbol{M}_{11} - \lambda \boldsymbol{I}_n$ is $\mathbf{O}(n)$ as well. Consider the main equation as

$$\det\left(\boldsymbol{M}_{22} - \lambda \boldsymbol{I}_m - \boldsymbol{M}_{21}(\boldsymbol{M}_{11} - \lambda \boldsymbol{I}_n)^{-1}\boldsymbol{M}_{12}\right) = 0 \tag{4.61}$$

After obtaining the value of $(\boldsymbol{M}_{11} - \lambda \boldsymbol{I}_n)^{-1}$ for a specific λ in an iteration, the complexity of the term $\boldsymbol{M}_{21}(\boldsymbol{M}_{11} - \lambda \boldsymbol{I}_n)^{-1}\boldsymbol{M}_{12}$ should be obtained. This is multiplication of three matrices of the computational complexity $\mathbf{O}(2m^2n^2)$. Since m is limited, the computational complexity would be $\mathbf{O}(n^2)$ that is the largest and dominant complexity in solving this equation. Thus, an efficient complexity is gained using the present method for the near-regular graphs/structures. This method is less complex than other conventional methods such as the method using a combination of householder and QR methods wherein the computational complexity is $\mathbf{O}(n^3)$.

Now, the present method is outlined in the following steps:

1. Numbering the near-regular graph/structure so that first the regular part is numbered and then the irregular part.
2. Forming the Laplacian/stiffness matrix for the graph/structure. The matrix will have the pattern of the two-block matrix $\boldsymbol{M}$ because of the mentioned numbering.
3. Forming Eq. (4.46) as the main equation.
4. Forming a Laplacian/stiffness matrix named $\boldsymbol{M}^*$ of the dimension $n + m$ (similar to the dimension of matrix $\boldsymbol{M}$) but with the pattern of matrix $\boldsymbol{M}_{11}$. In other words, forming a regular matrix corresponding to the near-regular matrix with the same dimension [see the Eq. (4.49)].
5. Finding the eigenvalues of the regular matrix $\boldsymbol{M}^*$ through a quick solution using the Eq. (2.16). These eigenvalues are used as the suitable initial guesses as they are close to the eigenvaluses of the near-regular graph. The eigenvalues are in an interval $[a, b]$ which is independent from the dimension of the matrix $\boldsymbol{M}^*$.

6. Solving Eq. (4.46) through an iterative bisection method without forming the characteristic equation. The eigenvalues found in step 5 are used as initial guesses in this step. Applying the iterative method in advance allows us to solve $(\boldsymbol{M}_{11} - \lambda \boldsymbol{I}_n)^{-1}$ quickly using the Eq. (4.47), decreasing the complexity of the method.

Previously, the inverse of the stiffness matrix of near-regular structures Type I was found. In Eq. (4.46), numbering is performed such that the matrix $\boldsymbol{M}_{11} - \lambda \boldsymbol{I}_n$ is a regular one and the condition $\boldsymbol{A}_1\boldsymbol{A}_2 = \boldsymbol{A}_2\boldsymbol{A}_1$ holds. Therefore, inverting this matrix using the eigenpairs and the Eq. (4.47) is possible. However, since the stiffness matrix of near-regular structures Type I can also be efficiently inverted, numbering of the graph/structure can be performed such that the block $\boldsymbol{M}_{11}$ is a near-regular Type I matrix. Thus, the matrix $\boldsymbol{M}_{11} - \lambda \boldsymbol{I}_n$ is swiftly inverted while the condition $\boldsymbol{A}_1\boldsymbol{A}_2 = \boldsymbol{A}_2\boldsymbol{A}_1$ does not hold. This way the present method becomes more generalized as the matrix $\boldsymbol{M}_{11}$ can be a regular or near-regular Type I one. The efficiency of the method is shown using some examples.

4.4.3 A Simple Illustrative Example

For the truss shown in the Fig. 4.22, the eigenvalues of the stiffness matrix are calculated. For all members the elastic modulus is considered as 210 kN/mm^2 and the cross section areas are taken as 15 cm^2.

Based on the present algorithm, first a proper ordering is performed. According to Fig. 4.23, the truss is composed of regular and irregular parts that are shown in green and red colors, respectively.

Ordering of the regular part is performed sector by sector as it is shown in Fig. 4.23. In this structure each five nodes form a sector, for example the first sector is composed of the nodes 1, 2, 3, 4 and 5. Thus, the matrix $\boldsymbol{M}_{11}$ contains 7 blocks (each sector represents a block) of dimension 10 (each sector consists of 5 nodes with each node having two degrees of freedom). The last regular sector does not belong to the matrix $\boldsymbol{M}_{11}$ since it has 2 nodes in common with the irregular part (the nodes 38 and 40) that causes this sector not to follow the pattern of other regular sectors. Consequently,

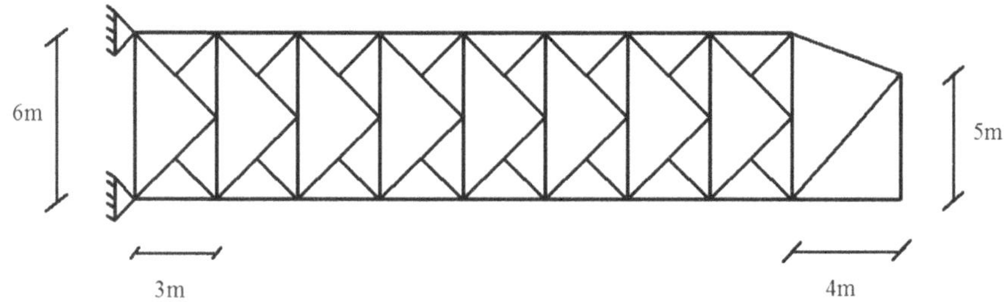

Fig. 4.22 The truss under study

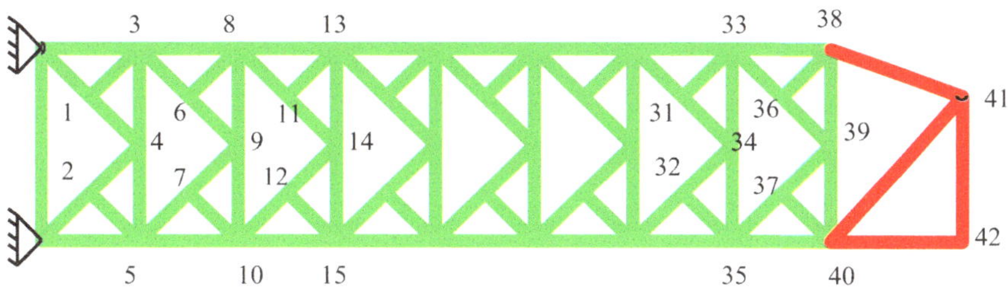

Fig. 4.23 The truss composed of regular and irregular parts with a suitable ordering

the sector number 8 is considered as a part of the matrix $\boldsymbol{M}_{22}$ as well as the irregular part shown in the red color. The matrix $\boldsymbol{M}_{22}$ contains 7 nodes and is of dimension 14. Using this partitioning, the reduced stiffness matrix will have the following form:

$$\boldsymbol{M} = \begin{bmatrix} \boldsymbol{M}_{11} & \boldsymbol{M}_{12} \\ \boldsymbol{M}_{21} & \boldsymbol{M}_{22} \end{bmatrix}_{84\times 84}$$

where $\boldsymbol{M}_{12}$, $\boldsymbol{M}_{21}$ and $\boldsymbol{M}_{22}$ are matrices of dimensions 70×14, 14×70 and 14×14, respectively, and

$$\boldsymbol{M}_{11} = \boldsymbol{F}_{70\times 70}(\boldsymbol{C}, \boldsymbol{B}, \boldsymbol{C}) = \boldsymbol{I} \otimes \boldsymbol{C} + \boldsymbol{T} \otimes \boldsymbol{B}$$

where $\boldsymbol{B}$ and $\boldsymbol{C}$ are matrices of the dimension 10.

For finding the initial guesses and the interval $[a, b]$ the matrix $\boldsymbol{M}^*$ should be formed. It was mentioned that this matrix should have dimension identical to that of the matrix $\boldsymbol{M}$. But, since the dimension of the matrix $\boldsymbol{M}$ is 84, the matrix $\boldsymbol{M}^*$ cannot be formed because it is composed of blocks of dimension 10. In other words, this matrix just can have dimensions of the multiples of 10 such as 80 or 90. To overcome the problem, 6 rows and columns filled with zeros are added to the matrix M to get the dimension 90. These zeros do not affect the eigenvalues of the matrix $\boldsymbol{M}$ and only add 6 additional zero eigenvalues to the final answer. This new matrix is named $\boldsymbol{M}'$.

$$\boldsymbol{M}' = \begin{bmatrix} \boldsymbol{M} & \vdots & 0_{84\times 6} \\ \cdots & & \cdots \\ 0_{6\times 84} & \vdots & 0_{6\times 6} \end{bmatrix}_{90\times 90}$$

The matrix $\boldsymbol{M}^*$ of the dimension 90 (9 blocks of the dimension 10) is formed as:

$$\boldsymbol{M}^* = \boldsymbol{F}(\boldsymbol{C}, \boldsymbol{B}, \boldsymbol{C}) = \boldsymbol{I} \otimes \boldsymbol{C} + \boldsymbol{T} \otimes \boldsymbol{B}$$

In fact the matrix $\boldsymbol{M}^*$ is the stiffness matrix of the regular structure of Fig. 4.24.

Fig. 4.24 A regular structure used for appropriate initial guesses

The supports in the right side are chosen so that the last sector is similar to the other sectors.

Now using the Eq. (2.16), the eigenvalues of the matrix $\boldsymbol{M}^*$ are obtained. There are 90 eigenvalues for the matrix $\boldsymbol{M}^*$. Typical eigenvalues λ_{11} to λ_{20} are written in the following:

$$\lambda_{M^*} = 10^6[1.7329, 1.8604, 2.0003, 2.0649, 2.4029, \\ 2.4228, 2.9563, 3.4094, 3.5748, 3.9336]$$

Since for the matrix $\boldsymbol{M}^*$, $\lambda_{\min} = \lambda_1 = 0.2951 \times 10^6$ and $\lambda_{\max} = \lambda_{90} = 63.461 \times 10^6$ and regarding the six zero eigenvalues added to the main structure, the interval used in the bisection method is chosen as $[0\ 65 \times 10^6]$.

The initial guesses and the interval are ready to be utilized in the bisection method. All terms in the equation

$$\det\left(\boldsymbol{M}_{22} - \lambda \boldsymbol{I}_m - \boldsymbol{M}_{21}(\boldsymbol{M}_{11} - \lambda \boldsymbol{I}_n)^{-1}\boldsymbol{M}_{12}\right) = 0$$

were calculated. Now, the equation is solved in each step using the bisection method. The eigenvalues λ_{11} to λ_{20} are found as

$$\lambda_M = 10^6[1.5934, 1.8151, 1.9138, 2.1311, 2.2158, \\ 2.7474, 3.1877, 3.3628, 3.5226, 3.9020]$$

It can be observed that $\lambda^*_{\mathbf{M}}$ is an efficient initial guess for $\lambda_{\mathbf{M}}$.

In order to obtain more precise answers, one can use Rayleigh-Ritz relationship by means of which having the approximate eigenvalues and eigenvectors, one can get closer to the eigenvalues and eigenvectors of the main problem.

4.4.4 Complementary Examples

In this section different examples are studied to show the efficiency of the present method.

Example 1 In this example, the frequencies of the free vibration of the double layer grid, shown in the Fig. 4.25, are calculated. A real case of such a grid can be seen in Fig. 4.26. For all members the elastic modulus is taken as 200 kN/mm^2 and the cross section areas are 10 cm^2. The mass is equal to 7.8 kg for the vertical and horizontal members and 11.0 kg for the diagonal ones. The plan of the structure and its lateral view can be seen in the Figs. 4.27 and 4.28, respectively.

To find the frequencies of the structure, the following equation should be solved:

$$\left(\boldsymbol{K}-\omega_i^2\boldsymbol{M}\right)\{\emptyset\}_i = \{0\}$$

Multiplying the two sides by $\boldsymbol{M}^{-1}$, we will have

$$\left(\boldsymbol{M}^{-1}\boldsymbol{K} - \omega_i^2\boldsymbol{I}\right)\{\emptyset\}_i = \{0\}$$

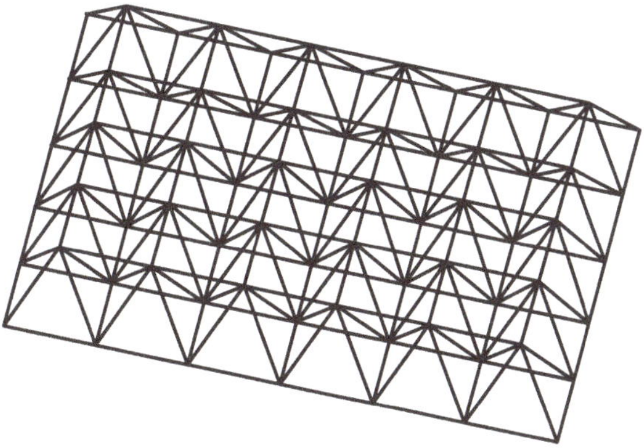

Fig. 4.25 A double layer grid with near regular pattern

Fig. 4.26 A real double layer grid

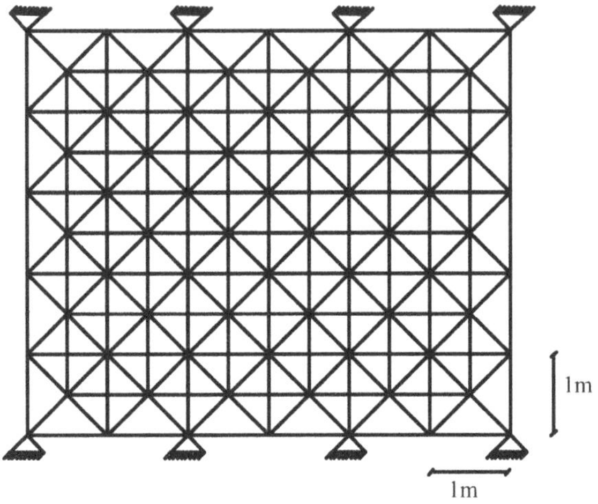

Fig. 4.27 The plan of the grid

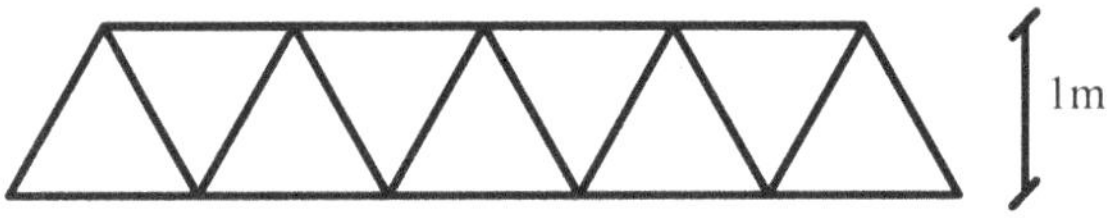

Fig. 4.28 lateral view of the grid

And then

$$\det\left(\boldsymbol{M}^{-1}\boldsymbol{K} - \omega_i^2\boldsymbol{I}\right) = 0$$

The structure is geometrically regular. However, it is considered as a near-regular structure mathematically using a suitable nodal ordering. In this paper, a graph or structure is considered as regular if the pattern of its Laplacian matrix or stiffness matrix follows graph product rules. A graph or structure with just a regular geometry is not considered as regular. The structure is composed of a central regular part shown in green and an irregular part shown in red, Fig. 4.29. Using a proper ordering and considering $\boldsymbol{M}^{-1}\boldsymbol{K} = \boldsymbol{S}$, the matrix $\boldsymbol{S}$ of the dimension 192 with the following pattern is obtained:

$$\boldsymbol{S} = \begin{bmatrix} \boldsymbol{S}_{11} & \boldsymbol{S}_{12} \\ \boldsymbol{S}_{21} & \boldsymbol{S}_{22} \end{bmatrix}$$

where $\boldsymbol{S}_{11}$ is the block of the regular green part, and $\boldsymbol{S}_{22}$ corresponds to the block of the irregular red part.

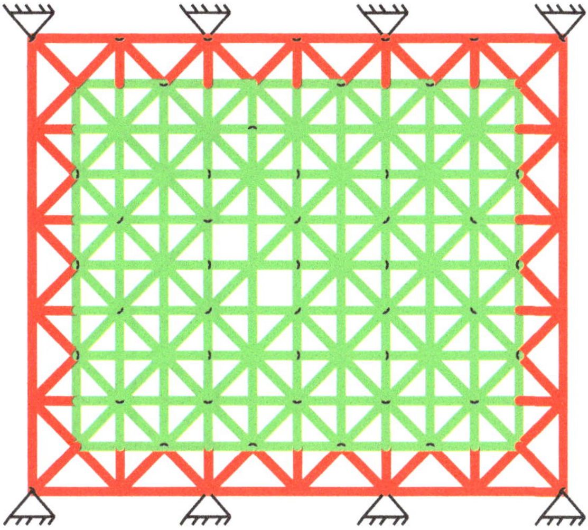

Fig. 4.29 The regular and irregular parts of a double layer grid

Using the blocks of the matrix, the main equation is formed

$$\det\left(\boldsymbol{S}_{22}-\omega^2\boldsymbol{I}_m-\boldsymbol{S}_{21}\left(\boldsymbol{S}_{11}-\omega^2\boldsymbol{I}_n\right)^{-1}\boldsymbol{S}_{12}\right)=0$$

The solution of this equation using the present algorithm results in finding the values of ω. The first 10 frequencies are obtained as

$$\omega=10^4[0.8791, 1.1253, 1.3324, 1.6811, 1.9371, 2.0916, 2.2983, 2.4577, 2.7348, 3.0060]$$

Example 2 An important practical application of the present method is in the analysis of structures with constructional imperfection. Consider the structure shown in Fig. 4.30. Because of constructional imperfection the nodes 31 and 35 are not in the right place as shown in Fig. 4.31. The amount of their deviation is seen in Fig. 4.31. For all members, the elastic modulus is taken as 210 kN/mm^2 and the cross section areas are 15 cm^2. The specific gravity of steel is equal to 7800 kg/m^3. The frequencies of the real structure are found.

The frequencies of the regular ideal structure shown in Fig. 4.30 are swiftly found using graph products. However, due to the constructional imperfection, the real structure changes into a near-regular one composed of regular and irregular parts as shown in Figs. 4.32 and 4.33.

To solve the real structure, the frequencies of the regular ideal structure can be used as an initial guess. The main equation is written as

$$\det\left(\boldsymbol{S}_{22}-\omega^2\boldsymbol{I}_m-\boldsymbol{S}_{21}\left(\boldsymbol{S}_{11}-\omega^2\boldsymbol{I}_n\right)^{-1}\boldsymbol{S}_{12}\right)=0$$

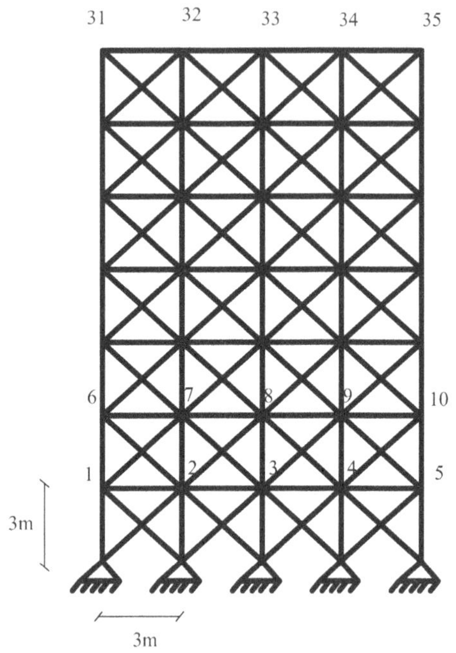

Fig. 4.30 The ideal form of the truss under study

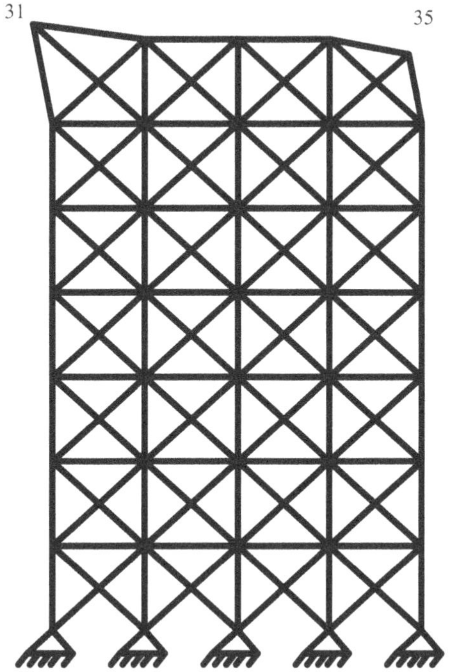

Fig. 4.31 The real form of the truss including constructional imperfection

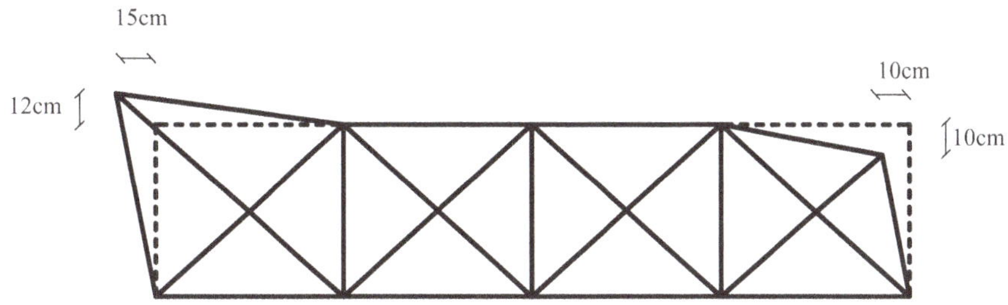

Fig. 4.32 Constructional imperfection of the last story

Fig. 4.33 A structure composed of regular and irregular parts

where the matrix $\boldsymbol{S}_{11}$ with dimension 50, represents the green regular part, and $\boldsymbol{S}_{22}$ with dimension 20, represents the red irregular part. Using Eq. (2.16), the frequencies of the ideal structure are calculated. The frequencies ω_{61} to ω_{70} are obtained as

$$\omega = 10^4[15.3324, 15.7053, 16.1053, 16.4158, 16.7371,\\ 17.0850, 17.3017, 17.5989, 17.7348, 17.9126]$$

Now, these frequencies are used as the initial guess in the bisection stage of the present algorithm. The frequencies ω_{61} to ω_{70} of the real structure are obtained as

$$\omega = 10^4[15.6824, 15.9358, 16.2489, 16.5298, 16.8626,\\ 17.2007, 17.4611, 17.7218, 17.9346, 18.1043]$$

References

1. Rahami, H., Kaveh, A., Shojaei, I., Gholipour, Y.: Analysis of irregular structures composed of regular and irregular parts using graph products. J. Comput. Civ. Eng. **28**(4) (2014)
2. Shojaei, I., Kaveh, A., Rahami, H.: Analysis of structures convertible to repeated structures using graph products. Comput. Struct. **125**, 153–163 (2013)
3. Kaveh, A.: Optimal Analysis of Structures by Concepts of Symmetry and Regularity. Springer, New York (2013)
4. Kaveh, A., Shojaei, I., Rahami, H.: New developments in the optimal analysis of regular and near-regular structures: decomposition, graph products, force method. Acta Mech. **226**(3), 665–681 (2015)
5. Kaveh, A.: Structural Mechanics: Graph and Matrix Methods, 3rd edn. Research Studies Press (John Wiley), UK (2006)
6. Kaveh, A., Rahami, H., Ardalan Asl, M., Mirghaderi, S.R.: Analysis of regular structures with member non-regularity using the equilibrium equations and singular valued decomposition. Eng. Comput. **30**, 21–48 (2013)
7. Pellegrino, S.: Structural computations with the singular value decomposition of the equilibrium matrix. Int. J. Solids Struct. **30**, 3025–3035 (1993)
8. Shojaei, I., Kaveh, A., Rahami, H.: A numerical method for eigensolution of near-regular structural and mechanical systems. Period. Polytech. Civ. Eng. **60**(2), 247–255 (2016)
9. Kaveh, A., Rahami, H.: An efficient analysis of repetitive structures generated by graph products. Int. J. Numer. Methods Eng. **84**(1), 108–126 (2010)
10. Kaveh, A., Rahami, H.: Block diagonalization of adjacency and laplacian matrices for graph product; applications in structural mechanics. Int. J. Numer. Methods Eng. **68**, 33–63 (2006)
11. Kaveh, A., Rahami, H.: Compound matrix block diagonalization for efficient solution of eigenproblems in structural mechanics. Acta Mech. **188**, 155–166 (2007)
12. Pothen, A., Simon, H., Liou, K.P.: Partitioning sparse matrices with eigenvectors of graphs. SIAM J. Matrix Anal. Appl. **11**, 430–452 (1990)
13. Yueh, WCh.: Eigenvalues of several tri-diagonal matrices. Appl. Math. **5**, 66–74 (2005)

Chapter 5
Static Analysis of Nearly Regular Continuous Domains

5.1 Introduction

During the past decades, numerical methods (e.g., finite element methods, mesh free methods) have extensively been developed and used to tackle structural and mechanical problems [1, 2]. The accuracy of numerical solutions, however, depends on the number of elements and nodes which highly affects the execution time and cost efficiency of the solution. Despite development of super computers for swift solution of complicated numerical problems, efficient algorithms are still of great interest because as computational capability increases new complicated problems arises. Regarding the style and elegance of new designs as well as growing application of prefabrication in structural and mechanical systems, many systems hold regular and near-regular geometrical patterns. Special methods for solving regular and near-regular structures were presented and discussed in Chaps. 2, 3, and 4. Because of natural repetition of elements and nodes in numerical methods and regularity of meshes, regular and near regular patterns are also observed in many numerical solution resulting in decomposable stiffness matrices. In this chapter, finite element and mesh free solutions of regular and near-regular structural and mechanical systems are developed and discussed.

5.2 Finite Element Solution of Regular and Near Regular Patterns

Finite element solution of many structural/mechanical systems and differential equations results in regular and near-regular patterns. Due to natural repetition of elements in FE models, regularity is very common in resultant stiffness matrices. This allows mathematical developments of graph product rules for efficient solution of FE models. Graph product rules are developed to decompose a regular pattern to its preliminary generating components, solving the problem using efficient divide and conquer

A. Kaveh et al., *Swift Analysis of Civil Engineering Structures Using Graph Theory Methods*, Studies in Systems, Decision and Control 290,
https://doi.org/10.1007/978-3-030-45549-1_5

approaches. Regularity is regarded as a specific condition mathematically; however, due to natural repetition of elements in FE models, regular/near-regular patterns are very common in these models. In this section, graph product mathematical formulations for FE models are developed and efficiency of the method is demonstrated using computational complexity. Comprehensiveness of the method is shown using diverse complementary examples.

5.2.1 Efficient Finite Element Solution Regular Patterns

Finite element (FE) solution of many differential equations results in regular decomposable patterns. Although in FE solutions utilizing more elements usually lead to more accurate results, computational complexity and dimension of the problems significantly increase. Alternatively, using the efficient formulations developed in Chap. 2, solution of the matrices and equations can be obtained.

Consider the following heat conduction in one dimension with $k(x)$ as a variable conductivity

$$(ku')' = f \quad u(0) = u(1) = 0 \tag{5.1}$$

Multiplying both sides of Eq. (5.1) by the function $v(x)$ with the condition $v(0) = v(1) = 0$ and integrating over the domain [0 1] lead to

$$\int_0^1 \left(k(x)u'(x)\right)' v(x)dx = \int_0^1 f(x)v(x)dx \tag{5.2}$$

Manipulating the left side results in

$$-\int_0^1 k(x)u'(x)v'(x)dx = \int_0^1 f(x)v(x)dx \tag{5.3}$$

$u(x)$ is approximated using the following relationship

$$u(x) = \sum_{j=1}^{m} c_j\phi_j(x) \quad \phi(0) = \phi(1) = 0 \tag{5.4}$$

Substituting Eq. (5.4) into Eq. (5.3) and using a set of m basis functions $\psi_i(x)$, we will have

$$\sum_{j=1}^{m} K_{ij}c_j = \int_0^1 f(x)\psi_i(x)dx \quad \text{where } K_{ij} = -\int_0^1 k(x)\phi_j'(x)\psi_i'(x)dx \tag{5.5}$$

And

$$\boldsymbol{Kc} = \boldsymbol{F} \quad \text{where } F_i = \int_0^1 f(x)\psi_i(x)dx \tag{5.6}$$

Considering the Galerkin method

$$\phi_j(x) = \begin{cases} (x - x_{j-1})/h & x_{j-1} \le x \le x_j \\ (x_{j+1} - x)/h & x_j \le x \le x_{j+1} \\ 0 & Otherwise \end{cases} \tag{5.7}$$

Finally,

$$\boldsymbol{K} = 1/h \begin{bmatrix} -2 & 1 & & & & \\ 1 & -2 & 1 & & & \\ & 1 & -2 & 1 & & \\ & & 1 & \ddots & \ddots & \\ & & & \ddots & -2 & 1 \\ & & & & 1 & -2 \end{bmatrix} = F\left(-\frac{2}{h}, \frac{1}{h}, -\frac{2}{h}\right) \tag{5.8}$$

where form $\boldsymbol{F}$ was introduced in Eq. (2.4), and the eigenpairs of matrix $\boldsymbol{F}(b, a, b)$ are calculated as

$$\lambda_k = b + 2a\cos\frac{k\pi}{n+1}, \quad k = 1, 2, \ldots, n \tag{5.9}$$

And

$$v_j^{(k)} = \sin\frac{kj\pi}{n+1}, \quad j = 1, 2, \ldots, n; \qquad \boldsymbol{v}^{(k)} = \left(v_1^{(k)}, v_2^{(k)}, \ldots, v_n^{(k)}\right)^t \tag{5.10}$$

Using Eqs. (5.9) and (5.10), components of the inverse of matrix $\boldsymbol{k}$ are readily found (see also Eq. (2.32)).

$$\boldsymbol{k}^{-1} = \boldsymbol{V\lambda}^{-1}\boldsymbol{V}^t \Rightarrow k_{ij}^{-1} = \sum_{k=1}^{n}\left(\frac{1}{\sum_{k=1}^{n}\sin\left[\frac{k\pi}{n+1}\right]^2}\sin\left[\frac{ik\pi}{n+1}\right]\sin\left[\frac{jk\pi}{n+1}\right]\frac{1}{\lambda_k}\right) \tag{5.11}$$

For higher order problems the solution will be more complicated. The Galerkin form of the two dimensional Poison problem would be as follows:

$$-\iint_{\Omega} \nabla u \cdot \nabla v dxdy = \iint_{\Omega} fvdxdy \tag{5.12}$$

And

$$u(x, y) = \sum_{j=1}^{n} cj\phi_j(x, y) \tag{5.13}$$

Therefore,

$$K_{ij} = -\iint_{\Omega} \nabla\phi_j \cdot \nabla\phi_i dxdy \tag{5.14}$$

Consider the following Poisson equation

$$\begin{aligned} -\left(u_{xx} + u_{yy}\right) = f(x, y) \qquad (x, y) \in [ab] \times [cd] \\ u(x, y)|_{\partial\Omega} = 0 \end{aligned} \tag{5.15}$$

Using a uniform triangulation mesh (Fig. 5.1) and piecewise continuous linear basis function defined at $(x_i, y_i) = (ih, jh)$, we will have

$$\phi_i\left(N_j\right) = \begin{cases} 1 & i = j \\ 0 & otherwise \end{cases} \tag{5.16}$$

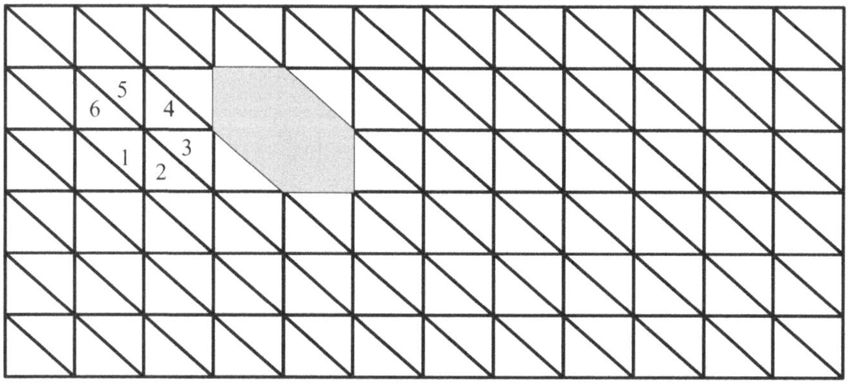

Fig. 5.1 A uniform triangulation mesh on a rectangular domain

And

$$\phi_{j(n-1)+i} = \begin{cases} \frac{x-(i-1)h+y-(j-1)h}{h} - 1 & \text{Region1} \\ \frac{y-(j-1)h}{h} & \text{Region2} \\ \frac{h-(x-ih)}{h} & \text{Region3} \\ 1 - \frac{x-ih+y-jh}{h} & \text{Region4} \\ \frac{h-(y-jh)}{h} & \text{Region5} \\ \frac{x-(i-1)h}{h} & \text{Region6} \\ 0 & \text{Otherwise} \end{cases} \tag{5.17}$$

Thus, the stiffness matrix will be defined as follows

$$\boldsymbol{K} = \boldsymbol{F}(\boldsymbol{B}, -\boldsymbol{I}, \boldsymbol{B}) = \boldsymbol{I} \otimes \boldsymbol{B} + \boldsymbol{T} \otimes -\boldsymbol{I} \tag{5.18}$$

where $\boldsymbol{I}$ is the identity matrix, $\boldsymbol{B} = \boldsymbol{F}(4, -1, 4)$, and $\boldsymbol{T} = \boldsymbol{F}(0, 1, 0)$.

Matrix $\boldsymbol{K}$ satisfies Eq. (2.12) and can be decomposed. Using eigenpairs of $\boldsymbol{K}$ the inverse of the matrix is calculate using Eq. (2.32).

The computational complexity of the method is $O(n)$ while the most efficient algorithm of inverting a matrix (of a general form) is of the order $O(n^{2.373})$ [3].

5.2.2 *Efficient Finite Element Solution of Near-Regular Patterns*

The concept of near-regularity for structures was introduced in Chap. 3. Similarly, FE formulation of many structural and mechanical systems may lead to repetitive patterns that are not readily decomposable, or may include both regular and irregular parts.

Consider FE solution of the plate element where the element forces and displacements are defined as

$$\boldsymbol{P}_i = \{P_{i1}, P_{i2}, P_{i3}\} = \{w_{zi}, M_{xi}, M_{yi}\} \quad i = 1:4 \tag{5.19}$$

$$\boldsymbol{q}_i = \{q_{i1}, q_{i2}, q_{i3}\} = \left\{w_i, \frac{\partial w_i}{\partial y}, \frac{\partial w_i}{\partial x}\right\} \quad i = 1:4 \tag{5.20}$$

Considering the following shape functions and operator $\bar{d}$, the strain-displacement matrix $\boldsymbol{B}$ is obtained

$$\begin{aligned} \boldsymbol{f}_i &= \{f_{i1}, f_{i2}, f_{i3}\} \\ \bar{d} &= \left\{\frac{\partial^2}{\partial x^2}, \frac{\partial^2}{\partial y^2}, \frac{2\partial^2}{\partial x \partial y}\right\} \end{aligned} \tag{5.21}$$

$$\overline{\boldsymbol{B}}_i = \bar{d}\boldsymbol{f}_i = \begin{bmatrix} f_{i1,xx} & f_{i2,xx} & f_{i3,xx} \\ f_{i1,yy} & f_{i2,yy} & f_{i3,yy} \\ 2f_{i1,xy} & 2f_{i2,xy} & 2f_{i3,xy} \end{bmatrix} \tag{5.22}$$

The stiffness matrix and nodal forces of the plate element are obtained as

$$\boldsymbol{K} = \int_A \overline{\boldsymbol{B}}^T \overline{\boldsymbol{E}}\overline{\boldsymbol{B}} dA$$
$$\boldsymbol{F} = \int_A \overline{\boldsymbol{B}}^T \overline{\boldsymbol{E}} \boldsymbol{\varphi}_0 dA \tag{5.23}$$

Consider the plate shown in Fig. 5.2.
The general canonical form of the assembled stiffness matrix will be as follows:

$$\boldsymbol{K} = \begin{bmatrix} \boldsymbol{A}_1 & \boldsymbol{B}_1^T & & & & \\ \boldsymbol{B}_1 & \boldsymbol{A} & \boldsymbol{B}^T & & & \\ & \boldsymbol{B} & \boldsymbol{A} & \boldsymbol{B}^T & & \\ & & \boldsymbol{B} & \ddots & \ddots & \\ & & & \ddots & \boldsymbol{A} & \boldsymbol{B}_2^T \\ & & & & \boldsymbol{B}_2 & \boldsymbol{A}_2 \end{bmatrix} \tag{5.24}$$

This near-regular pattern can be solved using the following regular pattern

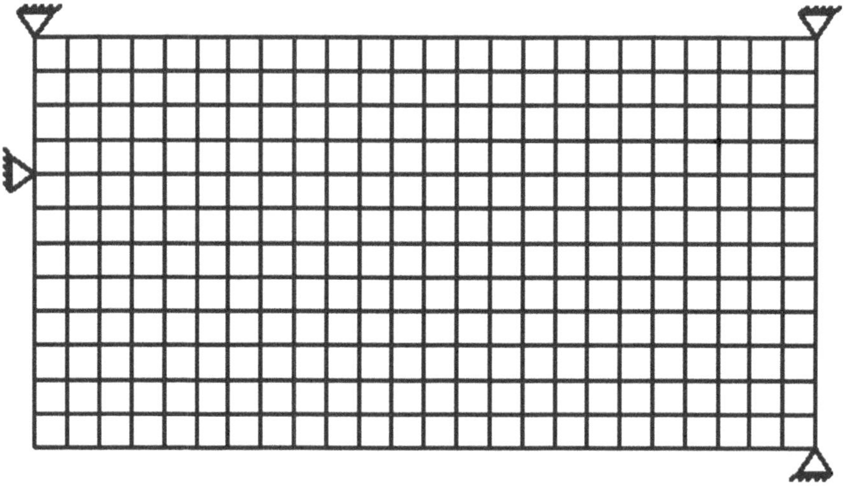

Fig. 5.2 A plate with near-regular pattern

$$K_1 = \begin{bmatrix} A & B^T & & & & B \\ B & A & B^T & & & \\ & B & A & B^T & & \\ & & B & \ddots & \ddots & \\ & & & \ddots & A & B^T \\ B^T & & & & B & A \end{bmatrix} \tag{5.25}$$

The regular pattern is decomposable because

$$K_1 = P_0 \otimes A + P_1 \otimes B + P_2 \otimes B^T \tag{5.26}$$

where P_0, P_1 and P_2 are permutation matrices with the property

$$P_i P_j = P_j P_i \tag{5.27}$$

The eigenpairs of the permutation matrix P_i are obtained as follows (also see Eq. (2.74))

$$\begin{aligned} &\lambda = \omega^k; \quad \omega = e^{\frac{2\pi i}{n}}; \quad i = \sqrt{-1}; \quad k = 0 : n-1 \\ &v_k = \left[1, \omega^k, \omega^{2k}, \ldots, \omega^{(n-1)k}\right]^t \end{aligned} \tag{5.28}$$

Accordingly, the inverse of matrix K_1 is readily obtained. Now, with the solution of the regular pattern in hand, the near-regular structure (i.e. the plate) is solved as follows:

$$\begin{bmatrix} F_1 \\ F_2 \\ F_3 \\ \vdots \\ F_{n-1} \\ F_n \end{bmatrix} = \begin{bmatrix} A_1 & B_1^T & & & & \\ B_1 & A & B^T & & & \\ & B & A & B^T & & \\ & & B & \ddots & \ddots & \\ & & & \ddots & A & B_2^T \\ & & & & B_2 & A_2 \end{bmatrix} \begin{bmatrix} \Delta_1 \\ \Delta_2 \\ \Delta_3 \\ \vdots \\ \Delta_{n-1} \\ \Delta_n \end{bmatrix} \tag{5.29}$$

Using the following manipulation (i.e. adding and subtracting B and B^T), we will have

$$\begin{bmatrix} \boldsymbol{F}_1 \\ \boldsymbol{F}_2 \\ \boldsymbol{F}_3 \\ \vdots \\ \boldsymbol{F}_{n-1} \\ \boldsymbol{F}_n \end{bmatrix} = \begin{bmatrix} \boldsymbol{A}_1 & \boldsymbol{B}_1^T & & & & \\ \boldsymbol{B}_1 & \boldsymbol{A} & \boldsymbol{B}^T & & \boldsymbol{B}-\boldsymbol{B} & \\ & \boldsymbol{B} & \boldsymbol{A} & \boldsymbol{B}^T & & \\ & & \boldsymbol{B} & \ddots & \ddots & \\ & \boldsymbol{B}^T-\boldsymbol{B}^T & & \ddots & \boldsymbol{A} & \boldsymbol{B}_2^T \\ & & & & \boldsymbol{B}_2 & \boldsymbol{A}_2 \end{bmatrix} \begin{bmatrix} \Delta_1 \\ \Delta_2 \\ \Delta_3 \\ \vdots \\ \Delta_{n-1} \\ \Delta_n \end{bmatrix} \tag{5.30}$$

Equation (5.30) can be rewritten as

$$\begin{bmatrix} \boldsymbol{F}_1 \\ \boldsymbol{F}_n \end{bmatrix} = \begin{bmatrix} \boldsymbol{A}_1 & \boldsymbol{B}_1^T & \boldsymbol{0} & \boldsymbol{0} \\ \boldsymbol{0} & \boldsymbol{0} & \boldsymbol{B}_2 & \boldsymbol{A}_2 \end{bmatrix} \begin{bmatrix} \Delta_1 \\ \Delta_2 \\ \Delta_{n-1} \\ \Delta_n \end{bmatrix}$$

And

$$\begin{bmatrix} \boldsymbol{F}_2 \\ \boldsymbol{F}_3 \\ \vdots \\ \boldsymbol{F}_{n-2} \\ \boldsymbol{F}_{n-1} \end{bmatrix} = \begin{bmatrix} \boldsymbol{A} & \boldsymbol{B}^T & & & \boldsymbol{B} \\ \boldsymbol{B} & \boldsymbol{A} & \boldsymbol{B}^T & & \\ & \boldsymbol{B} & \ddots & \ddots & \\ & & \ddots & \boldsymbol{A} & \boldsymbol{B}^T \\ \boldsymbol{B}^T & & & \boldsymbol{B} & \boldsymbol{A} \end{bmatrix} \begin{bmatrix} \Delta_2 \\ \Delta_3 \\ \vdots \\ \Delta_{n-2} \\ \Delta_{n-1} \end{bmatrix} + \begin{bmatrix} \boldsymbol{B}_1\Delta_1 - \boldsymbol{B}\Delta_{n-1} \\ \boldsymbol{0} \\ \vdots \\ \boldsymbol{0} \\ -\boldsymbol{B}^T\Delta_2 + \boldsymbol{B}_2^T\Delta_n \end{bmatrix} \tag{5.31}$$

Multiplying the two sides of the second equation by matrix $\boldsymbol{D}$ (i.e. the inverse of matrix $\boldsymbol{K}_1$ above)

$$\boldsymbol{D}\begin{bmatrix} \boldsymbol{F}_2 \\ \boldsymbol{F}_3 \\ \vdots \\ \boldsymbol{F}_{n-2} \\ \boldsymbol{F}_{n-1} \end{bmatrix} = \begin{bmatrix} \boldsymbol{F}_2^* \\ \boldsymbol{F}_3^* \\ \vdots \\ \boldsymbol{F}_{n-2}^* \\ \boldsymbol{F}_{n-1}^* \end{bmatrix} = \begin{bmatrix} \Delta_2 \\ \Delta_3 \\ \vdots \\ \Delta_{n-2} \\ \Delta_{n-1} \end{bmatrix}$$
$$+ \begin{bmatrix} \boldsymbol{D}_{11} & \boldsymbol{D}_{(n-2)1} \\ \boldsymbol{D}_{12} & \boldsymbol{D}_{(n-2)2} \\ \vdots & \vdots \\ \boldsymbol{D}_{1(n-3)} & \boldsymbol{D}_{(n-2)(n-3)} \\ \boldsymbol{D}_{1(n-2)} & \boldsymbol{D}_{(n-2)(n-2)} \end{bmatrix} \begin{bmatrix} \boldsymbol{B}_1 & \boldsymbol{0} & -\boldsymbol{B} & \boldsymbol{0} \\ \boldsymbol{0} & -\boldsymbol{B}^T & \boldsymbol{0} & \boldsymbol{B}_2^T \end{bmatrix} \begin{bmatrix} \Delta_1 \\ \Delta_2 \\ \Delta_{n-1} \\ \Delta_n \end{bmatrix} \tag{5.32}$$

Using the first and last rows in Eq. (5.32) and the first part in Eq. (5.31), we will have

$$\begin{bmatrix}\Delta_1\\ \Delta_2\\ \Delta_{n-1}\\ \Delta_n\end{bmatrix}=\begin{bmatrix}\boldsymbol{A}_1 & \boldsymbol{B}_1^T & \boldsymbol{0} & \boldsymbol{0}\\ \boldsymbol{D}_{11}\boldsymbol{B}_1 & \boldsymbol{I}-\boldsymbol{D}_{(n-2)1}\boldsymbol{B}^T & -\boldsymbol{D}_{11}\boldsymbol{B} & \boldsymbol{D}_{(n-2)1}\boldsymbol{B}_2^T\\ \boldsymbol{D}_{1(n-2)}\boldsymbol{B}_1 & -\boldsymbol{D}_{(n-2)(n-2)}\boldsymbol{B}^T & \boldsymbol{I}-\boldsymbol{D}_{1(n-2)}\boldsymbol{B} & \boldsymbol{D}_{(n-2)(n-2)}\boldsymbol{B}_2^T\\ \boldsymbol{0} & \boldsymbol{0} & \boldsymbol{B}_2 & \boldsymbol{A}_2\end{bmatrix}^{-1}\begin{bmatrix}\boldsymbol{F}_1\\ \boldsymbol{F}_2^*\\ \boldsymbol{F}_{n-1}^*\\ \boldsymbol{F}_n\end{bmatrix} \tag{5.33}$$

From Eq. (5.32), we will have

$$\begin{bmatrix}\boldsymbol{F}_3^*\\ \boldsymbol{F}_4^*\\ \vdots\\ \boldsymbol{F}_{n-3}^*\\ \boldsymbol{F}_{n-2}^*\end{bmatrix}=\begin{bmatrix}\Delta_3\\ \Delta_4\\ \vdots\\ \Delta_{n-3}\\ \Delta_{n-2}\end{bmatrix}+\begin{bmatrix}\boldsymbol{D}_{12} & \boldsymbol{D}_{(n-2)2}\\ \boldsymbol{D}_{13} & \boldsymbol{D}_{(n-2)3}\\ \vdots & \vdots\\ \boldsymbol{D}_{1(n-4)} & \boldsymbol{D}_{(n-2)(n-4)}\\ \boldsymbol{D}_{1(n-3)} & \boldsymbol{D}_{(n-2)(n-3)}\end{bmatrix}\begin{bmatrix}\boldsymbol{B}_1 & \boldsymbol{0} & -\boldsymbol{B} & \boldsymbol{0}\\ \boldsymbol{0} & -\boldsymbol{B}^T & \boldsymbol{0} & \boldsymbol{B}_2^T\end{bmatrix}\begin{bmatrix}\Delta_1\\ \Delta_2\\ \Delta_{n-1}\\ \Delta_n\end{bmatrix} \tag{5.34}$$

Finally, all unknowns are found in a closed form

$$\begin{bmatrix}\Delta_3\\ \Delta_4\\ \vdots\\ \Delta_{n-3}\\ \Delta_{n-2}\end{bmatrix}=\begin{bmatrix}\boldsymbol{F}_3^*\\ \boldsymbol{F}_4^*\\ \vdots\\ \boldsymbol{F}_{n-3}^*\\ \boldsymbol{F}_{n-2}^*\end{bmatrix}-\begin{bmatrix}\boldsymbol{D}_{12}\boldsymbol{B}_1 & -\boldsymbol{D}_{(n-2)2}\boldsymbol{B}^T & -\boldsymbol{D}_{12}\boldsymbol{B} & \boldsymbol{D}_{(n-2)2}\boldsymbol{B}_2^T\\ \boldsymbol{D}_{13}\boldsymbol{B}_1 & -\boldsymbol{D}_{(n-2)3}\boldsymbol{B}^T & -\boldsymbol{D}_{13}\boldsymbol{B} & \boldsymbol{D}_{(n-2)3}\boldsymbol{B}_2^T\\ \vdots & \vdots & \vdots & \vdots\\ \boldsymbol{D}_{1(n-4)}\boldsymbol{B}_1 & -\boldsymbol{D}_{(n-2)(n-4)}\boldsymbol{B}^T & -\boldsymbol{D}_{1(n-4)}\boldsymbol{B} & \boldsymbol{D}_{(n-2)(n-4)}\boldsymbol{B}_2^T\\ \boldsymbol{D}_{1(n-3)}\boldsymbol{B}_1 & -\boldsymbol{D}_{(n-2)(n-3)}\boldsymbol{B}^T & -\boldsymbol{D}_{1(n-3)}\boldsymbol{B} & \boldsymbol{D}_{(n-2)(n-3)}\boldsymbol{B}_2^T\end{bmatrix}\begin{bmatrix}\Delta_1\\ \Delta_2\\ \Delta_{n-1}\\ \Delta_n\end{bmatrix} \tag{5.35}$$

where

$$\begin{bmatrix}\Delta_1\\ \Delta_2\\ \Delta_{n-1}\\ \Delta_n\end{bmatrix}=\begin{bmatrix}\boldsymbol{A}_1 & \boldsymbol{B}_1^T & \boldsymbol{0} & \boldsymbol{0}\\ \boldsymbol{D}_{11}\boldsymbol{B}_1 & \boldsymbol{I}-\boldsymbol{D}_{(n-2)1}\boldsymbol{B}^T & -\boldsymbol{D}_{11}\boldsymbol{B} & \boldsymbol{D}_{(n-2)1}\boldsymbol{B}_2^T\\ \boldsymbol{D}_{1(n-2)}\boldsymbol{B}_1 & -\boldsymbol{D}_{(n-2)(n-2)}\boldsymbol{B}^T & \boldsymbol{I}-\boldsymbol{D}_{1(n-2)}\boldsymbol{B} & \boldsymbol{D}_{(n-2)(n-2)}\boldsymbol{B}_2^T\\ \boldsymbol{0} & \boldsymbol{0} & \boldsymbol{B}_2 & \boldsymbol{A}_2\end{bmatrix}^{-1}$$

$$\begin{bmatrix} \boldsymbol{F}_1 \\ \boldsymbol{F}_2^* \\ \boldsymbol{F}_{n-1}^* \\ \boldsymbol{F}_n \end{bmatrix}$$

While the computational complexity of inverting the matrix $\boldsymbol{K}$ in Eq. (5.29) is $O(n^3)$ in a schoolbook method (i.e. Gauss–Jordan elimination), $O(n^{2.807})$ in Strassen algorithm, $O(n^{2.376})$ in Coppersmith–Winograd algorithm, and $O(n^{2.373})$ in Williams algorithm [3], the dominant computational complexity of the current method in Eq. (5.35), using only a traditional schoolbook method, is $O(n^2)$. The computational complexity of inverting matrix $\boldsymbol{K}_1$ (i.e. finding matrix $\boldsymbol{D}$), which holds the regular pattern, is $O(n)$ [4].

The studied plate was a near-regular plate stiffness matrix of which was converted to the pattern of a regular plate (i.e. second part of Eq. 5.31) during the solution. Now (3.31), consider the near-regular plate in Fig. 5.3.

The plate is composed of a near-regular part (black) connected to an arbitrary part (red). This structure can also be readily solved. Similar to the algorithm proposed in Chap. 4 for the solution of near-regular structures Type II (Eq. (4.10)), the inverse of the stiffness matrix is calculated as

$$\boldsymbol{D} = \begin{bmatrix} \boldsymbol{C}_1 & \vdots & \boldsymbol{C}_2 & \vdots & \boldsymbol{C}_1\boldsymbol{B}^* \\ \dots & \vdots & \dots & \vdots & \dots \\ \boldsymbol{C}_3 & \vdots & \boldsymbol{C}_4 & \vdots & \boldsymbol{C}_3\boldsymbol{B}^* \\ \dots & \vdots & \dots & \vdots & \dots \\ \boldsymbol{C}_1^* & \vdots & \boldsymbol{C}_2^* & \vdots & \boldsymbol{C}_3^* \end{bmatrix} \tag{5.36}$$

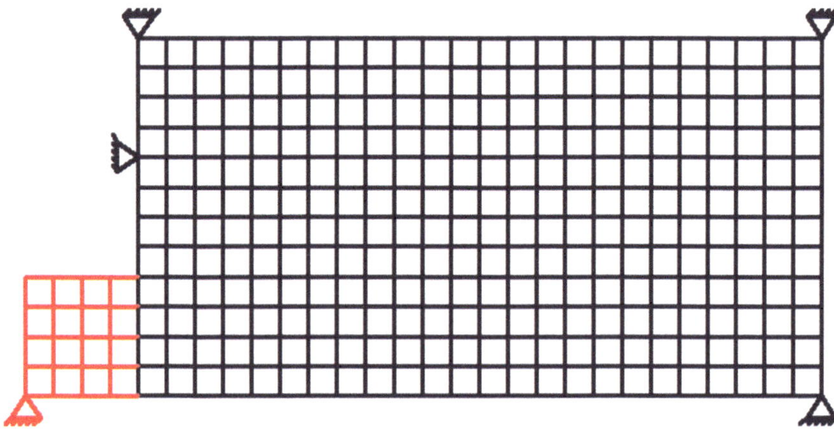

Fig. 5.3 A near-regular plate composed of two parts

The equations for finding the components of matrix $\boldsymbol{D}$ were presented in Chap. 4. The dominant computational complexity of the algorithm is $O(n^2)$ that belongs to the solution of the near-regular part (black).

5.2.3 Examples

Example 1 Consider the cantilever shown in Fig. 5.4 [2]. Using the meshes and elements introduced in Table 5.1 the problem was solved for four mesh conditions. The general pattern of the stiffness matrix will be as follows:

$$\boldsymbol{K}_{nm} = \begin{bmatrix} (\boldsymbol{A}_m)_1 & \boldsymbol{B}_m^T & & & & \\ \boldsymbol{B}_m & \boldsymbol{A}_m & \boldsymbol{B}_m^T & & & \\ & \boldsymbol{B}_m & \boldsymbol{A}_m & \boldsymbol{B}_m^T & & \\ & & \boldsymbol{B}_m & \ddots & \ddots & \\ & & & \ddots & \boldsymbol{A}_m & \boldsymbol{B}_m^T \\ & & & & \boldsymbol{B}_m & \boldsymbol{A}_m \end{bmatrix}_n$$

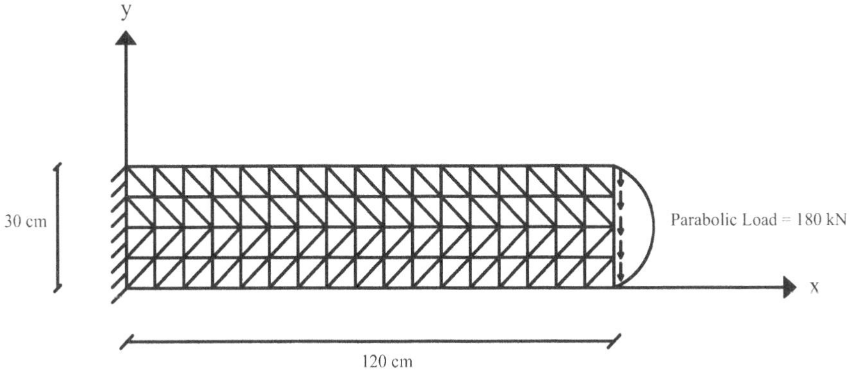

Fig. 5.4 A cantilever under parabolic load

Table 5.1 Different sets of meshes and elements used in the analysis of a cantilever

Mesh	Run	Number of nodes	Number of elements	Tip deflection	Stiffness matrix dimension
4 × 16 (M-1)	M-1	85	128CST	−0.75	160
8 × 32 (M-2)	M-2	297	512CST	−0.86	576
2 × 8 (M-3)	M-3	85	32LST	−0.85	160
4 × 16 (M-4)	M-4	297	128LST	−0.89	576

Table 5.2 The values of n and m in matrix $\boldsymbol{K}_{nm}$ for different elements and meshes

Mesh	Number of elements	n	m
4 × 16 (M-1)	128CST	16	10
8 × 32 (M-2)	512CST	32	18
2 × 8 (M-3)	32LST	8	20
4 × 16 (M-4)	128LST	16	36

This pattern is a simplified form of the pattern in Eq. (5.29) wherein $(\boldsymbol{A}_m)_2 = \boldsymbol{A}_m$ and $(\boldsymbol{B}_m)_1 = (\boldsymbol{B}_m)_2 = \boldsymbol{B}_m$. The block $(\boldsymbol{A}_m)_1$ is corresponding to the degrees of freedom on the free side. The values of n and m in matrix $\boldsymbol{K}_{nm}$ were defined in Table 5.2.

Using the closed-form relationships in Eq. (5.35) the displacements are found (see tip displacements in Table 5.1). For the mesh conditions M-1 and M-3, the time complexities of solving the problem using the present method and Williams algorithm are 2.5600e+004 and 1.6997e+005, respectively. For the mesh conditions M-2 and M-4, the time complexity of the present method is 3.3178e+005, compared to 3.5521e+006 in Williams algorithm.

Example 2 A square plate with a central hole is under the loads shown in Fig. 5.5. Due to symmetry of the structure and loading, one quarter of the plate can be modeled (Fig. 5.6). The plate is split into two parts (i.e., black and red parts in Fig. 5.7) to form a near-regular structure Type II in the cylindrical coordinate system. Using Eq. (5.36) and the relationships in Chap. 4 the plate structure is solved.

Stiffness matrix of the dominant black part holds the following pattern if one first numbers the outer blocks, corresponding to supports, and then numbers the inner blocks.

Fig. 5.5 A square plate with a central hole

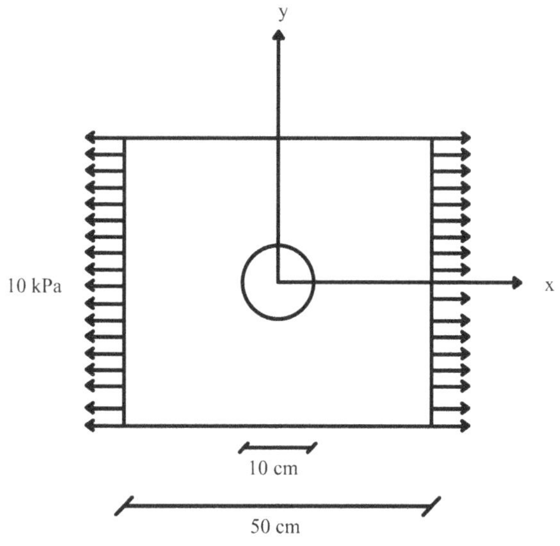

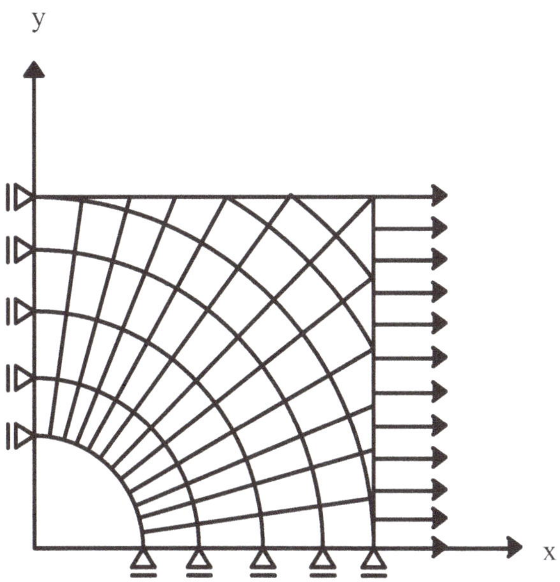

Fig. 5.6 One quarter of the square plate

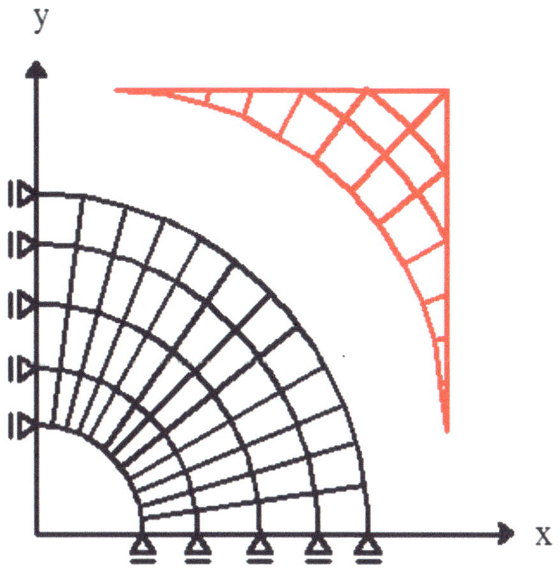

Fig. 5.7 The plate composed of two parts

$$\boldsymbol{K}_{nm} = \begin{bmatrix} (\boldsymbol{A}_m)_1 & (\boldsymbol{B}_m)_1^T & & & & \\ (\boldsymbol{B}_m)_1 & \boldsymbol{A}_m & \boldsymbol{B}_m^T & & & \\ & \boldsymbol{B}_m & \boldsymbol{A}_m & \boldsymbol{B}_m^T & & \\ & & \boldsymbol{B}_m & \ddots & \ddots & \\ & & & \ddots & \boldsymbol{A}_m & \boldsymbol{B}_m^T \\ & & & & \boldsymbol{B}_m & \boldsymbol{A}_m \end{bmatrix}_n$$

Similarly, this pattern is a simplified form of the pattern in Eq. (5.29) wherein $(\boldsymbol{A}_m)_2 = \boldsymbol{A}_m$, $(\boldsymbol{B}_m)_2 = \boldsymbol{B}_m$.

Example 3 A three-dimensional finite element model of spinal cord was presented in Fig. 5.8 [5]. The axial injury model of the spinal cord was simulated where all degrees of freedom at the bottom, top, and the two ends of the model were constrained. Geometry of the model is constructed through 70 times repetition of a sector, with 485 nodes, along the longitudinal axis, resulting in a stiffness matrix of dimension 101,850 (i.e., 70 × 485 × 3).[1] Using a sector by sector numbering along the longitudinal axis and considering the constraints at the two ends, the stiffness matrix will have the following pattern

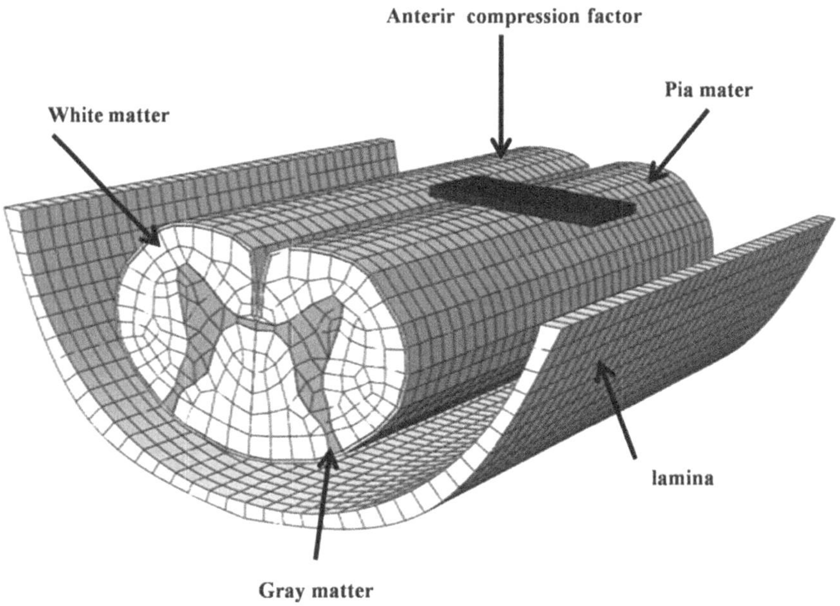

Fig. 5.8 Three-dimensional FE model of the spinal cord

[1] The modeled mesh includes more elements than the schematic Fig. 5.8.

$$\boldsymbol{K}_{nm} = \begin{bmatrix} \boldsymbol{A}_m & \boldsymbol{B}_m^T & & & & \\ \boldsymbol{B}_m & \boldsymbol{A}_m & \boldsymbol{B}_m^T & & & \\ & \boldsymbol{B}_m & \boldsymbol{A}_m & \boldsymbol{B}_m^T & & \\ & & \boldsymbol{B}_m & \ddots & \ddots & \\ & & & \ddots & \boldsymbol{A}_m & \boldsymbol{B}_m^T \\ & & & & \boldsymbol{B}_m & \boldsymbol{A}_m \end{bmatrix}_n$$

where $n = 70$ and $m = 1455$ (i.e., 485×3). In this near-regular pattern we have $(\boldsymbol{A}_m)_1 = (\boldsymbol{A}_m)_2 = \boldsymbol{A}_m$ and $(\boldsymbol{B}_m)_1 = (\boldsymbol{B}_m)_2 = \boldsymbol{B}_m$. Since the stiffness matrix $\boldsymbol{K}_{nm}$ is a specific form of Eq. (5.29), Eq. (5.35) can be more simplified. The corresponding regular pattern of the stiffness matrix $\boldsymbol{K}_{nm}$ will have the following pattern

$$(\boldsymbol{K}_{nm})_1 = \begin{bmatrix} \boldsymbol{A}_m & \boldsymbol{B}_m^T & & & & \boldsymbol{B}_m \\ \boldsymbol{B}_m & \boldsymbol{A}_m & \boldsymbol{B}_m^T & & & \\ & \boldsymbol{B}_m & \boldsymbol{A}_m & \boldsymbol{B}_m^T & & \\ & & \boldsymbol{B}_m & \ddots & \ddots & \\ & & & \ddots & \boldsymbol{A}_m & \boldsymbol{B}_m^T \\ \boldsymbol{B}_m^T & & & & \boldsymbol{B}_m & \boldsymbol{A}_m \end{bmatrix}_n$$

Matrix $(\boldsymbol{K}_{nm})_1$ can be decomposed as

$$(\boldsymbol{K}_{nm})_1 = \boldsymbol{P_0} \otimes \boldsymbol{A}_m + \boldsymbol{P_1} \otimes \boldsymbol{B}_m + \boldsymbol{P_2} \otimes \boldsymbol{B}_m^T$$

where $\boldsymbol{P}_0$ is identity matrix and $\boldsymbol{P}_1$ and $\boldsymbol{P}_2$ are the following permutation matrices

$$\boldsymbol{P}_1 = \begin{bmatrix} 0 & 0 & & & & 1 \\ 1 & 0 & 0 & & & \\ & 1 & 0 & 0 & & \\ & & 1 & \ddots & \ddots & \\ & & & \ddots & 0 & 0 \\ & & & & 1 & 0 \end{bmatrix}_n \qquad \boldsymbol{P}_2 = \begin{bmatrix} 0 & 1 & & & & \\ 0 & 0 & 1 & & & \\ & 0 & 0 & 1 & & \\ & & 0 & \ddots & \ddots & \\ & & & \ddots & 0 & 1 \\ 1 & & & & 0 & 0 \end{bmatrix}_n$$

Eigenpairs of the permutation matrices $\boldsymbol{P}_1$ and $\boldsymbol{P}_2$ are obtained as Eq. (2.74).

Thus, matrix $(\boldsymbol{K}_{nm})_1$ is decomposed into a block diagonal matrix. Instead of solving a matrix of dimension 101,850, matrices of dimension 1455 are solved that is much more efficient computationally (i.e. divide and conquer). The obtained components of matrix $\boldsymbol{D}$ (i.e., the inverse of matrix $(\boldsymbol{K}_{nm})_1$) are inserted in Eq. (5.35). The time complexity of solving the problem using the current method is 1.0373e+010, compared to 7.6541e+011 in Williams algorithm. It is notable that the efficiency of the present method is further pronounced in three-dimensional problems with substantial number of degrees of freedom.

5.3 Finite Element Solution of Problems with an Arbitrary Domain: A Hyper Pre-solved Element

In this section a finite element algorithm is presented using large pre-solved elements. Utilizing the largest rectangle/cuboid inside an arbitrary domain, a large hyper element is developed that is solved using graph product rules. This pre-solved hyper element is efficiently inserted into the finite element formulation of partial differential equations (PDE) and engineering problems to reduce the computational complexity and execution time of the solution. The general solution of the large pre-solved element for uniform mesh of triangular and rectangular elements is formulated for second order PDEs. The efficiency of the algorithm depends on the relative size of the large element and the domain; however the method remains as efficient as a classic method for even small relative sizes. The application of the method is demonstrated using various examples [6].

5.3.1 Largest Rectangle/Cuboid Inside an Arbitrary Polygon/Solid

Several algorithms for finding the largest rectangle inside an arbitrary polygon have been suggested in literature [7]. A recent algorithm solve efficiently the problem with a computational complexity only depending on the number of vertices on bounds of the polygon. A detailed description of the method can be found in [7]. Briefly, to obtain the largest rectangle inside the quasi-lattice polygon S, a square grid partition like G that includes all points of S is created (Fig. 5.9). Two sub algorithms, henceforth called radial algorithm and vector matrix $\boldsymbol{U}$ algorithm, are developed to define the position of any point p and any vector u in partition G. In the radial algorithm, using the edges of S as well as the angles formed by vertices $\boldsymbol{V} = \left[v_1 \ldots v_n \right]$ and the lines connecting the vertices to the point p, two possible solutions (i.e., 0 and $\pm 2\pi$) are obtained. The solution 0 corresponds to a point outside polygon S, whereas the solution $\pm 2\pi$ shows a point inside the polygon.

In vector matrix $\boldsymbol{U}$ algorithm, an upper triangular matrix $\boldsymbol{U}$, of the dimension N, is formed where N is the number of total nodes inside and on boundaries of polygon S.

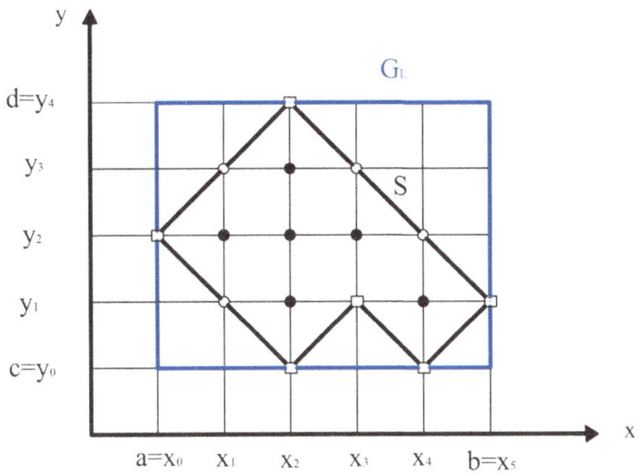

Fig. 5.9 A square grid partition G including all points of the quasi-lattice polygon S

$$
\boldsymbol{U} = \begin{bmatrix} \vec{0} & \vec{u}_{12} & \vec{u}_{13} & \dots & \vec{u}_{1N} \\ \vec{0} & \vec{0} & \vec{u}_{23} & \dots & \vec{u}_{2N} \\ \vdots & \vdots & \vdots & \vdots & \vdots \\ \vec{0} & \vec{0} & \vec{0} & \dots & \vec{u}_{N-1N} \\ \vec{0} & \vec{0} & \vec{0} & \dots & \vec{0} \end{bmatrix} \tag{5.37}
$$

This matrix consists of vectors from which the segments of the largest rectangle can be obtained. In this matrix a vector u_{ij} is non-zero if it completely lies inside the polygon S. For two given points p_i and p_j from partition G, it is controlled whether the vector $u_{ij} = p_{ij} = p_j - p_i$ crosses any edges of polygon S. If vector u_{ij} does not cross any edges, it lies completely either inside polygon S or outside polygon S.

Using an arbitrary point of vector u_{ij} and the radial algorithm above, the position of the vector is defined. Using these two sub algorithms, the matrix $\boldsymbol{U}$ is constructed in $O(n^2)$ time where n is the number of vertices of polygon S.

The main algorithm is now developed to obtain the coordinates of the largest rectangle enclosed with polygon S. For each edge, of a potential rectangle, made by points p_i and p_j, we need to have $u_{ij} \neq 0$ (i.e., the edge lies inside polygon S). Furthermore, four edges forming a rectangular have to be parallel and perpendicular two-by-two. As the algorithm runs for edges (i.e., u_{ij}) of polygon S, a found rectangle is saved and is updated when a better solution (i.e., a larger rectangular) is obtained. Computational complexity of the algorithm is $O(n^3)$ where n is the number of vertices of polygon S. In schematic Fig. 5.10, the largest rectangular inside a quasi-lattice polygon S is indicated. To obtain the largest rectangular inside a closed contour, the corresponding quasi-lattice polygon S inside the contour is obtained and then the rectangle inside the polygon is found (Fig. 5.11) [7].

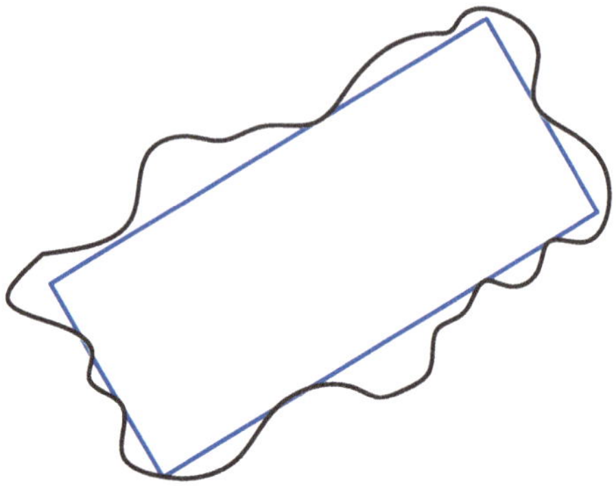

Fig. 5.10 The largest rectangular inside a closed contour

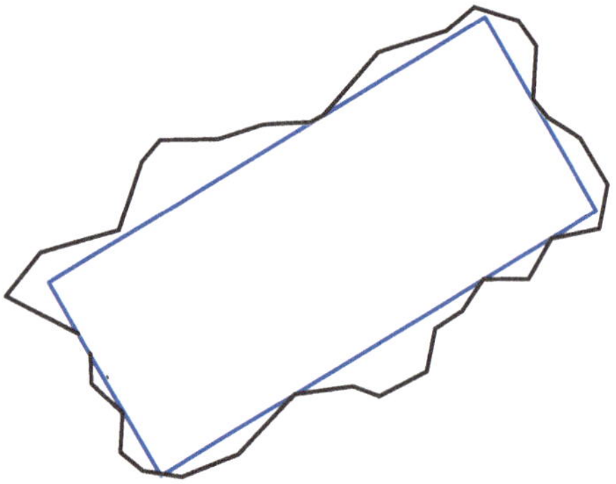

Fig. 5.11 The largest rectangular inside a quasi-lattice polygon S inside the contour

In the present study, in addition to finding the largest rectangle inside a polygon, finding the largest cuboid inside an arbitrary three-dimensional solid is of interest. Accordingly, the foregoing algorithm is extended to include an additional step wherein third edge of a cuboid is obtained using cross product of the other two edges. A found cuboid is saved and updated as the algorithm runs until the cuboid with the largest volume is obtained.

5.3.2 Developing a Hyper Pre-solved Element

Consider the following partial differential equation (PDE) with the given boundary condition (BC) on the 2D domain Ω

$$\mathrm{BVP}\begin{cases} Du = f & (PDE) \\ u|_{\partial\Omega} = g & (BC) \end{cases} \tag{5.38}$$

where D is a differential operator.

Finite element solution of the boundary value problem (BVP) using a weak form formulation is developed by multiplying the PDE by the test function $\vartheta(x, y)$ and integrating over the domain Ω [8].

$$\langle Du, v\rangle = \langle f, v\rangle \tag{5.39}$$

where scalar product $\langle ., .\rangle$ is the resulting integral.

By applying integration by parts on the left-hand side, we will have

$$\langle Du, v\rangle = a(u, v) \tag{5.40}$$

where $a(., .)$ is the bilinear form obtained in the procedure of integration by parts

$$a(u, v) = \langle f, v\rangle \tag{5.41}$$

Solutions for Eq. (5.41) are known as weak form solutions.

From a finite-dimensional function space an approximation of u (i.e. u_h) is sought and Eq. (5.41) is reformulated. Considering the solution of Eq. (5.41) in the form of $u_h = \sum_j u_j\phi_j$ and employing the Galerkin solution, we will have

$$a\left(\sum_j u_j\phi_j, \phi_i\right) = \langle f, \phi_i\rangle \tag{5.42}$$

Considering Eq. (5.42) in a matrix form

$$\boldsymbol{Mu} = \boldsymbol{f} \tag{5.43}$$

where the components of coefficient (i.e., stiffness) matrix $\boldsymbol{M}$ and vector $\boldsymbol{f}$ are obtained through

$$\boldsymbol{M}_{ij} = a(\phi_j, \phi_i), \quad \boldsymbol{f}_i = \langle f, \phi_i\rangle \tag{5.44}$$

To continue with a finite element approximation, consider a rectangular domain as follows

$$\Omega = \{(x, y)|x \in [ab], y \in [cd]\} \tag{5.45}$$

For a second order PDE, if a mesh of repetitive similar elements is used and a block by block numbering is conducted, the stiffness matrix will hold the pattern in Eq. (5.46). It is notable that the pattern of blocks $\boldsymbol{A}$ and $\boldsymbol{B}$ depends on the type of element used.

$$\boldsymbol{M}_{nm} = \begin{bmatrix} \boldsymbol{A}_m & \boldsymbol{B}_m^T & & & & \\ \boldsymbol{B}_m & \boldsymbol{A}_m & \boldsymbol{B}_m^T & & & \\ & \boldsymbol{B}_m & \boldsymbol{A}_m & \boldsymbol{B}_m^T & & \\ & & \boldsymbol{B}_m & \ddots & \ddots & \\ & & & \ddots & \boldsymbol{A}_m & \boldsymbol{B}_m^T \\ & & & & \boldsymbol{B}_m & \boldsymbol{A}_m \end{bmatrix}_n = \boldsymbol{F}_n\left(\boldsymbol{A}_m, \boldsymbol{B}_m, \boldsymbol{A}_m, \boldsymbol{B}_m^T\right) \tag{5.46}$$

Using methods similar to those in Eq. (5.24) through Eq. (5.35), matrix $\boldsymbol{M}_{nm}$ in Eq. (5.46) is solved. Consider the following decomposable pattern

$$(\boldsymbol{M}_1)_{mn} = \begin{bmatrix} \boldsymbol{A}_m & \boldsymbol{B}_m^T & & & & \boldsymbol{B}_m \\ \boldsymbol{B}_m & \boldsymbol{A}_m & \boldsymbol{B}_m^T & & & \\ & \boldsymbol{B}_m & \boldsymbol{A}_m & \boldsymbol{B}_m^T & & \\ & & \boldsymbol{B}_m & \ddots & \ddots & \\ & & & \ddots & \boldsymbol{A}_m & \boldsymbol{B}_m^T \\ \boldsymbol{B}_m^T & & & & \boldsymbol{B}_m & \boldsymbol{A}_m \end{bmatrix} = \boldsymbol{G}_n\left(\boldsymbol{A}_m, \boldsymbol{B}_m, \boldsymbol{A}_m, \boldsymbol{B}_m^T\right) \tag{5.47}$$

We can write

$$\boldsymbol{M}_1 = \boldsymbol{P}_0 \otimes \boldsymbol{A} + \boldsymbol{P}_1 \otimes \boldsymbol{B} + \boldsymbol{P}_2 \otimes \boldsymbol{B}^T \tag{5.48}$$

where $\boldsymbol{P}_0$, $\boldsymbol{P}_1$ and $\boldsymbol{P}_2$ are permutation matrices with the commutative property i.e.:

$$\boldsymbol{P}_i \boldsymbol{P}_j = \boldsymbol{P}_j \boldsymbol{P}_i \tag{5.49}$$

Matrix $\boldsymbol{M}_1$ is decomposable and can be inverted using the relationships in Chap. 2. Using matrix operations the solution of matrix $\boldsymbol{M}$ is obtained through the available inverse of matrix $\boldsymbol{M}_1$ (i.e. matrix $\boldsymbol{D}$)

$$\begin{bmatrix} \boldsymbol{u}_1 \\ \boldsymbol{u}_n \end{bmatrix} = \begin{bmatrix} \boldsymbol{I} - \boldsymbol{D}_{1n}\boldsymbol{B}^T & -\boldsymbol{D}_{11}\boldsymbol{B} \\ -\boldsymbol{D}_{nn}\boldsymbol{B}^T & \boldsymbol{I} - \boldsymbol{D}_{n1}\boldsymbol{B} \end{bmatrix}^{-1} \begin{bmatrix} \boldsymbol{D}_{11} & \boldsymbol{D}_{12} & \dots & \boldsymbol{D}_{1n-1} & \boldsymbol{D}_{1n} \\ \boldsymbol{D}_{n1} & \boldsymbol{D}_{n2} & \dots & \boldsymbol{D}_{nn-1} & \boldsymbol{D}_{nn} \end{bmatrix} \begin{bmatrix} \boldsymbol{f}_1 \\ \boldsymbol{f}_2 \\ \vdots \\ \boldsymbol{f}_{n-1} \\ \boldsymbol{f}_n \end{bmatrix}$$

And

$$\begin{bmatrix} u_2 \\ u_3 \\ \vdots \\ u_{n-2} \\ u_{n-1} \end{bmatrix} = \left(\begin{bmatrix} D_{21} & D_{22} & \cdots & D_{2n} \\ D_{31} & D_{32} & \cdots & D_{2n} \\ \vdots & \vdots & \ddots & \vdots \\ D_{(n-1)1} & D_{(n-1)2} & \cdots & D_{(n-1)n} \end{bmatrix} - \begin{bmatrix} -D_{2n}B^T & -D_{21}B \\ -D_{3n}B^T & -D_{31}B \\ \vdots & \vdots \\ -D_{(n-2)n}B^T & -D_{(n-2)1}B \\ -D_{(n-1)n}B^T & -D_{(n-1)1}B \end{bmatrix} \right.$$

$$\left. \begin{bmatrix} I - D_{1n}B^T & -D_{11}B \\ -D_{nn}B^T & I - D_{n1}B \end{bmatrix}^{-1} \begin{bmatrix} D_{11} & D_{12} & \cdots & D_{1n-1} & D_{1n} \\ D_{n1} & D_{n2} & \cdots & D_{nn-1} & D_{nn} \end{bmatrix} \right) \begin{bmatrix} f_1 \\ f_2 \\ \vdots \\ f_{n-1} \\ f_n \end{bmatrix} \tag{5.50}$$

Re-writing Eq. (5.50) in a compact form

$$\begin{bmatrix} u_2 \\ u_3 \\ \vdots \\ u_{n-2} \\ u_{n-1} \end{bmatrix} = D^*_{(n-2)\times n} \begin{bmatrix} f_1 \\ f_2 \\ \vdots \\ f_{n-1} \\ f_n \end{bmatrix}, \quad \begin{bmatrix} u_1 \\ u_n \end{bmatrix} = \begin{bmatrix} \overline{D}_1 \\ \overline{D}_2 \end{bmatrix}_{2\times n} \begin{bmatrix} f_1 \\ f_2 \\ \vdots \\ f_{n-1} \\ f_n \end{bmatrix} \tag{5.51}$$

and

$$\begin{bmatrix} u_1 \\ u_2 \\ \vdots \\ u_{n-1} \\ u_n \end{bmatrix} = \begin{bmatrix} \overline{D}_1 \\ D^* \\ \overline{D}_2 \end{bmatrix}_{n\times n} \begin{bmatrix} f_1 \\ f_2 \\ \vdots \\ f_{n-1} \\ f_n \end{bmatrix} \tag{5.52}$$

While the dominant computational complexity of the current method is $O(n^2)$, the computational complexity of solving the matrix equation $\boldsymbol{Mu} = \boldsymbol{f}$, using LU decomposition, is $O(n^{2.376})$ in Coppersmith–Winograd algorithm and $O(n^{2.373})$ in optimized CW-like algorithms [3].

a. **Uniform Triangulation Mesh**

Although a time-saving solution is achieved for a rectangular domain through Eqs. (5.47) to (5.52), the solution can become more simplified if incorporating element type in the formulation. Through partitioning the domain Ω using a uniform triangulation mesh, utilizing block by block numbering, and employing the piecewise continuous linear basis function, we will have (Figs. 5.12 and 5.13)

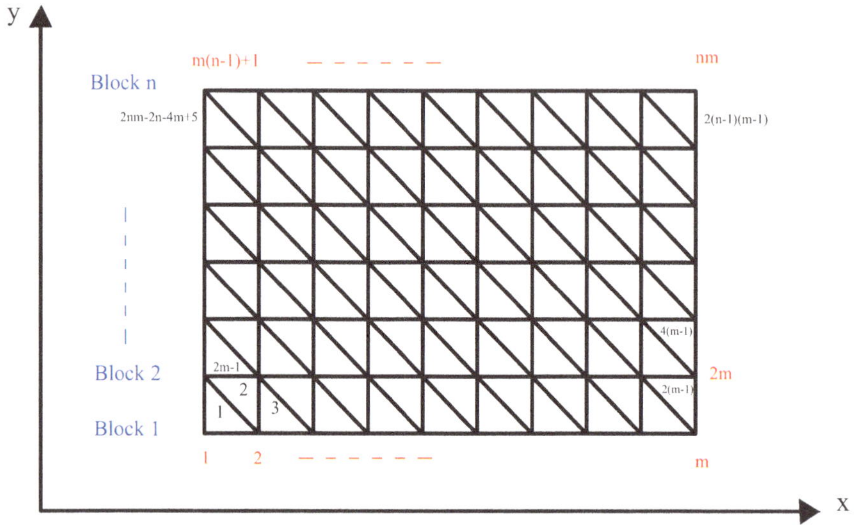

Fig. 5.12 Partitioning the domain Ω using a uniform triangulation mesh

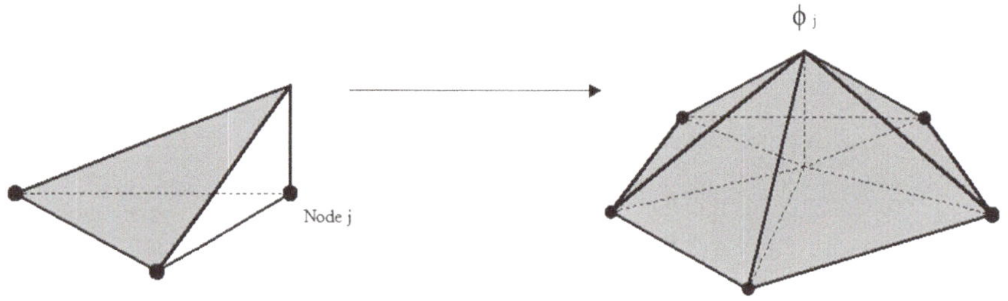

Fig. 5.13 The piecewise continuous linear basis function

$$\phi_i(N_j) = \begin{cases} 1 & i = j \\ 0 & otherwise \end{cases} \tag{5.53}$$

By applying the boundary conditions, the reduced stiffness matrix is obtained as follows:

$$\boldsymbol{M} = \boldsymbol{F}(\boldsymbol{A}, \boldsymbol{B}, \boldsymbol{A}, \boldsymbol{B}) = \boldsymbol{I} \otimes \boldsymbol{A} + \boldsymbol{T} \otimes \boldsymbol{B} \tag{5.54}$$

where $\boldsymbol{I}$ is identity matrix and $\boldsymbol{T} = \boldsymbol{F}(0, 1, 0, 1)$.

The matrix $\boldsymbol{M}$ is block diagonalized using the relationships in Chap. 2 (see also [9, 10] for details).

$$\boldsymbol{D}_M = \boldsymbol{D}_I \otimes \boldsymbol{A} + \boldsymbol{D}_T \otimes \boldsymbol{B}$$

$$
\boldsymbol{D}_M = \begin{bmatrix} 2\cos\frac{\pi}{n+1}\boldsymbol{B}+\boldsymbol{A} & & & & \\ & 2\cos\frac{2\pi}{n+1}\boldsymbol{B}+\boldsymbol{A} & & & \\ & & \ddots & & \\ & & & 2\cos\frac{(n-1)\pi}{n+1}\boldsymbol{B}+\boldsymbol{A} & \\ & & & & 2\cos\frac{n\pi}{n+1}\boldsymbol{B}+\boldsymbol{A} \end{bmatrix} \tag{5.55}
$$

The kth block in Eq. (5.55) will have the form in Eq. (5.56) and the matrices $\boldsymbol{A}$ and $\boldsymbol{B}$ are tri-diagonal matrices

$$
2\cos\frac{k\pi}{n+1}\boldsymbol{B}+\boldsymbol{A} = 2\cos\frac{k\pi}{n+1}\boldsymbol{F}(c,d,c)+\boldsymbol{F}(e,g,e) \tag{5.56}
$$

where the scalars c, d, e and g depend on the governed second order equation.

$$
\lambda_M = \lambda_{D_M} = \bigcup_{k=1}^{n-2}\bigcup_{l=1}^{m-2}\left[2\cos\frac{k\pi}{n+1}\left(2\cos\frac{l\pi}{m+1}d+c\right)+\left(2\cos\frac{l\pi}{m+1}g+e\right)\right] \tag{5.57}
$$

Finally, similar to what is performed in modal analysis (also see Eq. 2.31), the obtained eigenvalues are used to solve the Eq. (5.43):

$$
\begin{aligned}
&\boldsymbol{Mu}=\boldsymbol{f} \Rightarrow \{\varphi\}_j^T\boldsymbol{M}\{\varphi\}_j y_j = \lambda_j y_j = \{\varphi\}_j^T\boldsymbol{f}\\
&y_j = \frac{f_j}{\lambda_j} \Rightarrow \{u\}_n = \sum_{i=1}^{n}\{\varphi\}_i y_i = \sum_{i=1}^{n}\{\varphi\}_i\frac{f_i}{\lambda_i} = \sum_{i=1}^{n}\frac{\{\varphi\}_i\{\varphi\}_i^T}{\lambda_i}\boldsymbol{f}
\end{aligned} \tag{5.58}
$$

where λ_i and $\{\varphi\}_{\mathrm{i}}$ are the eigenpairs of the matrix $\boldsymbol{M}$.

For the case that the second order BVP is Poisson equation [i.e. $D=\nabla^2$ in Eq. (5.38)]

$$
\mathrm{BVP}\begin{cases}\nabla^2 u(x,y)=f(x,y)\\ u(x,y)|_{\partial\Omega}=0\end{cases} \tag{5.59}
$$

The stiffness matrix $\boldsymbol{M}$ and vector $\boldsymbol{f}$ are obtained as

$$
\boldsymbol{M}_{ij} = \iint\limits_{\Omega}\nabla\phi_j\cdot\nabla\phi_i dxdy,\quad \boldsymbol{f}_i = \iint\limits_{\Omega} f\phi_i dxdy \tag{5.60}
$$

The matrix $\boldsymbol{M}$ will have the following form

$$
\boldsymbol{M} = \boldsymbol{I}\otimes\boldsymbol{A}+\boldsymbol{T}\otimes-\boldsymbol{I} \tag{5.61}
$$

where $\boldsymbol{I}$ is the identity matrix, $\boldsymbol{A}$ is $\boldsymbol{F}(4, -1, 4, -1)$ and $\boldsymbol{T}$ is $\boldsymbol{F}(0, 1, 0, 1)$.

Since $\boldsymbol{IT} = \boldsymbol{TI}$, the matrix $\boldsymbol{M}$ can be block diagonalized and decomposed [10–12] as follows:

$$\lambda_{M} = \lambda_{D_M} = \bigcup_{k=1}^{n-2}\bigcup_{l=1}^{m-2}\left(4 - 2\cos\frac{k\pi}{n+1} - 2\cos\frac{l\pi}{n+1}\right) \tag{5.62}$$

And the solution is obtained using Eq. (5.58).

As an illustrative example, consider the Laplace equation with the following boundary condition:

$$\text{BVP}\begin{cases} \nabla^2 u(x, y) = 0 \\ u(x, 0) = \sin(\pi x) \\ u(x, 1) = \sin(\pi x)e^{-\pi} \\ u(0, y) = u(1, y) = 0 \end{cases} \quad \Omega = \{(x, y)|x, y \in [0\ 1]\} \tag{5.63}$$

Similar to Fig. 5.12, triangular elements are used to partition the rectangular domain. 200 elements in each direction are employed such that we will have a total of $101 \times 101 = 10{,}201$ degrees of freedom.

Instead of solving a matrix equation of dimension 9801 (i.e. $10{,}201 - 400 = 9801$), the problem is efficiently solved using Eqs. (5.62) and (5.58), and the displacement field $\boldsymbol{U}_1$ is obtained as shown in Fig. 5.14.

b. **Uniform Rectangular Mesh**

Consider the same problem in Eq. (5.38) but partitioned with rectangular elements. Consider an arbitrary element with the domain Ω^e and boundary $\partial\Omega^e$ within the domain Ω with boundary $\partial\Omega$ (Fig. 5.15):

Assume the following solution for an element with n nodes

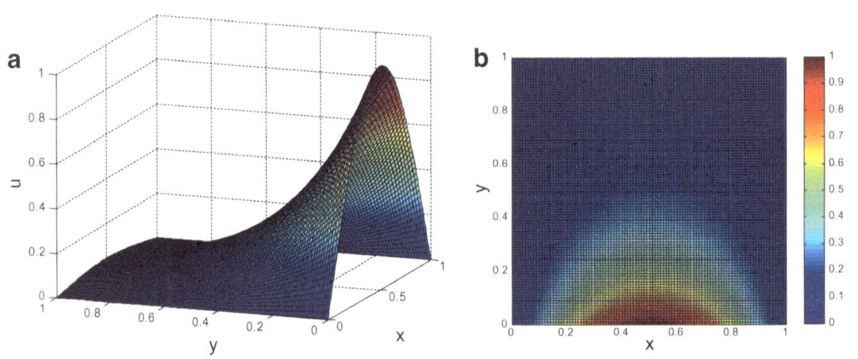

Fig. 5.14 **a** The 3D contour of displacement field. **b** The 2D contour of displacement field

Fig. 5.15 An arbitrary element with the domain $\boldsymbol{\Omega}^{\mathrm{e}}$ within the domain $\boldsymbol{\Omega}$

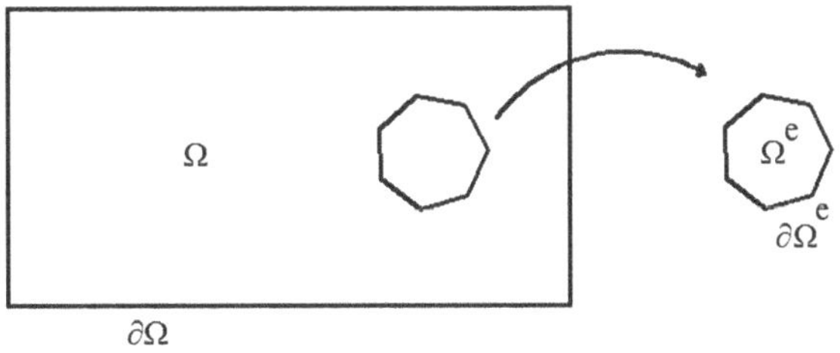

$$u_h^e(x, y) = \sum_{j=1}^{n} u_j^e \phi_j^e(x, y) \tag{5.64}$$

where u_j^e is the nodal value for $u(x, y)$ at the jth node of element and $\phi_j^e(x, y)$ is the interpolation function for $u(x, y)$ at node j within the element. Considering the Galerkin solution and substituting Eq. (5.64) into Eq. (5.41), we will have

$$a\left(\sum_j u_j^e \phi_j^e, \phi_i^e\right) = \langle f, \phi_i^e \rangle \tag{5.65}$$

Writing in a matrix form results in

$$\boldsymbol{M}^{\mathrm{e}} \boldsymbol{u}^{\mathrm{e}} = \boldsymbol{f}^{\mathrm{e}} \tag{5.66}$$

where the element stiffness matrix $\boldsymbol{M}^{\mathrm{e}}$ and vector $\boldsymbol{f}$ are obtained using Eq. (5.66)

$$\boldsymbol{M}_{ij}^e = a\left(\phi_j^e, \phi_i^e\right), \quad \boldsymbol{f}_i^{\mathrm{e}} = \langle f, \phi_i^e \rangle \tag{5.67}$$

Considering 4-node rectangular elements, interpolation functions are constructed using Lagrange interpolation polynomials

$$L_k(x) = \prod_{\substack{i=1 \\ i \neq j}}^{n} \frac{x - x_i}{x_k - x_i} = \phi_k^e(x); \quad L_k\left(x_j\right) = \delta_{jk}; \quad \sum_{k=1}^{n} L_k(x) = 1 \tag{5.68}$$

where $\prod$ represent a multiplication sign.

For the rectangular element with 4 nodes (Fig. 5.16), we will have

$$\begin{aligned} &\phi_1^e(x, y) = \left(1 - \frac{x}{a}\right)\left(1 - \frac{y}{b}\right) \qquad &&\phi_2^e(x, y) = \frac{x}{a}\left(1 - \frac{y}{b}\right) \\ &\phi_3^e(x, y) = \frac{x}{a}\frac{y}{b} &&\phi_4^e(x, y) = \left(1 - \frac{x}{a}\right)\frac{y}{b} \end{aligned} \tag{5.69}$$

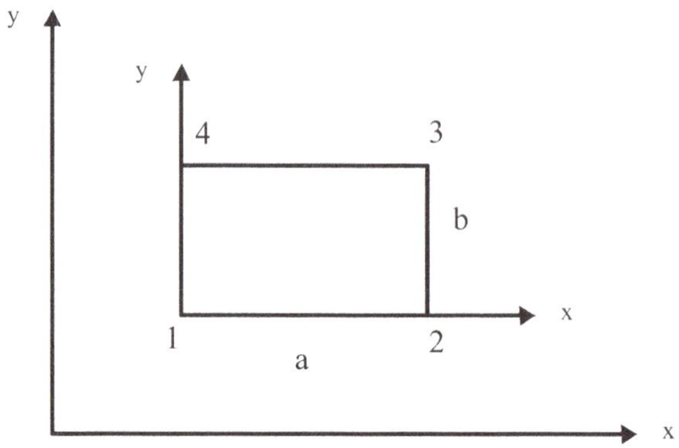

Fig. 5.16 The rectangular element with 4 nodes

Using the mesh and numbering shown in Fig. 5.17, the stiffness matrix is assembled

$$\boldsymbol{M} = \boldsymbol{F}(\boldsymbol{A}, \boldsymbol{B}, \boldsymbol{A}, \boldsymbol{B}) = \boldsymbol{I} \otimes \boldsymbol{A} + \boldsymbol{T} \otimes \boldsymbol{B} \tag{5.70}$$

And

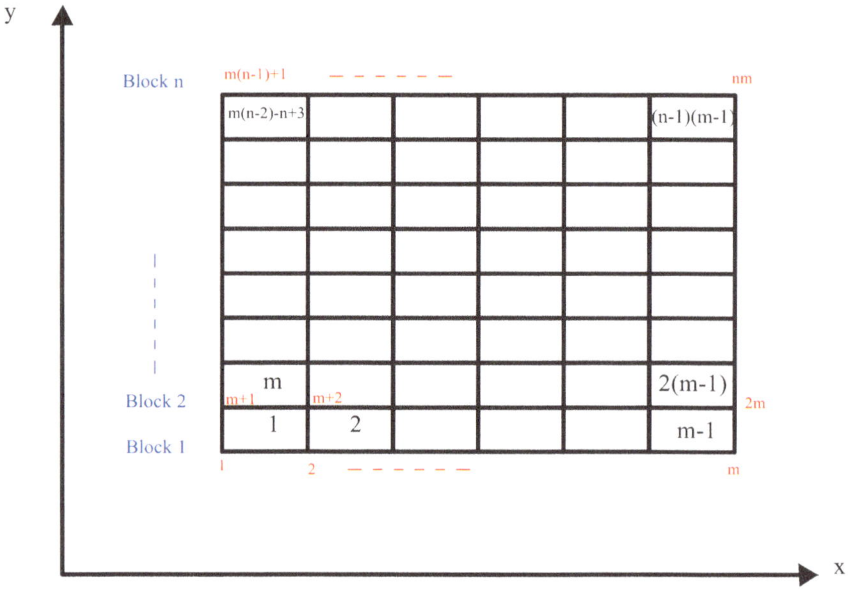

Fig. 5.17 Partitioning the domain **Ω** using a uniform rectangular mesh

$$\lambda_M = \lambda_{D_M} = \bigcup_{k=1}^{n-2}\bigcup_{l=1}^{m-2}\left[2\cos\frac{k\pi}{n+1}\left(2\cos\frac{l\pi}{m+1}d + c\right) + \left(2\cos\frac{l\pi}{m+1}g + e\right)\right] \tag{5.71}$$

where the scalars c, d, e and g depend on the governed second order equation.

Finally, similar to what is performed in modal analysis (Eq. 5.57), the obtained eigenpairs are used to solve Eq. (5.43).

For the case in which the second order BVP is Poisson equation [i.e. $D = \nabla^2$ in Eq. (5.38)], the element stiffness matrix $\boldsymbol{M}^{\mathrm{e}}$ and vectors and $\boldsymbol{f}^{\mathrm{e}}$ are obtained as follows:

$$\boldsymbol{M}_{ij}^{e} = \iint_{\Omega^{e}} \nabla\phi_j^e \cdot \nabla\phi_i^e dxdy, \quad \boldsymbol{f}_i^{\mathrm{e}} = \iint_{\Omega^{e}} f(x,y)\phi_i^e(x,y)dxdy \tag{5.72}$$

And the element stiffness matrix is constructed as:

$$\boldsymbol{M}^{\mathrm{e}} = \begin{bmatrix} 2(a^2+b^2) & a^2-2b^2 & -(a^2+b^2) & b^2-2a^2 \\ & 2(a^2+b^2) & b^2-2a^2 & -(a^2+b^2) \\ & & 2(a^2+b^2) & a^2-2b^2 \\ Sym & & & 2(a^2+b^2) \end{bmatrix} \tag{5.73}$$

The assembled stiffness matrix will have the pattern in Eq. (5.70), where the blocks $\boldsymbol{A}$, $\boldsymbol{B}$ and $\boldsymbol{C}$ are obtained as

$$\boldsymbol{A} = \boldsymbol{F}\left(8(a^2+b^2), a^2-2b^2, 8(a^2+b^2)\right) \tag{5.74}$$

And

$$\boldsymbol{B} = 2(b^2-2a^2); \quad \boldsymbol{C} = \boldsymbol{F}(1,1,1) \tag{5.75}$$

Using Eq. (5.70) a full decomposition is obtained and the eigenvalues of matrix $\boldsymbol{D}_M$ are calculated as:

$$\lambda_M = \lambda_{D_M} = \bigcup_{k=1}^{n-2}\bigcup_{l=1}^{m-2}\left(\left[4(b^2-2a^2)\cos\frac{k\pi}{n+1} + 8(a^2+b^2)\right] + 2\left[4(b^2-2a^2)\cos\frac{k\pi}{n+1} + a^2 - 2b^2\right]\cos\frac{l\pi}{n+1}\right) \tag{5.76}$$

Similarly, the rest of solution is obtained using Eq. (5.58).

Although we utilized triangular and 4-node rectangular elements, the sufficient condition for decomposing the stiffness matrix (Eq. 5.46) still holds for other types of elements. In a rectangular domain, as long as we use repetitive patterns of elements in one or two directions and number the degrees of freedom in a block by block order, the pattern in Eq. (5.46) is obtained.

Now, the entire rectangular domain Ω (Figs. 5.12 and 5.17) can be used as a large element since we could define an efficient pre-solved solution for the domain. Such a rectangular domain allows direct construction of stiffness matrix rather than dealing with individual elements and subsequent assembling procedures. The obtained solution for the rectangular domain through application of identical repetitive elements, block by block numbering, and graph product decomposition rules (see the relationships in Chap. 2) is regarded as a pre-solution for a large element (i.e. the entire rectangular domain). Such a solution will efficiently be inserted in the finite element formulation of an arbitrary domain wherein the large pre-solved element reduces the computational complexity of the problem. In the following section the method is presented in detail.

5.3.3 Finite Element Solution Using the Large Pre-solved Element

In this section finite element formulation for an arbitrary domain Ω is presented. The developed concepts in previous sections are combined and inserted into an algorithm to establish an efficient numerical solution. To get advantages of the results developed by graph product rules, finite element formulation, and block numbering for rectangular domain (see previous sections above), the domain Ω should be split into a rectangular domain plus the remaining parts. Although there is no limitation for the relative size of the rectangular and remaining parts, the larger the rectangular domain, the more efficient the algorithm. Obviously, for an extreme condition where the domain Ω approaches a rectangular domain, the unknowns are simply found using the results obtained in Eqs. (5.57), (5.58) and (5.71). For the arbitrary domain Ω to find the largest rectangle (or largest cuboid for a three-dimensional problem) the proposed algorithm in Sect. 5.3.1 is used. This rectangular domain, solved by an efficient solution from Sect. 5.3.2, is henceforth called "the large pre-solved element". Consider the following PDE on the arbitrary domain Ω partitioned by arbitrary triangular elements (Fig. 5.18)

$$\text{BVP}\begin{cases} Du = f & (PDE) \\ u|_{\partial\Omega} = g & (BC) \end{cases} \tag{5.77}$$

Consider u_h as the minimizer of the following functional [2, 8]

$$J(v) = \frac{1}{2}\langle Dv, v\rangle - \langle f, v\rangle \tag{5.78}$$

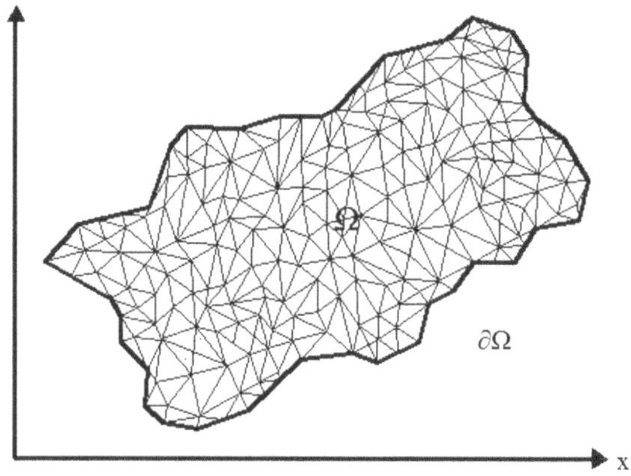

Fig. 5.18 The arbitrary domain Ω partitioned by arbitrary triangular elements

It is aimed to determine u such that

$$\frac{dJ(u_h)}{du_h} = 0 \tag{5.79}$$

Approximating u by $u_h = \sum_j u_j \phi_j$ results in

$$\left\langle D \sum_j \beta_j \phi_j, \phi_i \right\rangle - \langle f, \phi_i \rangle = 0 \tag{5.80}$$

For the case that BVP is the Poisson equation [i.e. $D = \nabla^2$ in Eq. (5.77)]

$$J(u_h) = \frac{1}{2} \int_\Omega |\nabla u_h(x, y)|^2 \mathrm{dxdy} - \int_\Omega f(x, y) u_h(x, y) \mathrm{dxdy} \tag{5.81}$$

Writing the minimization problem in a matrix form

$$\text{find } \boldsymbol{u} \text{ such that } \frac{1}{2}\boldsymbol{u}^{\mathrm{T}} \boldsymbol{M} \boldsymbol{u} - \boldsymbol{u}^{\mathrm{T}} \boldsymbol{f} \text{ is minimized} \tag{5.82}$$

where $\boldsymbol{M}$ is the global stiffness matrix and $\boldsymbol{f}$ is the global load vector.

Since the domain is arbitrary and triangles hold different shapes and geometries, a classic finite element method solves the problem for any triangle k in the domain Ω using the global coordinate system $\mathbf{r} = (\mathrm{x}, \mathrm{y})$, the local coordinate (ξ, η) system, and Jacobi matrix J for transformation

$$\mathbf{r} = (1 - \xi - \eta)\mathbf{r}_1 + \xi r_2 + \eta r_3 = \mathbf{r}_1 \varphi_1(\xi, \eta) + \mathbf{r}_2 \varphi_2(\xi, \eta) + \mathbf{r}_3 \varphi_3(\xi, \eta) \tag{5.83}$$

where the set $\{\varphi_1, \varphi_2, \varphi_3\}$ is the nodal basis.

The Jacobi matrix J for the transformation from (ξ, η) to **r** = (x, y) is defined by

$$|\mathrm{J}| = \left|\frac{\partial(x, y)}{\partial(\xi, \eta)}\right| = \det\begin{bmatrix} x_1 & y_1 & 1 \\ x_2 & y_2 & 1 \\ x_3 & y_3 & 1 \end{bmatrix} = 2A_{123} \tag{5.84}$$

where A_{123} is the area of the triangle k.

Using a classic finite element formulation the element stiffness matrix for kth element is obtained as

$$\boldsymbol{M}_k^e = \frac{1}{4A_{123}}\begin{bmatrix} |\mathbf{r}_2 - \mathbf{r}_3|^2 & (\mathbf{r}_2 - \mathbf{r}_3)\cdot(\mathbf{r}_3 - \mathbf{r}_1) & (\mathbf{r}_2 - \mathbf{r}_3)\cdot(\mathbf{r}_1 - \mathbf{r}_2) \\ & |\mathbf{r}_3 - \mathbf{r}_1|^2 & (\mathbf{r}_3 - \mathbf{r}_1)\cdot(\mathbf{r}_1 - \mathbf{r}_2) \\ \text{sym} & & |\mathbf{r}_1 - \mathbf{r}_2|^2 \end{bmatrix} \tag{5.85}$$

Relating the local numbering to the global numbering using the Boolean matrix L^k, the full stiffness matrix is defined as

$$\boldsymbol{M}^* = \sum_k L^k \boldsymbol{M}_k^e (L^k)^T \tag{5.86}$$

Finally, by removing the rows and columns corresponding to the known values of the vector $\boldsymbol{u}$ (i.e. boundary values), the global stiffness matrix $\boldsymbol{M}$ is defined and the matrix equation is obtained

$$\boldsymbol{M}\boldsymbol{u} = \boldsymbol{f} \tag{5.87}$$

In the classic method a general triangulation was generated wherein for each element a transformation from (ξ, η) coordinate system to **r** = (x, y) coordinate system was required because of arbitrary shape and geometry of the elements. Furthermore, the matrix equation in Eq. (5.87) should be solved using a conventional method as the matrix does not hold a desired pattern.

Alternatively, using the proposed method here, the problem can be solved with less computational efforts. After finding the largest rectangle inside the domain Ω (Fig. 5.19), the global coordinate system for the problem is chosen parallel to the sides of the rectangle. First, the domain Ω_2 is partitioned and solved using the classic method (Fig. 5.20).

The rectangular domain Ω_1 is then partitioned using the uniform triangulation mesh (or uniform rectangular mesh) and the stiffness matrix for the domain Ω_1 is constructed and inverted using Eq. (5.57) (or Eq. (5.71) for rectangular element) and Eq. (5.58). However, it is not required to partition the domain and construct the stiffness matrix since it was already done. The only information required is the type of element (i.e. triangular or rectangular) and the number of the blocks (i.e. the values m and n). The latter is decided in a way to be consistent with the size of elements in the border of Ω_1 and Ω_2. Then, the entire rectangular domain Ω_1 is considered as a large element with available stiffness matrix and pre-defined solution. In case we need to

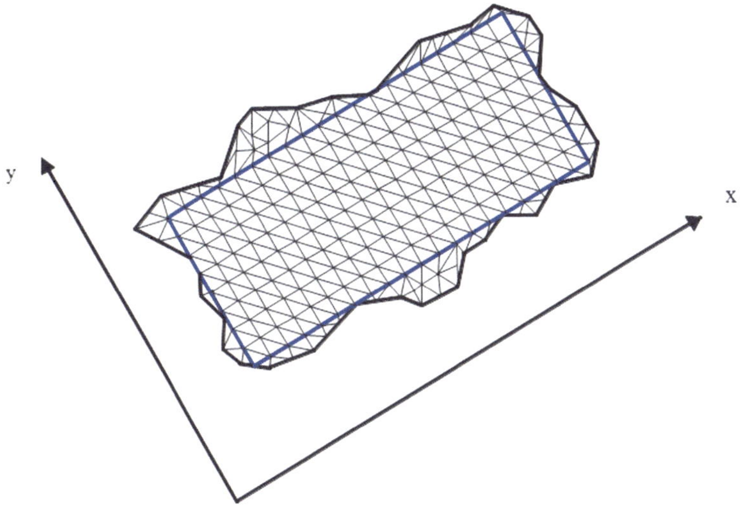

Fig. 5.19 The largest rectangle inside the domain Ω partitioned with uniform triangulation mesh

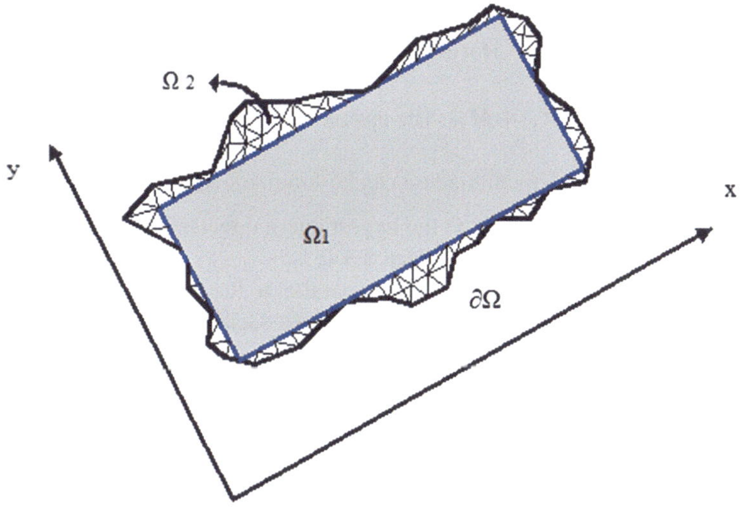

Fig. 5.20 The large element with available pre-defined solution and the remaining domain $\boldsymbol{\Omega_2}$

improve the accuracy of the solution, adaptive meshing is also possible. Specifically, regularity of the rectangular domain Ω_1 can be preserved after a mesh refinement (e.g., bisection and subdivision) so that the developed closed-form formulas for the large element stay valid. Finally, the stiffness matrix of the large element and the global stiffness matrix of the domain Ω_2 are assembled on the shared degrees of freedom such that the global stiffness matrix for the entire domain Ω is obtained.

Since we first number the inside nodes of the large element and then the boundary ones, assembling will not affect our desired pattern in Eq. (5.46) so that we still can use the pre-defined solution of the large element in the solution of the main problem. The shared degrees of freedom (i.e. the degrees of freedom on the boundary of the large element) are considered to be part of domain Ω_2. Stiffness matrix of the entire domain Ω is partitioned using the following block matrix

$$\boldsymbol{M} = \begin{bmatrix} \boldsymbol{F}(\boldsymbol{A}, \boldsymbol{B}, \boldsymbol{A}, \boldsymbol{B}) & \vdots & \boldsymbol{M}_{12} \\ \cdots & \cdots & \cdots \\ \boldsymbol{M}_{21} & \vdots & \boldsymbol{M}_{22} \end{bmatrix} = \begin{bmatrix} \boldsymbol{M}_{11} & \vdots & \boldsymbol{M}_{12} \\ \cdots & \cdots & \cdots \\ \boldsymbol{M}_{21} & \vdots & \boldsymbol{M}_{22} \end{bmatrix} \tag{5.88}$$

where block $\boldsymbol{M}_{11}$ is corresponding to the degrees of freedom inside the domain Ω_1, block $\boldsymbol{M}_{22}$ is corresponding to the degrees of freedom inside the domain Ω_2 plus the shared degrees of freedom, and $\boldsymbol{M}_{12}$ defines the way that these two domains are connected. The unknown values are calculated as follows:

$$\begin{bmatrix} \boldsymbol{U}_1 \\ \cdots \\ \boldsymbol{U}_2 \end{bmatrix} = \begin{bmatrix} \boldsymbol{M}_{11}^{-1}\left(\boldsymbol{I} + \boldsymbol{M}_{12}\boldsymbol{\Gamma}^{-1}\boldsymbol{M}_{21}\boldsymbol{M}_{11}^{-1}\right) & \vdots & -\boldsymbol{M}_{11}^{-1}\boldsymbol{M}_{12}\boldsymbol{\Gamma}^{-1} \\ \cdots & \cdots & \cdots \\ -\boldsymbol{\Gamma}^{-1}\boldsymbol{M}_{21}\boldsymbol{M}_{11}^{-1} & \vdots & -\boldsymbol{\Gamma}^{-1} \end{bmatrix} \begin{bmatrix} \boldsymbol{F}_1 \\ \cdots \\ \boldsymbol{F}_2 \end{bmatrix} \tag{5.89}$$

where $\boldsymbol{\Gamma} = -\boldsymbol{M}_{21}\boldsymbol{M}_{11}^{-1}\boldsymbol{M}_{12} + \boldsymbol{M}_{22}$. The inverse of matrix $\boldsymbol{M}_{11}$ was already obtained using Eq. (5.58).

The proposed finite element method can be summarized in the following steps:

1. For a given arbitrary domain Ω the largest internal rectangle/cuboid is obtained using the proposed algorithm in Sect. 5.3.1.
2. The global coordinate system is chosen parallel to the sides of the rectangle.
3. The type of element and number of blocks are decided for the domain Ω_1 and the solution of large element is obtained using Eqs. (5.57)/(5.71) and Eq. (5.58) for triangulation and rectangular meshes, respectively.
4. The remaining domain Ω_2, including the degrees of freedom inside the domain Ω_2 and on the boundary of the domain Ω_1, is partitioned using a classic finite element solution. Element and global stiffness matrices are constructed using the procedures shown in Eqs. (5.83) to (5.87).
5. The solution of the entire domain Ω is obtained using Eq. (5.89).

5.3.4 Complementary Examples

Example 1 Optical imaging technique is vastly used for body molecular imaging (Fig. 5.21). The photon propagation in heterogeneous tissues has been accepted to be governed by diffusion equation:

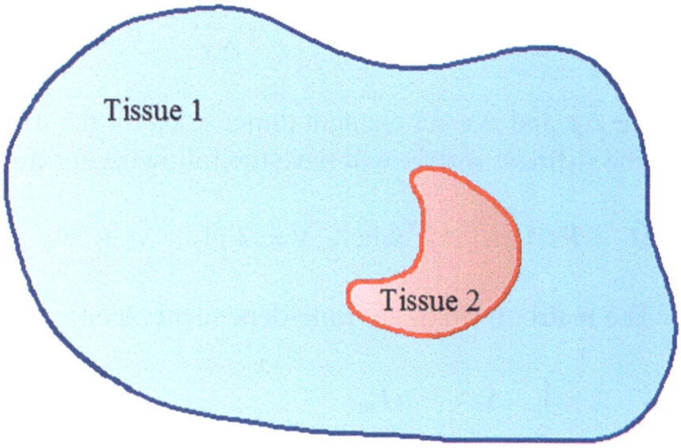

Fig. 5.21 Heterogeneous tissues used for optical imaging

$$\frac{\partial(x, y, t)}{\partial t} = \nabla \cdot (D(u(x, y, t), r)\nabla u(r, t))$$

where $D(u(r, t), r)$ denotes the collective diffusion coefficient.

For the condition that the diffusion coefficient does not depend on the density, D would be a constant value (i.e. thermal diffusivity).

$$\frac{\partial(x, y, t)}{\partial t} = D\nabla^2 u(r, t)$$

By utilizing thermal diffusivity of tissue 1 (i.e. D_1) for the whole domain Ω_1 and using a uniform triangulation mesh for this domain, the time-dependent dynamic equation for domain Ω_1 is obtained (Figs. 5.21 and 5.22)

$$-a\left(u_{i+1j}^{n+1} + u_{i-1j}^{n+1}\right) + (1 + 2a + 2b)u_{ij}^{n+1} - b\left(u_{ij+1}^{n+1} + u_{ij-1}^{n+1}\right) = u_{ij}^{n}$$

where

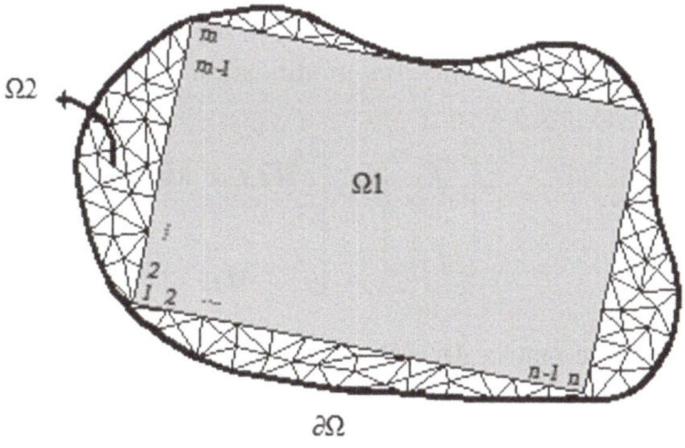

Fig. 5.22 The large element with available pre-defined solution and the remaining domain $\boldsymbol{\Omega_2}$

$$a = D_1 \frac{\Delta t}{\Delta x^2}, \quad b = D_1 \frac{\Delta t}{\Delta y^2}$$

where Δx and Δy are element dimensions for the triangular element.

The stiffness matrix will have the following form:

$$\boldsymbol{M} = \boldsymbol{F}(\boldsymbol{A}, \boldsymbol{B}, \boldsymbol{A}) \quad \text{where } \boldsymbol{A} = \boldsymbol{F}(1 + 2a + 2b, -b, 1 + 2a + 2b), \boldsymbol{B} = -a\boldsymbol{I}$$

The matrix form of the time-dependent equation for time t is obtained as

$$\begin{bmatrix} \boldsymbol{A}_m & -a\boldsymbol{I}_m & & & & \\ -a\boldsymbol{I}_m & \boldsymbol{A}_m & -a\boldsymbol{I}_m & & & \\ & -a\boldsymbol{I}_m & \boldsymbol{A}_m & \ddots & & \\ & & \ddots & \ddots & -a\boldsymbol{I}_m & \\ & & & -a\boldsymbol{I}_m & \boldsymbol{A}_m & -a\boldsymbol{I}_m \\ & & & & -a\boldsymbol{I}_m & \boldsymbol{A}_m \end{bmatrix} \begin{bmatrix} (\boldsymbol{u}_1^t)_m \\ (\boldsymbol{u}_2^t)_m \\ (\boldsymbol{u}_3^t)_m \\ \vdots \\ \left(\boldsymbol{u}_{(n-1)}^t\right)_m \\ (\boldsymbol{u}_n^t)_m \end{bmatrix}$$

$$= \begin{bmatrix} (\boldsymbol{u}_1^{t-1})_m + (g_1(a, b, u_b^t))_m \\ (\boldsymbol{u}_2^{t-1})_m + (g_2(a, b, u_b^t))_m \\ (\boldsymbol{u}_3^{t-1})_m + (g_3(a, b, u_b^t))_m \\ \vdots \\ \left(\boldsymbol{u}_{(n-1)}^{t-1}\right)_m + (g_{(n-1)}(a, b, u_b^t))_m \\ (\boldsymbol{u}_n^{t-1})_m + (g_n(a, b, u_b^t))_m \end{bmatrix}$$

The eigenvalues are obtained as follows:

$$\lambda_{\boldsymbol{M}} = \lambda_{\boldsymbol{D}_M} = \bigcup_{k=1}^{n-2} \bigcup_{l=1}^{m-2} \left[-2a \cos \frac{k\pi}{n+1} + \left(-2b \cos \frac{l\pi}{m+1} + (1 + 2a + 2b) \right) \right]$$

And the inverse of matrix $\boldsymbol{M}$ is found using Eq. (5.58).

To account for the differences in the diffusion coefficient in the two tissues (i.e., D_1 and D_2), the following modification is applied to the stiffness matrix $\boldsymbol{M}$ [4, 13]:

$$\begin{bmatrix} \boldsymbol{f}_I \\ \cdots \\ \boldsymbol{f}_{II} \end{bmatrix} = \begin{bmatrix} \boldsymbol{M}_{I,I} + \boldsymbol{M}'_{I,I} & \vdots & \boldsymbol{M}_{I,II} \\ \cdots & \cdots & \cdots \\ \boldsymbol{M}_{II,I} & \vdots & \boldsymbol{M}_{II,II} \end{bmatrix} \begin{bmatrix} \boldsymbol{u}_I \\ \cdots \\ \boldsymbol{u}_{II} \end{bmatrix}$$

where the matrix $\boldsymbol{M}$ is the stiffness matrix for the whole rectangular domain using the diffusion coefficient D_1 and the matrix $\boldsymbol{M}'$ is the effect of the difference between D_1 and D_2. The size of the block including $\boldsymbol{M}$ is equal to the number of degrees of freedom inside the boundary of tissue 2. It was previously shown the inverse of matrix above is obtained as follows [13]:

$$\begin{bmatrix} u'_I \\ \cdots \\ u'_{II} \end{bmatrix} = \begin{bmatrix} \overline{N}_{I,I} N_{I,I} & \vdots & \overline{N}_{I,I} N_{I,II} \\ \cdots & \cdots & \cdots \\ -N_{II,I} M'_{I,I} \overline{N}_{I,I} N_{I,I} + N_{II,I} & \vdots & -N_{II,I} M'_{I,I} \overline{N}_{I,I} N_{I,II} + N_{II,II} \end{bmatrix} \begin{bmatrix} f_I \\ \cdots \\ f_{II} \end{bmatrix}$$

where N is the inverse of matrix M which was efficiently solved using Eq. (5.58) and $\overline{N}_{I,I} = \left[I + N_{II,I} M'_{I,I}\right]^{-1}$. This means the required effort for finding the inverse of the rectangular domain including tissue 1 and tissue 2 is equal to the finding the inverse of a rectangular homogenous domain (i.e., the thermal diffusivity D_1 for the whole rectangular domain) using the Eq. (5.58) and finding the small matrix $\overline{N}_{I,I}$ of the dimension equal to the degrees of freedom inside the boundary of tissue 2. For the rectangular domain including 200 elements in each direction, we will have a total 10,201 degrees of freedom. While the time complexity for solving the rectangular domain using the current method is 9.5113e+008, the LU decomposition solves the matrix equation of the domain (partitioned by classic mesh) with the time complexity 3.2547e+009 using the swift optimized CW-like algorithms.

The solution of the rectangular domain from the equation above will be the matrix M_{11}^{-1} in Eq. (5.89). Now, the rest of the domain Ω (i.e. domain Ω_2) is partitioned and solved using steps 4 and 5 in the suggested algorithm. For the whole domain including 15,700 degrees of freedom the time complexity using the current method is 5.2974e+009, while this number will be 9.0546e+009 in a classic solution. In this example the area of the largest rectangular domain Ω_1 is ~65% of the whole domain Ω. The effect of the area of the rectangular domain on the time complexity of the problem is shown in Fig. 5.23. The exiting partial current j^+ as a function of the density of the diffusing material $u(r, t)$, ranges from 0.1 to 0.8 nW/mm^2.

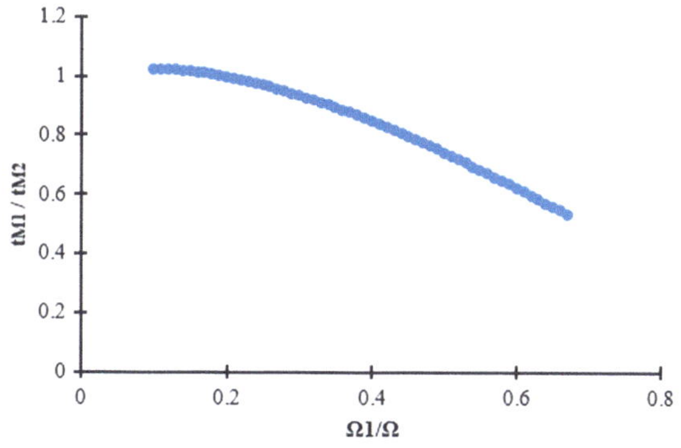

Fig. 5.23 The relative time complexity of the current method (i.e. tM1) and the classic method (i.e. tM2) versus the relative size of the rectangular domain and the whole domain

Example 2 Consider a typical lumbar spine vertebra composed of the vertebral body and the posterior neural arch (Fig. 5.24). Under the effect of gravity, loads carried by upper limbs (e.g. hands), and the response from muscles a compression force will be exerted on the vertebral column. Each vertebra is composed of an outer sell cortical bone and an internal vertical trabecular bone (Fig. 5.25).

The cortical bone is not strong enough to resist the compression forces. However, the vertical trabecular structures brace the cortical shell and transmit the compression forces. While the trabecular anisotropy is specimen-specific, the cortical anisotropy is almost generic [14]. Therefore, it has been suggested that orthotropic material properties obtained from micro-computed tomography data, and generic-anisotropic material properties are used for trabecular bone material modeling and cortical bone material modeling, respectively [14]. Axial compression (AC) and anterior bending (AB) are applied to the rigid plate (Fig. 5.26). For the finite element analysis using

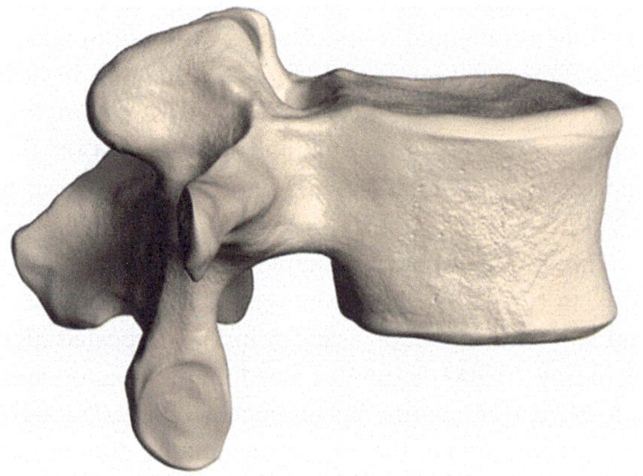

Fig. 5.24 A typical lumbar spine vertebra including the vertebral body and the posterior neural arch

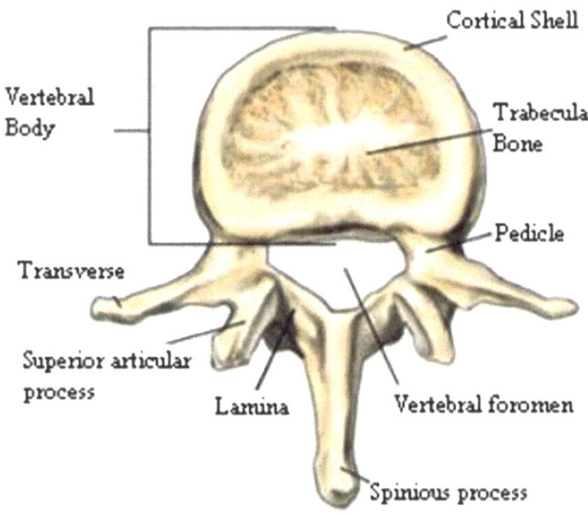

Fig. 5.25 The vertebra composed of an outer sell cortical bone and an internal vertical trabecular bone

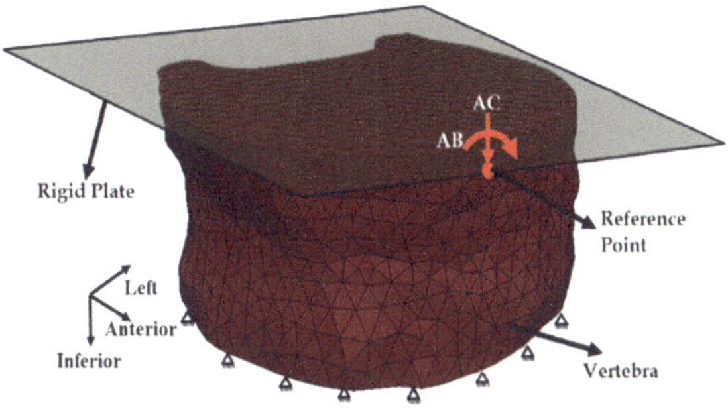

Fig. 5.26 Axial compression (AC) and anterior bending (AB) applied to the vertebra [14]

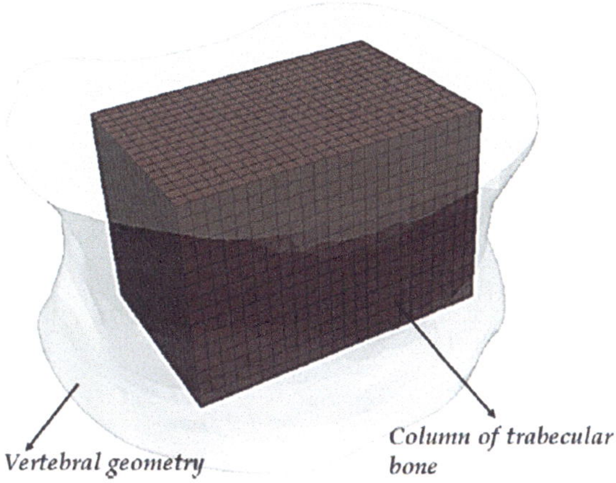

Fig. 5.27 The trabecular column considered as the large pre-solved element [14]

the present method, the trabecular column is considered as the large pre-solved element attached to the remainder part of the vertebra (Fig. 5.27). In the case that the whole vertebra is modeled with generic-anisotropic material properties, the largest cuboid inside the vertebra can be considered as the pre-solved element. The trabecular column and the remainder cortical part are partitioned using 8-node hexahedral elements and quadratic, tetrahedral elements respectively (Figs. 5.26 and 5.27).

It should be mentioned that there are some limitations associated with modeling of the vertebra and, in general, with biomaterial. The assumptions for modeling are borrowed from the relevant literature. Such simplified assumptions are, however, coming from the fact that

1. Vertebra properties are subject-specific;
2. In vivo study of the vertebra involves technical and ethical limitations;
3. Cadaver studies may not provide actual properties of the vertebra.

Now we formulate the problem. Using a block by block numbering along the larger edge of the trabecular cuboid followed by numbering the cortical part, we will have

$$\boldsymbol{M} = \begin{bmatrix} \boldsymbol{F}(\boldsymbol{A}, \boldsymbol{B}, \boldsymbol{A}, \boldsymbol{B}) & \vdots & \boldsymbol{M}_{12} \\ \cdots & \cdots & \cdots \\ \boldsymbol{M}_{21} & \vdots & \boldsymbol{M}_{22} \end{bmatrix}$$

For the trabecular cuboid we can write

$$\boldsymbol{M}_c = \boldsymbol{F}(\boldsymbol{A}, \boldsymbol{B}, \boldsymbol{A}, \boldsymbol{B}) = \boldsymbol{I} \otimes \boldsymbol{A} + \boldsymbol{T} \otimes \boldsymbol{B}$$

And

$$\boldsymbol{D}_{\boldsymbol{M}_c} = \begin{bmatrix} 2\cos\frac{\pi}{n+1}\boldsymbol{B} + \boldsymbol{A} & & & & \\ & 2\cos\frac{2\pi}{n+1}\boldsymbol{B} + \boldsymbol{A} & & & \\ & & \ddots & & \\ & & & 2\cos\frac{(n-1)\pi}{n+1}\boldsymbol{B} + \boldsymbol{A} & \\ & & & & 2\cos\frac{n\pi}{n+1}\boldsymbol{B} + \boldsymbol{A} \end{bmatrix}$$

The matrices $\boldsymbol{A}$ and $\boldsymbol{B}$ are tri-diagonal block matrices themselves.

$$\begin{aligned} 2\cos\frac{k\pi}{n+1}\boldsymbol{B} + \boldsymbol{A} &= 2\cos\frac{k\pi}{n+1}\boldsymbol{F}(\boldsymbol{C}, \boldsymbol{D}, \boldsymbol{C}) + \boldsymbol{F}(\boldsymbol{E}, \boldsymbol{G}, \boldsymbol{E}) \\ &= 2\cos\frac{k\pi}{n+1}(\boldsymbol{I} \otimes \boldsymbol{C} + \boldsymbol{T} \otimes \boldsymbol{D}) + (\boldsymbol{I} \otimes \boldsymbol{E} + \boldsymbol{T} \otimes \boldsymbol{G}) \end{aligned}$$

Thus,

$$\lambda_{\boldsymbol{M}_c} = \bigcup_1^k \bigcup_1^l \left[2\cos\frac{k\pi}{n+1}\left(2\cos\frac{l\pi}{m+1}\boldsymbol{D} + \boldsymbol{C}\right) + \left(2\cos\frac{l\pi}{m+1}\boldsymbol{G} + \boldsymbol{E}\right)\right]$$

Again, the matrices $\boldsymbol{C}$, $\boldsymbol{D}$, $\boldsymbol{E}$ and $\boldsymbol{G}$ are tri-diagonal block matrices themselves.

$$\begin{aligned} \lambda_{\boldsymbol{M}_c} = \bigcup_1^k \bigcup_1^l \bigcup_1^j \Bigg[& 2\cos\frac{k\pi}{n+1}\left(2\cos\frac{l\pi}{m+1}\left(2\cos\frac{j\pi}{s+1}\boldsymbol{Z} + \boldsymbol{X}\right)\right. \\ & \left.+\left(2\cos\frac{j\pi}{s+1}\boldsymbol{V} + \boldsymbol{N}\right)\right) + \left(2\cos\frac{l\pi}{m+1}\left(2\cos\frac{j\pi}{s+1}\boldsymbol{M} + \boldsymbol{P}\right)\right. \\ & \left.+\left(2\cos\frac{j\pi}{s+1}\boldsymbol{O} + \boldsymbol{W}\right)\right)\Bigg] \end{aligned}$$

where $\boldsymbol{Z}, \boldsymbol{X}, \boldsymbol{V}, \boldsymbol{N}, \boldsymbol{M}, \boldsymbol{P}, \boldsymbol{O}$ and $\boldsymbol{W}$ are matrices of dimension 3 that are obtained using the element stiffness matrix of the 8-node hexahedral element.

Using the obtained eigenvalues, the inverse of the stiffness matrix for the trabecular cuboid is obtained (Eq. 5.58). The time complexity for the solution of the cuboid

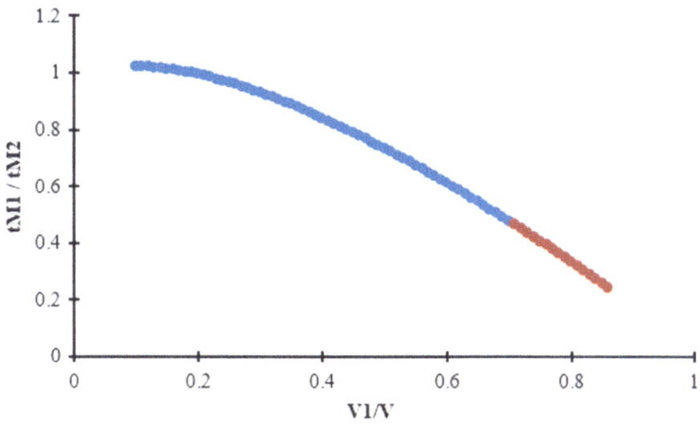

Fig. 5.28 The relative time complexity of the present method (tM1) and the classic method (tM2) versus the relative size of the cuboid domain (V1) and the whole domain (V)

partitioned by a mesh of dimension 100 × 50 × 50 is 6.2110e+011, while this number will be 9.8303e+013 in a classic finite element method. The rest of the problem (i.e. cortical part) is partitioned using quadratic, tetrahedral elements and solved through steps 4 and 5 in the proposed algorithm. Given a value of 70% for the volume of trabecular cuboid relative to the whole vertebral body, the time complexity for solving the whole model using the proposed and the classic methods will be 1.0977e+14 and 2.2920e+14, respectively. For the case that all materials model with generic-anisotropic material properties the largest cuboid will be 85% of the total volume. The relative decrease in time complexity form 70–85% is shown using red line (Fig. 5.28). The distribution of minimum principal strain under the axial compression (1000 N) and the anterior bending (30 Nm) ranges from −0.0075 to 0.001 and −0.02 to 0.00, respectively.

5.4 Mesh Free Solution of Problems with an Arbitrary Domain

In this section a mesh free algorithm using large pre-solved domain is developed. Using the largest rectangle/cuboid inside an arbitrary domain, a pre-solved rectangular domain is established using Kronecker product and graph theory rules. This pre-solved domain is efficiently inserted into the mesh free formulation of partial differential equations (PDEs) and engineering problems to reduce the computational complexity and execution time of the solution. The general solution of the pre-solved rectangular domain is formulated for second order shape functions. The efficiency of the present algorithm depends on the relative size of the large rectangular domain and the main domain; however, the method remains as efficient as a standard method for even small relative sizes. The application of the method is demonstrated using two examples [15].

5.4.1 Mesh Free Formulation and Computational Complexity

Consider the following differential equation with the given boundary condition on the domain Ω

$$\begin{cases} Du = f & \text{in } \Omega \\ Eu = g & \text{in } \Gamma_g \\ u = u_p & \text{in } \Gamma_u \end{cases} \tag{5.90}$$

where D and E are differential operators, and Γ_g and Γ_u are Neumann and Dirichlet boundaries, respectively.

A common method for numerical solution of Eq. (5.90) is the weighted residual method [16]. Function u_h approximates the solution of the differential equation as follows

$$\int_{\Omega} w_1(Du_h - f)d\Omega + \int_{\Gamma_g} w_2(Eu_h - g)d\Gamma + \int_{\Gamma_u} w_3(u_h - u_p)d\Gamma = 0 \tag{5.91}$$

where w_1, w_2 and w_3 are weighting functions.

To preserve the local features of the problem, the function u is approximated over a set of n nodes $\{x_i, i = 1 : n, x_i \in \overline{\Omega}\}$ in the interpolation domain $\overline{\Omega}$ (i.e. a cloud defined by star point[2] i) as follows (Figs. 5.29 and 5.30) [16].

$$u(\boldsymbol{x}) \cong u_h(\boldsymbol{x}) = \sum_{i=1}^{k} p_i(\boldsymbol{x})a_i = \boldsymbol{p}(\boldsymbol{x})^T\boldsymbol{a} \qquad \boldsymbol{x} \in \overline{\Omega} \tag{5.92}$$

where the vector $\boldsymbol{p}(x)^T$ includes a complete monomial basis (order k) and $\boldsymbol{a} = [a_1, a_2, a_3, \ldots, a_k]^T$.

For a two-dimensional domain, one can write

$$\begin{aligned} &\text{Linear } \boldsymbol{p}(\boldsymbol{x}) = [1, x, y]^T, \quad k = 3 \\ &\text{Quadratic } \boldsymbol{p}(\boldsymbol{x}) = \left[1, x, y, x^2, xy, y^2\right]^T, \quad k = 6 \\ &\text{etc.} \end{aligned} \tag{5.93}$$

To define the unknown vector $\boldsymbol{a}$ using the moving least square (MLS) approximation, the functional J is minimized

$$J = \sum_{j=1}^{n} \varphi(\boldsymbol{x} - \boldsymbol{x}_j)\left[u(\boldsymbol{x}_j) - \boldsymbol{p}^T(\boldsymbol{x}_j)\boldsymbol{a}\right]^2 \tag{5.94}$$

[2]The star point of a cloud is the node for which the function u (e.g., $u_h(\boldsymbol{x})$) is calculated. A cloud is known with its star point and the maximum value of the weight function in a cloud occurs in the star point.

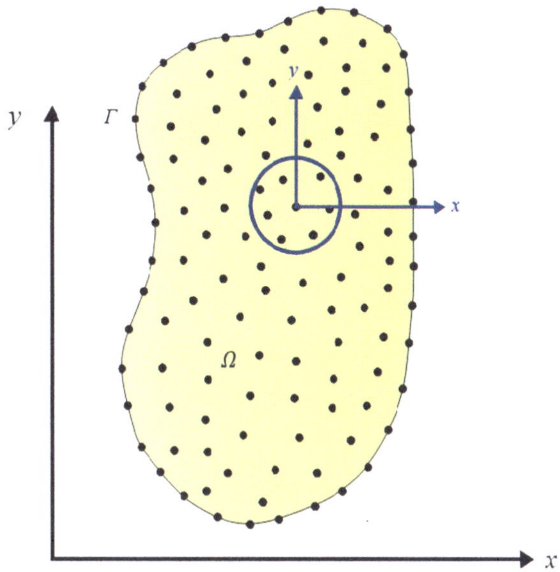

Fig. 5.29 The arbitrary domain Ω including a set of distributed points

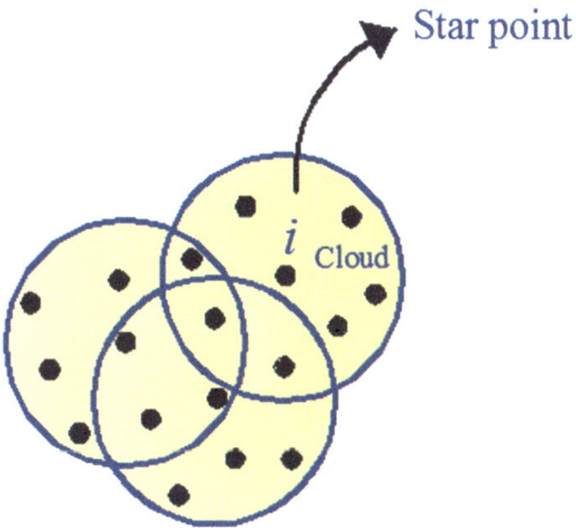

Fig. 5.30 Local cloud i including the domain $\overline{\Omega}$

where φ is the weight function.

Minimizing the functional J with respect to the parameters a_i leads to

$$\boldsymbol{a} = \left(\boldsymbol{P}^T \varphi \boldsymbol{P}\right)^{-1} \boldsymbol{P}^T \varphi \boldsymbol{u} = \boldsymbol{A}^{-1} \boldsymbol{B} \boldsymbol{u} \tag{5.95}$$

where

$$\boldsymbol{P} = \begin{bmatrix} \boldsymbol{p}^T(\boldsymbol{x}_1) \\ \boldsymbol{p}^T(\boldsymbol{x}_2) \\ \vdots \\ \boldsymbol{p}^T(\boldsymbol{x}_n) \end{bmatrix}, \boldsymbol{\varphi} = \begin{bmatrix} \varphi(\boldsymbol{x}-\boldsymbol{x}_1) & & & \\ & \varphi(\boldsymbol{x}-\boldsymbol{x}_2) & & \\ & & \ddots & \\ & & & \varphi(\boldsymbol{x}-\boldsymbol{x}_n) \end{bmatrix}, \boldsymbol{u} = \begin{bmatrix} u(\boldsymbol{x}_1) \\ u(\boldsymbol{x}_2) \\ \vdots \\ u(\boldsymbol{x}_n) \end{bmatrix} \tag{5.96}$$

Substituting $\boldsymbol{a}$ from the Eq. (5.95) into the Eq. (5.92) results in

$$u_h(\boldsymbol{x}) = \boldsymbol{N}^T(\boldsymbol{x})\boldsymbol{u} = \sum_{j=1}^{n} N_j(x)u_j \qquad \boldsymbol{x} \in \overline{\Omega} \tag{5.97}$$

where

$$\boldsymbol{N}^T(\boldsymbol{x}) = \boldsymbol{p}^T(\boldsymbol{x})\boldsymbol{A}^{-1}\boldsymbol{B}$$

and

$$N_j(\boldsymbol{x}) = \sum_{i=1}^{k} p_i(\boldsymbol{x})(\boldsymbol{A}^{-1}\boldsymbol{B})_{ij} \tag{5.98}$$

The first order partial derivative of $u_h(\boldsymbol{x})$ with respect to x_l is obtained as

$$(u_h)_{,l}(\boldsymbol{x}) = \sum_{j=1}^{n} N_{j,l}(\boldsymbol{x})u_j \quad \boldsymbol{x} \in \overline{\Omega} \tag{5.99}$$

and

$$N_{j,l}(x) = \sum_{i=1}^{k} p_{i,l}(x)\left(\boldsymbol{A}^{-1}\boldsymbol{B}\right)_{ij} + p_i(x)\left(\boldsymbol{A}^{-1}\boldsymbol{B}_{,l} + \left(\boldsymbol{A}_{,l}^{-1}\boldsymbol{B}\right)_{ij}\right. \tag{5.100}$$

where

$$\boldsymbol{A}_{,l}^{-1} = \left(\boldsymbol{A}^{-1}\right)_{,l} = -\boldsymbol{A}^{-1}\boldsymbol{A}_{,l}\boldsymbol{A}^{-1} \tag{5.101}$$

Repeating the same procedure, higher order derivatives are obtained.

To solve the problem using finite point method (i.e. point collocation), we set $w_1 = w_2 = w_3 = \delta$ in Eq. (5.91), where δ is Dirac delta. This leads to a set of equations as follows:

$$\begin{aligned} (Du_h)_i - f_i &= 0 \quad \text{in } \Omega \\ (Eu_h)_i - g_i &= 0 \quad \text{in } \Gamma_g \\ (u_h)_i - u_p &= 0 \quad \text{in } \Gamma_u \end{aligned} \tag{5.102}$$

The components of the coefficient or stiffness matrix $\boldsymbol{M}$ are found as follows:

$$\boldsymbol{M}_{ij} = \left[D(N_j)\right]_i + \left[E(N_j)\right]_i \tag{5.103}$$

And the matrix equation is obtained as:

$$\boldsymbol{M}\boldsymbol{u} = \boldsymbol{b} \tag{5.104}$$

where $\boldsymbol{u}$ is the vector of unknown displacements and $\boldsymbol{b}$ is the vector of known forces at the points. The computational complexity of solving the matrix equation using LU decomposition is $O(n^{2.373})$ in the Optimized CW-like algorithms [3].

However, if one can re-write Eq. (5.104) using the following decomposed form, the computational complexity will be reduced:

$$\boldsymbol{u} = \boldsymbol{V}\boldsymbol{\lambda}^{-1}\boldsymbol{V}^T\boldsymbol{b} \tag{5.105}$$

where the matrices $\boldsymbol{V}$ and $\boldsymbol{\lambda}$ are the eigenvectors and eigenvalues of the matrix $\boldsymbol{M}$, respectively.

Using a proper multiplication order, the computational complexity of calculating $\boldsymbol{V}\boldsymbol{\lambda}^{-1}\boldsymbol{V}^T\boldsymbol{b}$ would be $O(n^2)$. However, we have not considered the required effort for finding the decomposed form yet. The computational complexity for eigensolution of a matrix using the QR algorithm is of the order $O(n^3)$ that is higher than computational complexity in the direct solution. Alternatively, we can use Kronecker products rules along with block representation of matrices to efficiently decompose matrix $\boldsymbol{M}$. To do so, we first need to re-construct matrix $\boldsymbol{M}$ in our desired form that, in turn, requires re-formulation of the mesh free method.

5.4.2 Decomposition of Mesh Free Matrices on Rectangular Domains

Suppose matrix $\boldsymbol{M}$ of the dimension $n = l \times m$ holds the following block form

$$\boldsymbol{M}_n = \begin{bmatrix} \boldsymbol{A}_m & \boldsymbol{B}_m & & & & \\ \boldsymbol{B}_m & \boldsymbol{A}_m & \boldsymbol{B}_m & & & \\ & \boldsymbol{B}_m & \boldsymbol{A}_m & \boldsymbol{B}_m & & \\ & & \boldsymbol{B}_m & \ddots & \ddots & \\ & & & \ddots & \boldsymbol{A}_m & \boldsymbol{B}_m \\ & & & & \boldsymbol{B}_m & \boldsymbol{A}_m \end{bmatrix}_l = \boldsymbol{F}_l(\boldsymbol{A}_m, \boldsymbol{B}_m, \boldsymbol{A}_m)$$
$$= \boldsymbol{I}_l \otimes \boldsymbol{A}_m + \boldsymbol{T}_l \otimes \boldsymbol{B}_m \tag{5.106}$$

where $\boldsymbol{I}$ is identity matrix and $\boldsymbol{T} = \boldsymbol{F}(0, 1, 0)$ [12].

Using the following lemma matrix $\boldsymbol{M}$ can be decomposed:

Sufficient conditions for converting Hermitian matrices $\boldsymbol{A}_1$ and $\boldsymbol{A}_2$ into upper triangular ones using one orthogonal matrix are [11]:

$$\boldsymbol{A}_1\boldsymbol{A}_2 = \boldsymbol{A}_2\boldsymbol{A}_1 \quad \text{or} \quad \boldsymbol{A}_1^2 = \boldsymbol{A}_2^2 \tag{5.107}$$

If one of the conditions in Eq. (5.107) holds, the eigenpairs of the block matrix $\boldsymbol{M} = \boldsymbol{A}_1 \otimes \boldsymbol{B}_1 + \boldsymbol{A}_2 \otimes \boldsymbol{B}_2$ are obtained as follows:

$$\lambda_M = \bigcup_{i=1}^{l} eig(\boldsymbol{M}_i);\ \boldsymbol{M}_i = \lambda_i(\boldsymbol{A}_1)\boldsymbol{B}_1 + \lambda_i(\boldsymbol{A}_2)\boldsymbol{B}_2 \tag{5.108}$$

Since $\boldsymbol{IT} = \boldsymbol{TI}$ in Eq. (5.106), we will have

$$\lambda_M = \bigcup_{k=1}^{l} eig\left(2\cos\frac{k\pi}{l+1}\boldsymbol{B} + \boldsymbol{A}\right) \tag{5.109}$$

In the suggested mesh free formulation (see below) matrices $\boldsymbol{A}$ and $\boldsymbol{B}$ will have tri-diagonal patterns as follows

$$2\cos\frac{k\pi}{l+1}\boldsymbol{B} + \boldsymbol{A} = 2\cos\frac{k\pi}{l+1}\boldsymbol{F}(c, d, c) + \boldsymbol{F}(e, g, e) \tag{5.110}$$

where c, d, e and g are scalars values

$$\lambda_M = \bigcup_{k=1}^{l}\bigcup_{j=1}^{m} 2\cos\frac{k\pi}{l+1}\left(c + 2d\cos\frac{j\pi}{m+1}\right) + \left(e + 2g\cos\frac{j\pi}{m+1}\right) \tag{5.111}$$

The eigenvalues of matrix $\boldsymbol{M}$ are obtained using $n = l \times m$ operations that is $O(n)$. Eigenvectors of matrix $\boldsymbol{M}$ are calculated with the same computational complexity, Kaveh [9]. By reducing the computational complexity of eigensolution problem form the order $O(n^3)$ to the order $O(n)$, we can efficiently solve Eq. (5.104) using the decomposed form in Eq. (5.105).

Generating the desired pattern for matrix $\boldsymbol{M}$ [i.e. the pattern in Eq. (5.106)] is possible on a rectangular domain Ω. To do so, the clouds in the domain should include the same number of points with the same symmetric distribution. These properties leads to the generation of blocks $\boldsymbol{A}_m$ and $\boldsymbol{B}_m$. Furthermore, to generate the repetitive pattern of these two blocks (Eq. 5.106), a block by block numbering should be performed.

Consider the rectangular domain Ω in Fig. 5.31. For a linear vector $\boldsymbol{p}(\boldsymbol{x}) = [1, x, y]^T$, the number of points in a cloud (i.e. n_c) should be greater than the order

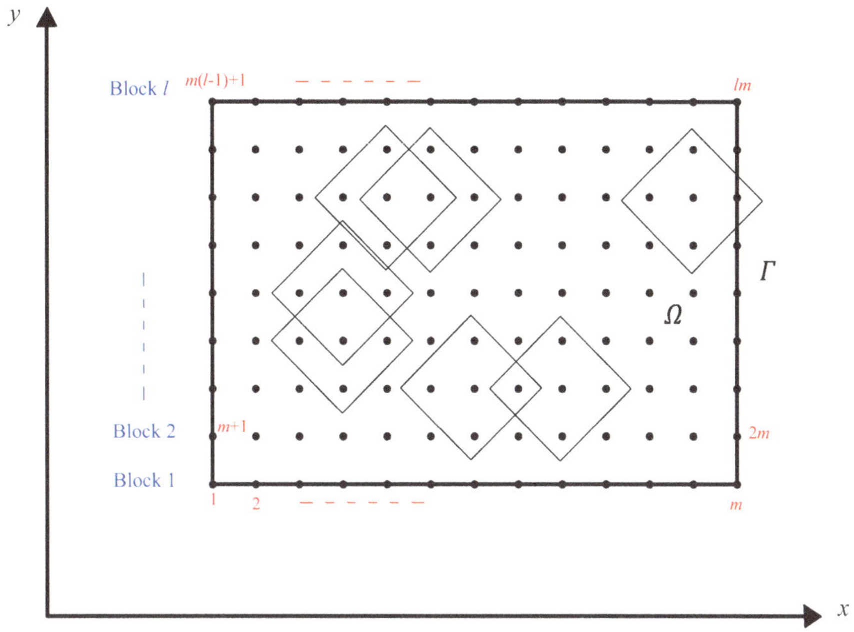

Fig. 5.31 The rectangular domain including horizontal and vertical blocks of points with similar 5-points clouds

of vector $\boldsymbol{p}(\boldsymbol{x})$ (i.e. $k = 3$). The domain is localized using square clouds ($n_p = 5$), and all points in the domain are numbered in a block by block order.

This result in the coefficient matrix in Eq. (5.106) with $\boldsymbol{B}_m = \boldsymbol{I}_m$

$$\boldsymbol{M}_n = \boldsymbol{F}_l(\boldsymbol{A}_m, \boldsymbol{I}_m, \boldsymbol{A}_m) = \boldsymbol{I}_l \otimes \boldsymbol{A}_m + \boldsymbol{T}_l \otimes \boldsymbol{I}_m \tag{5.112}$$

where $\boldsymbol{A}_m$ is a tri-diagonal matrix.

Since most partial differential equations in physics are of the second order, the solution of matrix $\boldsymbol{M}$ for the quadratic vector $\boldsymbol{p}(\boldsymbol{x}) = \left[1, x, y, x^2, xy, y^2\right]^T$ is presented in details. The rectangular domain Ω is localized using square clouds ($n_p = 9 > k = 6$) and all points in the domain are numbered in a block by block order (Fig. 5.32).

Consider the ith cloud including 9 points with equal horizontal and vertical distances h. Coordinates of the points (i.e. the set P) are calculated using the coordinate of the star point (x_i, y_i):

$$\begin{aligned} P = \{&(x_i - h, y_i + h), (x_i, y_i + h), (x_i + h, y_i + h), \\ &(x_i - h, y_i), (x_i, y_i), (x_i + h, y_i), (x_i - h, y_i - h), \\ &(x_i, y_i - h), (x_i + h, y_i - h)\} \end{aligned} \tag{5.113}$$

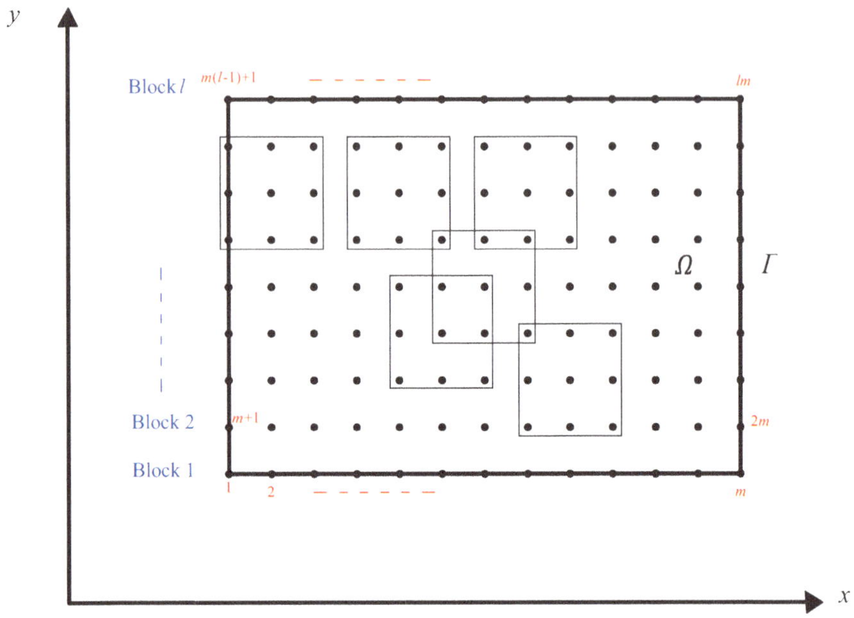

Fig. 5.32 The rectangular domain including horizontal and vertical blocks of points with similar 9-points clouds

The quartic spline weight function is employed as follows [17]:

$$\varphi(r) = \begin{cases} 1 - 6r^2 + 8r^3 - 3r^4 & r \le 1 \\ 0 & r \ge 1 \end{cases} \tag{5.114}$$

where $r = \|\boldsymbol{x} - \boldsymbol{x}i\|/di$ and di is the support size of node i.

Using the set P and setting $di = 1.2 \times \max(\|\boldsymbol{x} - \boldsymbol{x}i\|)$, vector $\boldsymbol{r}$ is calculated

$$\boldsymbol{r} = \left[\frac{\sqrt{2}h}{1.2\left(\sqrt{2}h\right)}, \frac{h}{1.2\left(\sqrt{2}h\right)}, \frac{\sqrt{2}h}{1.2\left(\sqrt{2}h\right)}, \frac{h}{1.2\left(\sqrt{2}h\right)}, 0, \frac{h}{1.2\left(\sqrt{2}h\right)}, \frac{\sqrt{2}h}{1.2\left(\sqrt{2}h\right)}, \frac{h}{1.2\left(\sqrt{2}h\right)}, \frac{\sqrt{2}h}{1.2\left(\sqrt{2}h\right)}\right] \tag{5.115}$$

A fixed least square (FLS) approximation is employed and the matrix of weight functions (i.e. the diagonal matrix $\boldsymbol{\varphi}$) is calculated using Eqs. (5.114) and (5.115). The following set includes diagonal components of matrix $\boldsymbol{\varphi}$:

$$\mathrm{diag}(\boldsymbol{\varphi}) = \{0.0162\ 0.1918\ 0.0162\ 0.1918\ 1.0000\ 0.1918\ 0.0162\ 0.1918\ 0.0162\} \tag{5.116}$$

Hereafter, we will deal with matrix calculations (i.e. multiplication, inverting etc.). Thus, it would be difficult to continue with coordinate parameters (i.e. (x_i, y_i) and h). Due to local and relative features of the points in the cloud, the global value of the star point (x_i, y_i) is not important. Furthermore, instead of changing the value of h for different problems or for a given problem with different accuracies of the solution, the dimension of the problem and given functions are scaled in a way to be compatible with the constant value of h. As such, we assume $(x_i, y_i) = (0.5, 0.5)$ and $h = 0.25$.

$$\boldsymbol{P} = \begin{bmatrix} \boldsymbol{p}^T(\boldsymbol{x}_1) \\ \boldsymbol{p}^T(\boldsymbol{x}_2) \\ \vdots \\ \boldsymbol{p}^T(\boldsymbol{x}_9) \end{bmatrix} = \begin{bmatrix} 1\ 0.25\ 0.25\ 0.06\ 0.06\ 0.06 \\ 1\ 0.25\ 0.25\ 0.25\ 0.12\ 0.06 \\ 1\ 0.25\ 0.25\ 0.56\ 0.18\ 0.06 \\ 1\ 0.50\ 0.50\ 0.06\ 0.12\ 0.25 \\ 1\ 0.50\ 0.50\ 0.25\ 0.25\ 0.25 \\ 1\ 0.50\ 0.50\ 0.56\ 0.37\ 0.25 \\ 1\ 0.75\ 0.75\ 0.06\ 0.18\ 0.56 \\ 1\ 0.75\ 0.75\ 0.25\ 0.37\ 0.56 \\ 1\ 0.75\ 0.75\ 0.56\ 0.56\ 0.56 \end{bmatrix} \tag{5.117}$$

Substituting the matrices $\boldsymbol{P}$ and $\boldsymbol{\varphi}$ into the Eqs. (5.95) and (5.97) results in

$$u_h(\boldsymbol{x}) = \boldsymbol{p}^T(\boldsymbol{x})\boldsymbol{A}^{-1}\boldsymbol{B}\boldsymbol{u}$$
$$= \begin{bmatrix} 1 \\ x \\ y \\ x^2 \\ xy \\ y^2 \end{bmatrix}^{T} \begin{bmatrix} 1.46 & 2.21 & -0.67 & 2.21 & -5.71 & 0.50 & -0.67 & 0.50 & 1.17 \\ -2.81 & 1.33 & 1.47 & -8.37 & 13.33 & -4.95 & 1.18 & 1.33 & -2.52 \\ -2.81 & -8.37 & 1.18 & 1.33 & 13.33 & 1.33 & 1.47 & -4.95 & -2.52 \\ 0.66 & -1.33 & 0.66 & 6.67 & -13.33 & 6.66 & 0.66 & -1.33 & 0.66 \\ 4.00 & 0.00 & -4.00 & 0.00 & 0.00 & 0.00 & -4.00 & 0.00 & 4.00 \\ 0.66 & 6.67 & 0.66 & -1.33 & -13.33 & -1.33 & 0.66 & 6.66 & 0.66 \end{bmatrix}$$
$$\begin{bmatrix} u_1 \\ u_2 \\ u_3 \\ \vdots \\ u_8 \\ u_9 \end{bmatrix} \tag{5.118}$$

Equation (5.118), repressing the displacement field for ith cloud, will have the same pattern for all other clouds because same clouds with the same point distributions were used throughout the domain (Figs. 5.32 and 5.33). Also, this pattern would be the same for all PDE problems. We reduced the mathematical computations through utilizing the same clouds (i.e. Eq. 5.118 represents any cloud in the domain). However, since the computational complexity of the problem is governed by the solution of Eq. (5.104), the main goal of the manipulations above is obtaining the desired pattern in Eq. (5.106) and then solving Eq. (5.103) via Eq. (5.105).

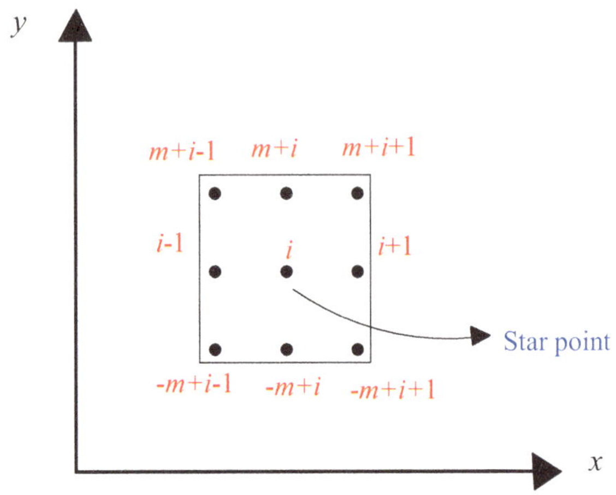

Fig. 5.33 The local 9-points cloud with repetitive numbering

Accordingly, we apply the local displacement field (Eq. 5.118) to a PDE to construct the global matrix equation (Eq. 5.103). Consider Laplace equation with the following boundary conditions

$$\text{BVP}\begin{cases} \nabla^2 u(x, y) = 0 \\ u(x, 0) = \sin(2\pi x) \\ u(x, 0.5) = \sin(2\pi x)e^{-\pi} \\ u(0, y) = u(0.5, y) = 0 \end{cases} \quad \Omega = \{(x, y)|x, y \in [0\ 0.5]\} \tag{5.119}$$

25 points are distributed in the domain and on the boundary with the same distances. This results in $h = 0.125$; however, since we developed Eq. (5.118) for $h = 0.25$, the dimensions and functions are scaled as follows

$$\text{BVP}\begin{cases} \nabla^2 u(X, Y) = 0 \\ u(X, 0) = \sin(\pi X) \\ u(X, 1) = \sin(\pi X)e^{-\pi} \\ u(0, Y) = u(1, Y) = 0 \end{cases} \quad \Omega = \{(X, Y)|X, Y \in [0\ 1]\} \tag{5.120}$$

where X, Y are the scaled variables.

By solving the Eq. (5.120) for 9 internal points (i.e. degrees of freedom) and substituting the boundary values for the 16 boundary points, the load vector $\boldsymbol{b}$ and stiffness matrix $\boldsymbol{M}$ are obtained as follows

$$\boldsymbol{b} = -[0.44, 0.62, 0.44, 0.00, 0.00, 0.00, 10.20, 14.43, 10.20] \tag{5.121}$$

And

$$\boldsymbol{M}_9 = \boldsymbol{F}_3(\boldsymbol{A}_3, \boldsymbol{B}_3, \boldsymbol{A}_3) = \boldsymbol{I}_3 \otimes \boldsymbol{A}_3 + \boldsymbol{T}_3 \otimes \boldsymbol{B}_3 \tag{5.122}$$

where

$$\boldsymbol{A} = \begin{bmatrix} -53.33 & 10.66 & 0 \\ 10.66 & -53.33 & 10.66 \\ 0 & 10.66 & -53.33 \end{bmatrix} \quad \boldsymbol{B} = \begin{bmatrix} 10.66 & 2.66 & 0 \\ 2.66 & 10.66 & 2.66 \\ 0 & 2.66 & 10.66 \end{bmatrix} \tag{5.123}$$

Matrix $\boldsymbol{M}$ holds the desired pattern in Eq. (5.106) with tri-diagonal matrices $\boldsymbol{A}$ and $\boldsymbol{B}$.

Therefore, the eigenvalues of matrix $\boldsymbol{M}$ are found using Eq. (5.111) with $c = 10.66$, $d = 2.66$, $e = -53.33$ and $g = 10.66$.

$$\lambda = \{-17.83, -78.17, -53.33, -38.24, -38.24, -68.41, -68.41, -58.66, -58.66\} \tag{5.124}$$

Using Eq. (5.105) the vector $\boldsymbol{u}$ is efficiently found

$$\boldsymbol{u} = [0.0670, 0.0948, 0.0670, 0.1470, 0.2079, 0.1470, 0.3224, 0.4560, 0.3224] \tag{5.125}$$

Justifying via the analytical solution in Eq. (5.126), the components of vector $\boldsymbol{u}$ are accurate up to 4 digits.

$$u(x, y) = \sin(\pi x)\exp(-\pi y) \tag{5.126}$$

For the general solution of Laplace equation using n points (i.e. n degrees of freedom) inside a domain including $m \times l$ blocks, we will have

$$\boldsymbol{M}_n = \boldsymbol{F}_l(\boldsymbol{A}_m, \boldsymbol{B}_m, \boldsymbol{A}_m) = \boldsymbol{I}_l \otimes \boldsymbol{A}_m + \boldsymbol{T}_l \otimes \boldsymbol{B}_m \tag{5.127}$$

where

$$\boldsymbol{A}_m = \boldsymbol{F}_m(-53.33, 10.66, -53.33) \quad \boldsymbol{B}_m = \boldsymbol{F}_m(10.66, 2.66, 10.66) \tag{5.128}$$

And finally the general equation for eigenvalues of the stiffness matrix $\boldsymbol{M}$ is obtained

$$\lambda_M = \bigcup_{k=1}^{l} \bigcup_{j=1}^{m} 2\cos\frac{k\pi}{l+1}\left(10.66 + 2(2.66)\cos\frac{j\pi}{m+1}\right) + \left(-53.33 + 2(10.66)\cos\frac{j\pi}{m+1}\right) \tag{5.129}$$

For a domain including 10,000 degrees of freedom (i.e. $m = 100, l = 100$) the contour of solution is obtained as follows (Fig. 5.34).

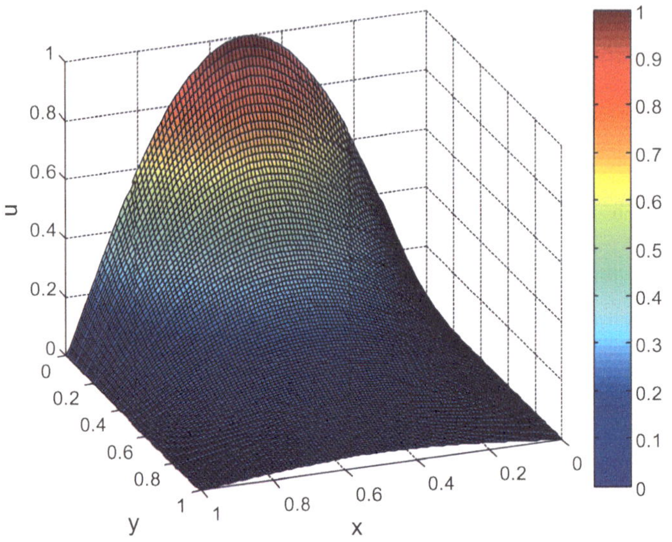

Fig. 5.34 The contour of displacement field

The formulation obtained via application of repetitive identical clouds, block by block numbering, and Kronecker product rules is considered as a pre-defined solution for the rectangular domain. Such a solution will be inserted in the mesh free formulation of an arbitrary domain to reduce the computational complexity of the problem.

5.4.3 Mesh Free Solution Using the Hyper Pre-solved Domain

Developing an efficient mesh free formulation for an arbitrary domain Ω is aimed here. The concepts developed above are combined and inserted into an algorithm to establish a numerical solution. To utilize the obtained results for rectangular domain in the solution of the arbitrary domain Ω, the domain is split into a rectangular domain plus the remaining part. Although there is no limitation for the relative size of the rectangular and remaining parts, the larger the rectangular domain, the more efficient the algorithm. Obviously, for an extreme condition where the domain Ω approaches a rectangular domain, the unknowns of problem are simply found using the results obtained in Eqs. (5.105), (5.111) and (5.118). The rectangular domain, solved via Kronecker product rules, is henceforth called the large pre-solved domain.

Consider the PDE in Eq. (5.90) on the arbitrary domain Ω (Fig. 5.29). In a standard mesh free formulation, the final matrix equation in Eq. (5.104) should be solved. The newly-developed algorithms [3] have improved the time complexity of LU decompositions. The proposed method in this part can help further reduce computational complexity of the solution. As the first step, the largest rectangle inside the domain Ω (Fig. 5.35), is found using the earlier proposed algorithm.

The global coordinate system for the problem is chosen parallel to the sides of the obtained rectangle. A block by block point distribution is used in the rectangular domain Ω_1, and the stiffness matrix for the domain Ω_1 is constructed and decomposed. It is not required to partition the domain and construct the stiffness matrix as it was already conducted using Eqs. (5.106), (5.111) and (5.118) (Fig. 5.11). The only thing one needs to know is the number of the blocks (i.e. the values of m and l) in Eq. (5.111). The whole rectangular domain Ω_1 is considered as a large domain with pre-defined solution whereas the remaining domain Ω_2 is formulated and solved using the standard mesh free method (Fig. 5.36). Finally, the stiffness matrices of the rectangular domain Ω_1 and the domain Ω_2 are assembled such that the global stiffness matrix for the whole domain Ω is obtained. Since we first number the points of the rectangular domain (i.e. block by block numbering) and then those of the domain Ω_2, assembling will not affect our desired pattern in Eq. (5.106). As such, one can use the pre-defined solution of the large rectangular domain in the solution of the main problem. The stiffness matrix of the entire domain Ω is partitioned using the following block matrix as:

$$\boldsymbol{M} = \begin{bmatrix} \boldsymbol{F}(\boldsymbol{A}, \boldsymbol{B}, \boldsymbol{A}) & \vdots & \boldsymbol{M}_{12} \\ \cdots & \cdots & \cdots \\ \boldsymbol{M}_{21} & \vdots & \boldsymbol{M}_{22} \end{bmatrix} = \begin{bmatrix} \boldsymbol{M}_{11} & \vdots & \boldsymbol{M}_{12} \\ \cdots & \cdots & \cdots \\ \boldsymbol{M}_{21} & \vdots & \boldsymbol{M}_{22} \end{bmatrix} \tag{5.130}$$

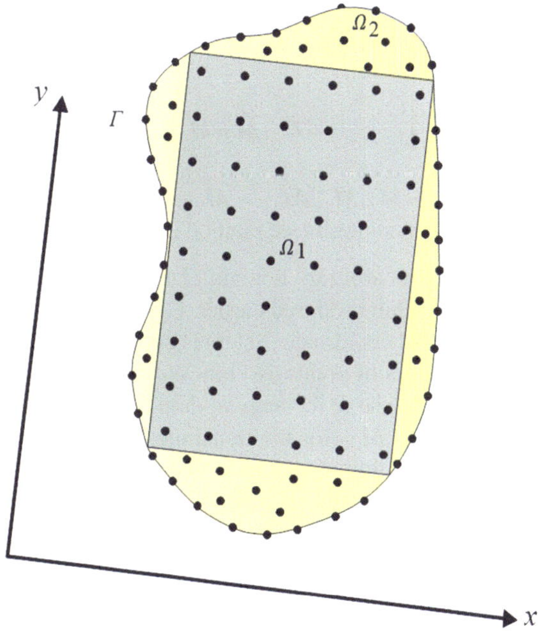

Fig. 5.35 The largest rectangle Ω_1 and the remaining part Ω_2 inside the arbitrary domain Ω including the distributed points

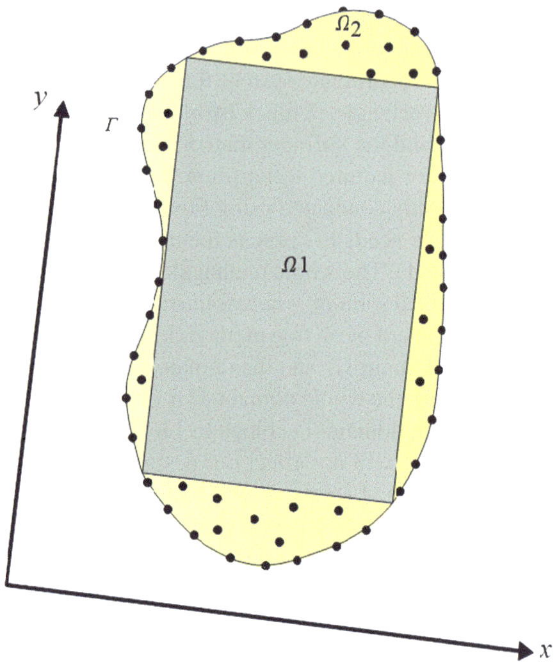

Fig. 5.36 A large domain Ω_1 with available pre-defined solution and the remaining domain Ω_2

where the blocks $\boldsymbol{M}_{11}$ and $\boldsymbol{M}_{22}$ correspond to the degrees of freedom inside the domains Ω_1 and Ω_2, respectively. The unknowns are obtained as follows

$$\begin{bmatrix} \boldsymbol{U}_1 \\ \cdots \\ \boldsymbol{U}_2 \end{bmatrix} = \begin{bmatrix} \boldsymbol{M}_{11}^{-1}(\boldsymbol{I} + \boldsymbol{M}_{12}\boldsymbol{\Gamma}^{-1}\boldsymbol{M}_{21}\boldsymbol{M}_{11}^{-1}) & \vdots & -\boldsymbol{M}_{11}^{-1}\boldsymbol{M}_{12}\boldsymbol{\Gamma}^{-1} \\ \cdots & \cdots & \cdots \\ -\boldsymbol{\Gamma}^{-1}\boldsymbol{M}_{21}\boldsymbol{M}_{11}^{-1} & \vdots & -\boldsymbol{\Gamma}^{-1} \end{bmatrix} \begin{bmatrix} \boldsymbol{F}_1 \\ \cdots \\ \boldsymbol{F}_2 \end{bmatrix} \quad (5.131)$$

where $\boldsymbol{\Gamma} = -\boldsymbol{M}_{21}\boldsymbol{M}_{11}^{-1}\boldsymbol{M}_{12} + \boldsymbol{M}_{22}$ and $\boldsymbol{M}_{11}^{-1} = \boldsymbol{V}\boldsymbol{\lambda}^{-1}\boldsymbol{V}^T$.

The proposed mesh free method is summarized in the following steps:

1. For a given arbitrary domain Ω the largest internal rectangle is obtained using the algorithm proposed earlier
2. The global coordinate system is chosen parallel to sides of the rectangle
3. The number of points and blocks are decided for the domain Ω_1 and the matrix $\boldsymbol{M}_{11}$ for the large rectangular domain is constructed.
4. The matrix $\boldsymbol{M}_{22}$ for the remaining domain Ω_2 is constructed using a standard mesh free solution and the connecting matrices $\boldsymbol{M}_{12}$ and $\boldsymbol{M}_{21}$ are obtained.
5. The solution of the whole domain Ω is found using Eq. (5.131) wherein $\boldsymbol{M}_{11}^{-1}$ is calculated using Eqs. (5.105), (5.111) and (5.118).

5.4.4 Complementary Examples

In this section two real-life problems are investigated using the proposed method.

Example 1 The head (u) of the water flowing into the soil with anisotropic material satisfies the following second order PDE:

$$k_x \frac{\partial^2 u}{\partial x^2} + k_y \frac{\partial^2 u}{\partial y^2} = 0$$

where k_x and k_y are permeability coefficients in x and y directions, respectively.

Consider the soil with the given boundary conditions in Fig. 5.37 and $k_x = 4$ and $k_y = 1$. First, the mixed boundary condition on the upper layer is converted into the pure Dirichlet boundary condition.

$$30 \le x \le 60 : 4u_{xx} + u_{yy} = 0 \xrightarrow{u_y=0 \to u_{yy}=0} u_{xx} = 0 \to u = Ax + B$$

$$\begin{cases} x = 30 \to u = 10 \\ x = 60 \to u = 2 \end{cases} \to \begin{cases} A = -4/15 \\ B = 18 \end{cases} \to u(x,50) = \begin{cases} 10 & 0 \le x \le 30 \\ 18 - \frac{4x}{15} & 30 \le x \le 60 \\ 2 & 60 \le x \le 100 \end{cases}$$

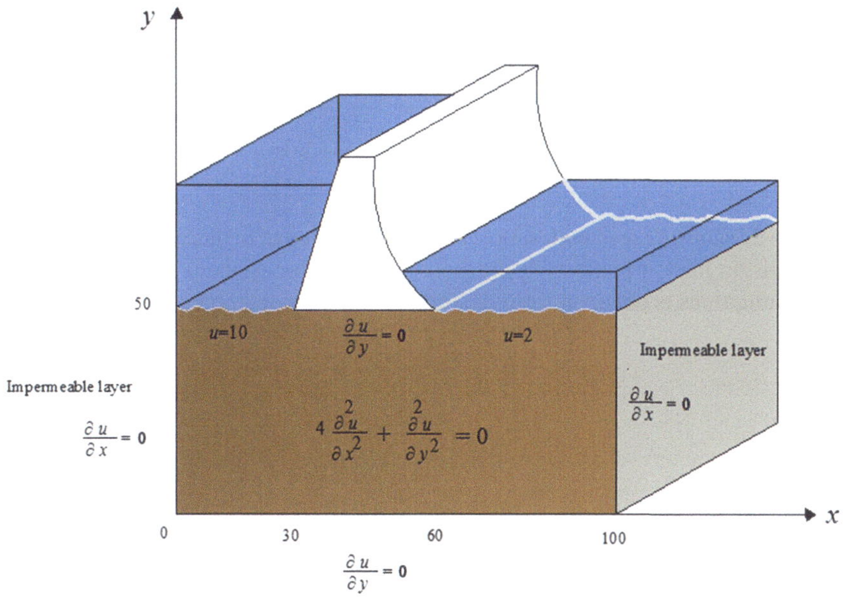

Fig. 5.37 The water flowing through the soil under the dam

By substituting the local displacement field (Eq. 5.118) into the PDE and applying the boundary conditions, we will have

$$\boldsymbol{M}_n = \boldsymbol{F}_l(\boldsymbol{A}_m, \boldsymbol{B}_m, \boldsymbol{A}_m) = \boldsymbol{I}_l \otimes \boldsymbol{A}_m + \boldsymbol{T}_l \otimes \boldsymbol{B}_m$$

where

$$\boldsymbol{A}_m = \boldsymbol{F}_m(-82.66, 50.66, -133.33), \quad \boldsymbol{B}_m = \boldsymbol{F}_m(9.33, 6.66, 2.66)$$

These two matrices are of the form $\boldsymbol{F}_m(a, b, c)$ with the condition $a = b + c$. The eigenvalues of this form are obtained using the equation $\lambda_k = c + 2b\cos((k-1)\pi)/m$. Accordingly, the equation for eigenvalues of the stiffness matrix $\boldsymbol{M}$ is obtained

$$\lambda_{\boldsymbol{M}} = \bigcup_{k=1}^{l}\bigcup_{j=1}^{m} 2\cos\frac{k\pi}{l+1}\left(2.66 + 2(6.66)\cos\frac{(j-1)\pi}{m+1}\right) + \left(-133.33 + 2(50.66)\cos\frac{(j-1)\pi}{m+1}\right)$$

Setting $l = 50$ and $m = 100$ above, the heads of the water for 5000 points are obtained using Eq. (5.105). The time complexity for solving the problem using the present method is 7.500e+007, compared to the value 6.2432e+008, for the same number of points, in the Optimized CW-like algorithms [3]. The contour of solution is depicted in Fig. 5.38 that is accurate up to 4 decimal digits relative to the analytical solution below:

$$u(x, y) = \frac{28}{5} + \frac{160}{3\pi^2}\sum_{n=1}^{\infty}\frac{\cos\left(\frac{3\pi n}{10}\right) - \cos\left(\frac{3\pi n}{5}\right)}{n^2\cosh(n\pi)}\cos\left(\frac{n\pi x}{100}\right)\cosh\left(\frac{n\pi y}{50}\right)$$

Example 2 A mobile roof with simple supports is designed for a stadium (Fig. 5.39). For a uniform load of intensity P, the theoretical solution of the problem using thin plate assumptions is obtained from the following fourth order equation:

$$\nabla^4 w = \frac{\partial^4 w}{\partial x^4} + \frac{2\partial^4 w}{\partial x^2 \partial y^2} + \frac{\partial^4 w}{\partial y^4} = \frac{P}{D}$$

where the parameters w and D are the lateral deflection and the flexural rigidity per unit width of plate, respectively.

This equation can be converted into two second order equations [18] as

$$\nabla^2 M = \frac{\partial^2 M}{\partial x^2} + \frac{\partial^2 M}{\partial y^2} = -P, \quad \nabla^2 w = \frac{\partial^2 w}{\partial x^2} + \frac{\partial^2 w}{\partial y^2} = -\frac{M}{D}$$

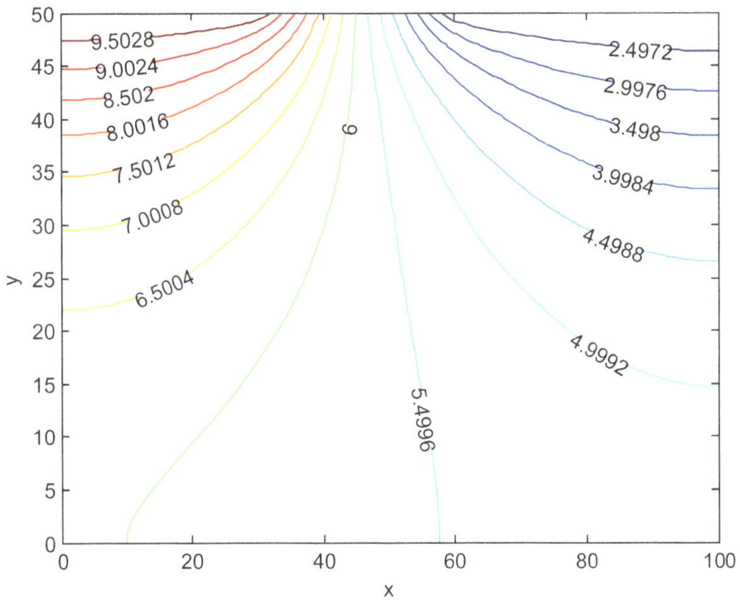

Fig. 5.38 The contour of solution including the lines with the same head of water

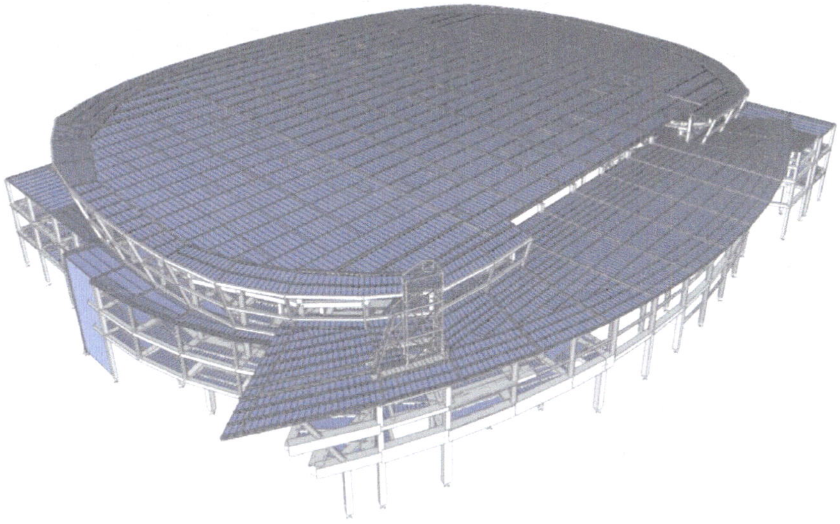

Fig. 5.39 The stadium with a simply supported mobile roof

From the moment (M) equation, the bending moment is found in term of the external load, and in the next equation this moment is entered as the known load in order to calculate the deflection. To generate the matrix pattern in Eq. (5.130) and solve the problem using Eq. (5.131), the largest rectangle inside the domain of the problem is defined (with a time complexity equals to 6.4e+007) as shown in Figs. 5.40 and 5.41.

Efficient solutions of the equations on the rectangular domain are pre-defined using graph product rules. Thus, only the remaining domain is solved using distributed points. The solution of the rectangular part will form the matrix $\boldsymbol{M}_{11}^{-1}$ in Eq. (5.131). The matrix $\boldsymbol{\Gamma}$ is of the dimension equal to that of matrix $\boldsymbol{M}_{22}$ (i.e. the degrees of freedom on the domain Ω_2). The area of the domain Ω_1 is ~75% of the entire domain. The number of distributed points in each domain is assumed to correspond to the area of the domain. Specifically, for a set of 10,000 points including 7500 points in the rectangular domain Ω_1 (i.e. 75 horizontal blocks, each block including 100 points) and 2500 points in the domain Ω_2, the time complexity of solving the

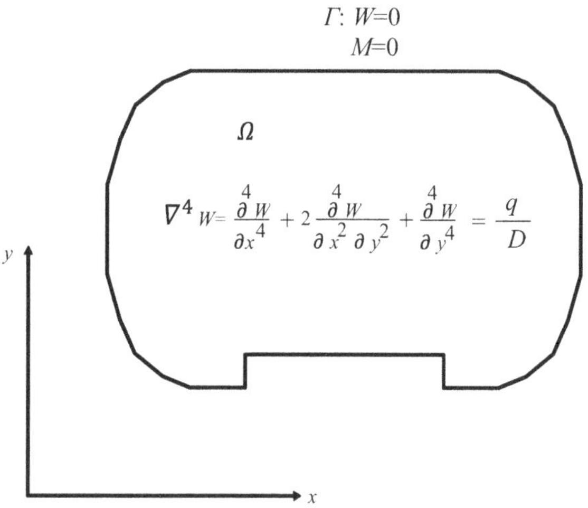

Fig. 5.40 The roof domain Ω governed by the fourth order differential equation

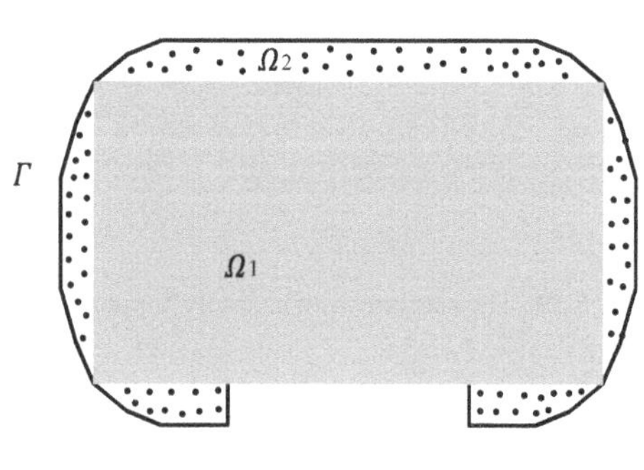

Fig. 5.41 The domain $\boldsymbol{\Omega}$ partitioned by the largest rectangle $\boldsymbol{\Omega}_1$ and the remaining part $\boldsymbol{\Omega}_2$

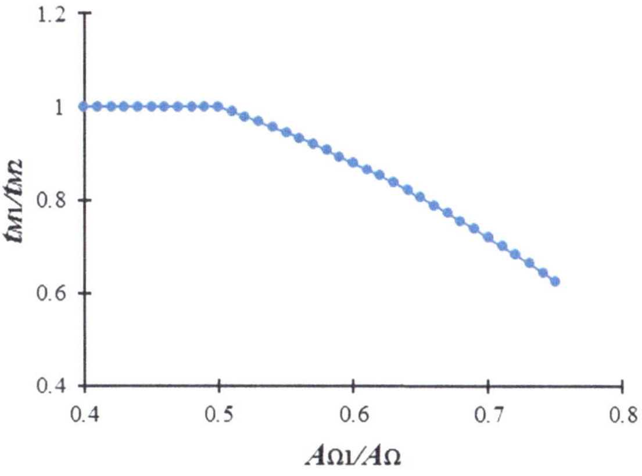

Fig. 5.42 Time complexity of the present method relative to that of a swift LU decomposition solution as a function of the relative size of the largest rectangle inside the domain

problem using Eq. (5.131) (i.e. t_{M1}) will be 6.2548e+011, whereas the time complexity using LU decomposition in the Optimized CW-like algorithms [3] (i.e. t_{M2}) is 1.0001e+012 for the same number of points. The efficiency of the solution using Eq. (5.131) depends on the area of the rectangular domain Ω_1 (i.e. $A\Omega_1$) such that for rectangles with areas equal or less than 50% of the whole domain (i.e. $A\Omega$), the present method would be as efficient as the swift LU decomposition method (Fig. 5.42).

References

1. Shojaei, I., Rahami, H., Kaveh, A.: Efficient finite element solution of regular and near-regular systems using graph products. Acta Mech. **226**(7), 2393–2405 (2015)
2. Logan, D.: A First Course in the Finite Element Method. Cengage Learning, Boston (2011)
3. Williams, V.V.: Breaking the Coppersmith-Winograd Barrier (2011)
4. Rahami, H., Kaveh, A., Shojaei, I., Gholipour, Y.: Analysis of irregular structures composed of regular and irregular parts using graph products. J. Comput. Civil Eng. **28**(4) (2014)
5. Nishida, N., Kanchiku, T., Kato, Y., Imajo, Y., Kawano, S., Taguchi, T.: Biomechanical analysis of the spinal cord in Brown-Séquard syndrome. Exp. Ther. Med. **6**, 1184–1188 (2013)
6. Shojaei, I., Kaveh, A., Rahami, H.: An efficient finite element solution using a large pre-solved regular element. Acta Mech. (2016). http://doi.org/10.1007/s00707-015-1552-7
7. Molano, R., Rodríguez, P.G., Caro, A., Durán, M.L.: Finding the largest area rectangle of arbitrary orientation in a closed contour. Appl. Math. Comput. **218**, 9866–9874 (2012)
8. Johnson, C.: Numerical Solution of Partial Differential Equations by the Finite Element Method. Courier Corporation, USA (2012)
9. Kaveh, A.: Optimal Structural Analysis, 2nd edn. Wiley, UK (2014)
10. Kaveh, A., Rahami, H.: An efficient analysis of repetitive structures generated by graph products. Int. J. Numer. Methods Eng. **84**, 108–126 (2010)
11. Kaveh, A., Shojaei, I., Rahami, H.: New developments in the optimal analysis of regular and near-regular structures: decomposition, graph products, force method. Acta Mech. **226**, 665–681 (2015)
12. Kaveh, A., Rahami, H.: Factorization for efficient solution of eigenproblems of adjacency and Laplacian matrices for graph products. Int. J. Numer. Methods Eng. **75**, 58–82 (2008)

13. Shojaei, I., Kaveh, A., Rahami, H.: Analysis of structures convertible to repeated structures using graph products. Comput. Struct. **125**, 153–163 (2013)
14. Unnikrishnan, G.U., Barest, G.D., Berry, D.B., Hussein, A.I., Morgan, E.F.: Effect of specimen-specific anisotropic material properties in quantitative computed tomography-based finite element analysis of the vertebra. J. Biomech. Eng. **135**, 101007 (2013)
15. Shojaei, I., Rahami, H., Kaveh, A.: A mesh free method using rectangular pre-solved domains using Kronecker products. Mech. Based Des. Struct. Mach. (2016). https://doi.org/10.1080/15397734.2016.1146147
16. Onate, E., Idelsohn, S., Zienkiewicz, O., Taylor, R., Sacco, C.: A stabilized finite point method for analysis of fluid mechanics problems. Comput. Methods Appl. Mech. Eng. **139**, 315–346 (1996)
17. Belytschko, T., Krongauz, Y., Organ, D., Fleming, M., Krysl, P.: Meshless methods: an overview and recent developments. Comput. Methods Appl. Mech. Eng. **139**, 3–47 (1996)
18. Szilard, R.: Theories and Applications of Plate Analysis: Classical Numerical and Engineering Methods. Wiley, Hoboken (2004)

Chapter 6
Dynamic Analysis of Near-Regular Structures

6.1 Introduction

In previous chapters methods for eigensolution and static analysis of regular and near-regular skeletal and continuum structures were presented. Application of Kronecker products and developed decomposition methods is not limited to regular and near-regular structures in which a geometrical regularity is present. The methods can be used in any formulations wherein forms of repetition exist. One of such formulations is the finite difference method in which the governing differential equation in each point depends on the point itself and the two neighboring points and this pattern is repeated for all points. Consequently, this repetition is seen in the obtained tri-diagonal or block tri-diagonal matrix which can efficiently be solved. In this chapter methods are developed for efficient dynamic analysis of structures using the principles presented in Chapter 2. The proposed methods are based on converting initial-value problems (e.g., dynamic analysis of single DOF systems) into boundary-value problems and taking advantages of the properties of the latter problems in efficient solution of the former ones. Finite difference formulation of boundary-value problems leads to matrices with repetitive tri-diagonal and block tri-diagonal patterns whose eigensolution and matrix inversion are obtained using the relationships in Chap. 2. The formulation for dynamic analysis is first developed for the single DOF systems and then generalized to multi DOF systems utilizing the modal analysis [1].

A. Kaveh et al., *Swift Analysis of Civil Engineering Structures Using Graph Theory Methods*, Studies in Systems, Decision and Control 290,
https://doi.org/10.1007/978-3-030-45549-1_6

6.2 Swift Solution of Boundary-Value Problems Using a Combined Finite Difference and Graph Product Methods

Finite difference method is a known numerical method for solving boundary-value problems. Like other numerical methods, by increasing the dimension of the problem or reducing the step size (to decrease the error) the amount of computations grows. However, using the solution of tri-diagonal matrices and block tri-diagonal matrices, presented in the references [2, 3] and combining these with finite difference method, the solution of boundary-value problems can be obtained. In the present method, due to finding closed-form formulas, increasing the dimension and step size of the problem lead to additional scalar summation operations. These operations are much less complicated than matrix operations with increased dimension and step size.

In the solution of those problems which leads to ordinary differential equations with boundary conditions, if the coefficients of the equations are constant values, then for second order equations one obtains numerical tri-diagonal matrices and for fourth order equations penta-diagonal matrices can be obtained. In case an equation is a differential equation of dimension 2 or 3, the corresponding matrices will be block tri-diagonal matrices. The deformation of beams and plates, buckling of frames and plates, free vibration and forced vibrations are some examples of this kind which are thoroughly discussed in [2]. As an example, in the simplest case when we have an ordinary differential equation of order 2 with some boundary conditions (e.g. deformation of a beam), the final matrix will be in the following form:

$$Ax = B; \quad A = F_n(b, a, b) \tag{6.1}$$

where

$$F_n(a, b, c) = \begin{bmatrix} a & b & & & & \\ b & c & b & & & \\ & b & \ddots & \ddots & & \\ & & \ddots & c & b & \\ & & & b & c & b \\ & & & & b & a \end{bmatrix}_{n\times n} \tag{6.2}$$

The eigenvalues and eigenvectors of the following matrix can be obtained via Eqs. (6.3) and (6.5), respectively:

$$\lambda_k = b + 2a\cos\frac{k\pi}{n+1}, k = 1, 2, \ldots, n \tag{6.3}$$

$$v_j^{(k)} = \sin\frac{kj\pi}{n+1}, j = 1, 2, \ldots, n \tag{6.4}$$

$$v^{(k)} = (v_1^{(k)}, v_2^{(k)}, \ldots, v_n^{(k)})^t \tag{6.5}$$

And also we have

$$A^{-1} = V\lambda^{-1}V^t \tag{6.6}$$

Now, using Eqs. (6.3) to (6.6) the components of the inverse of matrix A are found

$$A_{ij}^{-1} = \sum_{k=1}^{n}\left(\frac{1}{\sum_{k=1}^{n}\sin\left(\frac{k\pi}{n+1}\right)^2}\sin\left(\frac{ik\pi}{n+1}\right)\sin\left(\frac{jk\pi}{n+1}\right)\frac{1}{\lambda_k}\right) \tag{6.7}$$

where i and j indicate row and column, respectively.

In the boundary-value problems with two or three dimensional domains, the final matrix obtained from the finite difference method will have the block tri-diagonal form below:

$$M = F_n(A_m, B_m, C_m) = \begin{bmatrix} A_m & B_m & & & & \\ B_m & C_m & B_m & & & \\ & B_m & \ddots & \ddots & & \\ & & \ddots & C_m & B_m & \\ & & & B_m & C_m & B_m \\ & & & & B_m & A_m \end{bmatrix}_{nm\times nm} \tag{6.8}$$

The eigenpairs and inverse of the matrix M were previously obtained using graph product rules as:

$$M = A_1 \otimes B_1 + A_2 \otimes B_2 \tag{6.9}$$

$$if\, A_1A_2 = A_2A_1 \Rightarrow \lambda_M = \bigcup_{i=1}^{n} eig(M_i);\ M_i = \lambda_i(A_1)B_1 + \lambda_i(A_2)B_2 \tag{6.10}$$

$$(A_1 \otimes B_1 + A_2 \otimes B_2)(u \otimes v) = u \otimes (B_1 + \lambda B_2)v = \mu(u \otimes v) \tag{6.11}$$

Equation (6.11) shows that $u \otimes v$ are the eigenvectors of the matrix M, Kaveh [2]. Using Eqs. (6.9) and (6.11) the eigenpairs of matrix M are obtained, and then using Eq. (6.6) the matrix M is inverted.

The algorithm for solving boundary-value problems using finite difference method and graph product rules is summarized as follows:

(1) Constructing matrix form (i.e. $Ax = B$) of the differential equation using finite difference method
(2) Finding eigenpairs of matrix A using graph products:

(a) Eq. (6.3) to (6.5) for one-dimensional problems
(b) Eq. (6.10) and (6.11) for two/three-dimensional problems

(3) Inverting matrix A using Eq. (6.6). For one-dimensional problems this leads to the closed-form solution in Eq. (6.7)
(4) Calculating the unknowns via $x = A^{-1}B$.

Solving the governing differential equation of plates is a known example of the application of the combined finite difference method and graph products.

Solution of boundary-value problems were obtained using the features of tri-diagonal and block tri-diagonal matrices. In the next section we will study the more complicated initial-value problems.

6.3 Analysis of Systems with Single DOF by Transforming an Initial-Value Problem to a Boundary-Value Problem

The solution of initial-value problems via finite difference method leads to a recursive solution where for obtaining the function value of a specific point, all previous points should be solved in advance. While, in boundary-value problems, as seen previously, by inverting a matrix the function values of all points are gained simultaneously.

Considering the importance of initial-value problems in structural engineering; especially in the dynamic solution of structures under seismic loads, an efficient method for the analysis of these differential equations is presented. The method works based on converting the initial-value problems into the boundary-value problems and consequently utilizing the features of boundary-value problems in the efficient solution. Consider the following equation:

$$\ddot{u}(t) + u(t) = p(t); \quad u(0) = u_s; \quad \dot{u}(0) = v_s; \quad t \geq 0 \tag{6.12}$$

This holds the form of the dynamic equation:

$$m\ddot{u}(t) + c\dot{u} + ku(t) = p(t) \tag{6.13}$$

Consider the following equations:

$$\ddot{u}_i = \frac{1}{h^2}(u_{i-1} - 2u_i + u_{i+1}); \quad \dot{u}_i = \frac{1}{2h}(u_{i+1} - u_{i-1}) \tag{6.14}$$

where h is the step size. By substituting in the governing equation we will have

$$\frac{1}{h^2}(u_{i-1} - 2u_i + u_{i+1}) + u_i = p_i \tag{6.15}$$

Suppose we want to find function values in the interval [0, T]. Consider the n points used in the finite difference method. Suppose $u(0) = u_s$ and $u(T) = u_f$, then we have:

$$i=2:\ \frac{1}{h^2}\left(\underbrace{u_1}_{u_s} - 2u_2 + u_3\right) + u_2 = p_2, \ldots, i = n-1$$

$$:\ \frac{1}{h^2}\left(u_{n-2} - 2u_{n-1} + \underbrace{u_n}_{u_f}\right) + u_{n-1} = p_{n-1}$$

$$\underbrace{\begin{bmatrix} 1-\frac{2}{h^2} & \frac{1}{h^2} & & & & & & \\ \frac{1}{h^2} & 1-\frac{2}{h^2} & \frac{1}{h^2} & & & & & \\ & \frac{1}{h^2} & 1-\frac{2}{h^2} & \frac{1}{h^2} & & & & \\ & & \frac{1}{h^2} & \ddots & \ddots & & & \\ & & & \ddots & 1-\frac{2}{h^2} & \frac{1}{h^2} & & \\ & & & & \frac{1}{h^2} & 1-\frac{2}{h^2} & \frac{1}{h^2} & \\ & & & & & \frac{1}{h^2} & 1-\frac{2}{h^2} & \frac{1}{h^2} \\ & & & & & & \frac{1}{h^2} & 1-\frac{2}{h^2} \end{bmatrix}}_{K_{(n-2)\times(n-2)}} \underbrace{\begin{bmatrix} u_2 \\ u_3 \\ u_4 \\ \vdots \\ u_{n-4} \\ u_{n-3} \\ u_{n-2} \\ u_{n-1} \end{bmatrix}}_{u(n-2)} = \underbrace{\begin{bmatrix} p_2 - \frac{u_s}{h^2} \\ p_3 \\ p_4 \\ \vdots \\ p_{n-4} \\ p_{n-3} \\ p_{n-2} \\ p_{n-1} - \frac{u_f}{h^2} \end{bmatrix}}_{p(n-2)} \tag{6.16}$$

Now, the matrix K has the form as in Eq. (6.1) and therefore it can be inverted using Eq. (6.7). However, the vector U cannot be found since the value u_f in the last term of the vector P is not defined. Now, we want to define u_f

$$U = K^{-1}P = DP \tag{6.17}$$

$$u_2 = \underbrace{D_{11}\left(p_2 - \frac{u_s}{h^2}\right)}_{a_1} + \underbrace{D_{12}p_3}_{a_2} + \underbrace{D_{13}p_4}_{a_3} + \cdots + D_{1(n-2)}\left(p_{n-1} - \frac{u_f}{h^2}\right) \tag{6.18}$$

Defining a virtual node at $i = 1$ we can write:

$$\dot{u}_1 = \frac{1}{2h}(u_2 - u_0) = \dot{u}(0) = v_s \Rightarrow u_0 = u_2 - 2hv_s \tag{6.19}$$

The governing equation at $i = 1$ can be written as:

$$\frac{1}{h^2}(u_0 - 2u_1 + u_2) + u_1 = p_1 \tag{6.20}$$

Substituting Eq. (6.19) in (6.20), we have

$$\frac{1}{h^2}(u_2 - 2hv_s - 2u_1 + u_2) + u_1 = p_1 \tag{6.21}$$

Rewriting Eq. (6.18)

$$u_2 = \underbrace{a_1 + a_2 + a_3 + \cdots}_{a} + D_{1(n-2)}\left(p_{n-1} - \frac{u_f}{h^2}\right)$$

$$u_2 = a + D_{1(n-2)}\left(p_{n-1} - \frac{u_f}{h^2}\right) \tag{6.22}$$

Substituting u_2 from Eq. (6.22) into (6.21) results in

$$p_1 h^2 - u_1 h^2 + 2hv_s + 2u_1 = 2\left[a + D_{1(n-2)}\left(p_{n-1} - \frac{u_f}{h^2}\right)\right] \tag{6.23}$$

Therefore

$$p_{n-1} - \frac{u_f}{h^2} = \frac{p_1 h^2 - u_1 h^2 + 2hv_s + 2u_1 - 2a}{2D_{1(n-2)}}$$

$$p_{n-1} - \frac{u_f}{h^2} = \frac{p_1 h^2 - u_s h^2 + 2hv_s + 2u_s - 2a}{2D_{1(n-2)}} \tag{6.24}$$

All terms in the right hand side of this equation are numerical values and thus the value of right hand of Eq. (6.24) can be inserted in the last term of the vector P in Eq. (6.16). Therefore, using Eq. (6.17) all function values are calculated simultaneously.

$$U = K^{-1}P = DP$$

And the value of u_f is obtained utilizing Eq. (6.24) as

$$u_f = h^2 p_{n-1} - \frac{(p_1 - u_s)h^4 + 2h^3 v_s + 2u_s h^2 - 2ah^2}{2D_{1(n-2)}} \tag{6.25}$$

Now, consider Eq. (6.13)

$$m\ddot{u}(t) + c\dot{u} + ku(t) = p(t)$$

The general solution for this equation will result in

$$
\underbrace{\begin{bmatrix}
k-\frac{2m}{h^2} & \frac{m}{h^2}+\frac{c}{2h} & & & & & & \\
\frac{m}{h^2}-\frac{c}{2h} & k-\frac{2m}{h^2} & \frac{m}{h^2}+\frac{c}{2h} & & & & & \\
 & \frac{m}{h^2}-\frac{c}{2h} & k-\frac{2m}{h^2} & \frac{m}{h^2}+\frac{c}{2h} & & & & \\
 & \frac{m}{h^2}-\frac{c}{2h} & \ddots & \ddots & & & & \\
 & & \ddots & & k-\frac{2m}{h^2} & \frac{m}{h^2}+\frac{c}{2h} & & \\
 & & & & \frac{m}{h^2}-\frac{c}{2h} & k-\frac{2m}{h^2} & \frac{m}{h^2}+\frac{c}{2h} & \\
 & & & & & \frac{m}{h^2}-\frac{c}{2h} & k-\frac{2m}{h^2} & \frac{m}{h^2}+\frac{c}{2h} \\
 & & & & & & \frac{m}{h^2}-\frac{c}{2h} & \frac{m}{h^2}+\frac{c}{2h}
\end{bmatrix}}_{k_{(n-2)\times(n-2)}}
\underbrace{\begin{bmatrix} u_2 \\ u_3 \\ u_4 \\ \vdots \\ u_{n-4} \\ u_{n-3} \\ u_{n-2} \\ u_{n-1} \end{bmatrix}}_{u_{(n-2)}}
= \underbrace{\begin{bmatrix} p_2-\left(\frac{m}{h^2}-\frac{c}{2h}\right)u_s \\ p_3 \\ p_4 \\ \vdots \\ p_{n-4} \\ p_{n-3} \\ p_{n-2} \\ p_{n-1}-\left(\frac{m}{h^2}+\frac{c}{2h}\right)u_f \end{bmatrix}}_{p_{(n-2)}} \tag{6.26}
$$

And

$$
p_{n-1}-\left(\frac{m}{h^2}+\frac{c}{2l}\right)u_f=\frac{p_1-\left(k-\frac{2m}{h^2}\right)u_s-\left(\frac{2m}{h^2}\right)a+2hv_s\left(\frac{m}{h^2}-\frac{c}{2h}\right)}{\left(\frac{2m}{h^2}\right)D_{1(n-2)}} \tag{6.27}
$$

where

$$
a=\underbrace{D_{11}\left[p_2-\left(\frac{m}{h^2}-\frac{c}{2h}\right)u_s\right]}_{a_1}+\underbrace{D_{12}p_3}_{a_2}+\underbrace{D_{13}p_4}_{a_3}+\cdots+\underbrace{D_{1(n-3)}p_{(n-2)}}_{a_{(n-3)}} \tag{6.28}
$$

Therefore

$$
U=K^{-1}P=DP
$$

Now, the only point is obtaining the inverse of matrix K. This matrix holds the pattern of a tri-diagonal matrix with lower diagonal a, main diagonal b and upper diagonal c, therefore

$$\lambda_k = b + 2\sqrt{ac}\cos\frac{k\pi}{n+1}, k = 1, 2, \ldots, n \tag{6.29}$$

And

$$v_j^{(k)} = \left(\sqrt{\frac{a}{c}}\right)^{j-1} \sin\frac{kj\pi}{n+1}, j = 1, 2, \ldots, n \tag{6.30}$$

And

$$v^{(k)} = \left(v_1^{(k)}, v_2^{(k)}, \ldots, v_n^{(k)}\right) \tag{6.31}$$

Substituting Eqs. (6.29) and (6.31) into (6.6) result in:

$$A_{ij}^{-1} = \sum_{k=1}^{n}\left(\frac{(\sqrt{\frac{a}{c}})^{i-j}}{\sum_{k=1}^{n}\sin\left(\frac{k\pi}{n+1}\right)^2}\sin\left(\frac{ik\pi}{n+1}\right)\sin\left(\frac{jk\pi}{n+1}\right)\frac{1}{\lambda_k}\right) \tag{6.32}$$

where i and j indicate row and column, respectively. Multiplying the vector P by Eq. (6.32) results in the displacements as

$$u_i = \sum_{j=1}^{n}\sum_{k=1}^{n}\left(\frac{(\sqrt{\frac{a}{c}})^{i-j}}{\sum_{k=1}^{n}\sin\left(\frac{k\pi}{n+1}\right)^2}\sin\left(\frac{ik\pi}{n+1}\right)\sin\left(\frac{jk\pi}{n+1}\right)\frac{1}{\lambda_k}p_k\right) \tag{6.33}$$

Since the matrices in Eq. (6.26) are of dimension $n-2$, Eq. (6.33) changes to

$$u_i = \sum_{j=1}^{n-2}\sum_{k=1}^{n-2}\left(\frac{\left(\sqrt{\frac{\frac{m}{h^2}-\frac{c}{2h}}{\frac{m}{h^2}+\frac{c}{2h}}}\right)^{i-j}}{\sum_{k=1}^{n}\sin\left(\frac{k\pi}{n-1}\right)^2}\sin\left(\frac{ik\pi}{n-1}\right)\sin\left(\frac{jk\pi}{n-1}\right)\frac{1}{\lambda_k}p_{k+1}\right) \tag{6.34}$$

where

$$p_{k+1} = \begin{cases} p_2 - \left(\frac{m}{h^2} - \frac{c}{2h}\right)u_s & k = 1 \\ p_{k+1} & 1 < k < n-2 \\ p_{n-1} - \left(\frac{m}{h^2} + \frac{c}{2h}\right)u_f & k = n-2 \end{cases} \tag{6.35}$$

The algorithm for the solution of initial-value problems using conversion, finite difference method and graph product rules is summarized as follows:

(1) Constructing matrix form of the problem such in Eq. (6.26) using finite difference method
(2) Defining the unknown (last) term in the right side of Eq. (6.26) using (6.27) that is conversion of the initial-value problem into a boundary-value one
(3) Inverting matrix K in Eq. (6.26) using the closed-form formula in Eq. (6.32)

(4) Calculating the components of unknown vector U in Eq. (6.26) using (6.34)

An efficient solution for boundary-value problems was presented using the solution of tri-diagonal matrices and block tri-diagonal matrices solved by graph product rules. Then, the approach was applied to the initial-value problems by converting these equations to the boundary-value ones. The conversion was possible because although we lacked a boundary condition at the end of the interval, we had an initial value instead. Using additional equations at $i = 1$ and utilizing the virtual point u_0 situated out the interval and its omission later, the conversion was completed.

6.4 Dynamic Analysis of Structures Using an Efficient Modal Solution

In the previous sections an approach was presented for solving initial-value problems. Moreover, closed-form formulas were obtained for the eigenpairs of tri-diagonal and block tri-diagonal matrices. Now, we want to utilize the eigenpairs and initial-value solution in the seismic analysis of structures. In the modal analysis, a structure with n DOF system is decomposed into n single DOF systems. This decomposition leads to solution of an eigenvalue problem which is laborious for large structures. After the decomposition, n differential equations should be solved. Because of seismic loads in the right hand of these equations, the equations are solved using numerical methods. These numerical methods are recursive methods with step by step solutions and again are time-consuming for large structures. In this section, using the eigenpairs and initial-value solution, we want to ease the two difficult parts of modal analysis. The governing equations in the forced vibration of a structure in the modal analysis can be written as [4]

$$M\ddot{x} + C\dot{x} + Kx = P(t) \tag{6.36}$$

Changing the coordinate system leads to

$$x = \Phi u,\ \Phi = \left[\{\varphi\}_1, \ldots, \{\varphi\}_n\right] \tag{6.37}$$

When damping exists, by considering C as a combination of K and M, we will have

$$C = \alpha K + \beta M \Rightarrow \ddot{u}_i + 2\xi_i\omega_i\dot{u}_i + \omega_i^2 u_i = \varphi_i^t P(t) \tag{6.38}$$

Equation (6.38) shows the importance of calculating the eigenpairs.

Consider the shear building shown in Fig. 6.1.

For $k_i = k$ and $m_i = m$, the eigenvalues are obtained as follows

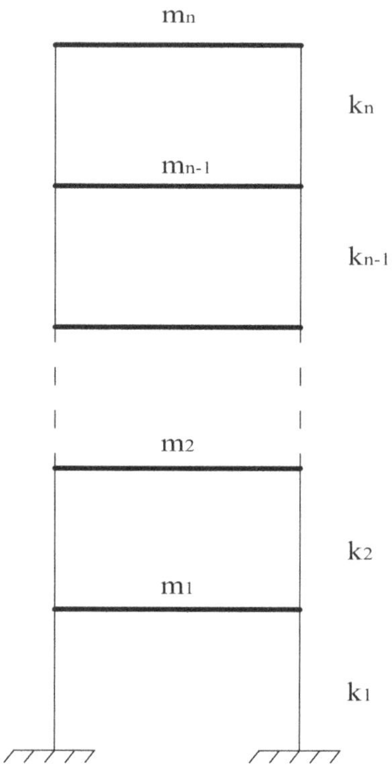

Fig. 6.1 Schematic of the shear building under study

$$\omega_i = 2\cos\frac{(n-i+1)\pi}{2n+1}\sqrt{\frac{k}{m}} \tag{6.39}$$

Now the eigenvectors are obtained by considering identical heights for the stories

$$\varphi_{ij} = \sin\left(\frac{(2i-1)j\pi}{2n+1}\right)\frac{2}{\sqrt{m(2n+1)}} \quad i = 1,\ldots,n,\ j = 1,\ldots,n \tag{6.40}$$

After obtaining closed-form formulas for the eigenpairs, the modal loading is found as

$$P_i = \left[\sin\left(\frac{(2i-1)\pi}{2n+1}\right), \sin\left(\frac{2(2i-1)\pi}{2n+1}\right), \ldots, \sin\left(\frac{n(2i-1)\pi}{2n+1}\right)\right]\frac{2P(t)}{\sqrt{m(2n+1)}} \tag{6.41}$$

When the structure is under seismic motions, we will have

$$P(t) = P_{eff}(t) = -M\{1\}\ddot{v}_g(t) \tag{6.42}$$

where $\{1\}$ is a column of ones and $\ddot{v}_g(t)$ is the applied acceleration at the base of the structure. Since $m_i = m$, we will have

$$P(t) = P_{eff}(t) = -m\{1\}\ddot{v}_g(t) \tag{6.43}$$

and

$$P_i = \left[\sin\left(\frac{(2i-1)\pi}{2n+1}\right), \sin\left(\frac{2(2i-1)\pi}{2n+1}\right), \ldots, \sin\left(\frac{n(2i-1)\pi}{2n+1}\right)\right]\frac{-2m\{1\}\ddot{v}_g(t)}{\sqrt{m(2n+1)}} \tag{6.44}$$

and

$$P_i = \left[\sin\left(\frac{(2i-1)\pi}{2n+1}\right) + \sin\left(\frac{2(2i-1)\pi}{2n+1}\right) + \cdots + \sin\left(\frac{n(2i-1)\pi}{2n+1}\right)\right]\frac{-2m\ddot{v}_g(t)}{\sqrt{m(2n+1)}} \tag{6.45}$$

We have

$$S = \sum_{j=1}^{n} \sin(jk) = \text{Im}(z);\; z = e^{ik} + e^{i2k} + \cdots + e^{ink};\; z = \frac{(2i-1)\pi}{2n+1} \tag{6.46}$$

and

$$z = \frac{e^{ink} - 1 - e^{i(n+1)k} + e^{ik}}{2 - 2\cos(k)} \tag{6.47}$$

Now the imaginary part of z will be

$$\text{Im}(z) = \frac{\sin(k) + \sin(nk) - \sin((n+1)k)}{2 - 2\cos(k)} \tag{6.48}$$

Therefore

$$S = \text{Im}(6.z) = \frac{\sin\left(\frac{(2i-1)\pi}{2n+1}\right) + \sin\left(\frac{n(2i-1)\pi}{2n+1}\right) - \sin\left(\frac{(n+1)(2i-1)\pi}{2n+1}\right)}{2 - 2\cos\left(\frac{(2i-1)\pi}{2n+1}\right)} \tag{6.49}$$

Consequently

$$P_i = \left[\frac{\sin\left(\frac{(2i-1)\pi}{2n+1}\right) + \sin\left(\frac{n(2i-1)\pi}{2n+1}\right) - \sin\left(\frac{(n+1)(2i-1)\pi}{2n+1}\right)}{2 - 2\cos\left(\frac{(2i-1)\pi}{2n+1}\right)}\right]\frac{-2m\ddot{v}_g(t)}{\sqrt{m(2n+1)}} = \alpha\ddot{v}_g(t)$$

$$\alpha = \left[\frac{\sin\left(\frac{(2i-1)\pi}{2n+1}\right) + \sin\left(\frac{n(2i-1)\pi}{2n+1}\right) - \sin\left(\frac{(n+1)(2i-1)\pi}{2n+1}\right)}{2 - 2\cos\left(\frac{(2i-1)\pi}{2n+1}\right)} \right] \frac{-2m}{\sqrt{m(2n+1)}} \tag{6.50}$$

According to Eq. (6.38), for the ith mode we will have

$$\ddot{u}_i + 2\xi_i \omega_i \dot{u}_i + \omega_i^2 u_i = P_i$$

According to Eq. (6.26), the general solution for this equation will result in

$$\underbrace{\begin{bmatrix} \omega_i^2 - \frac{2}{h^2} & \frac{1}{h^2} + \frac{2\xi_i\omega_i}{2h} & & & & & & \\ \frac{1}{h^2} - \frac{2\xi_i\omega_i}{2h} & \omega_i^2 - \frac{2}{h^2} & \frac{1}{h^2} + \frac{2\xi_i\omega_i}{2h} & & & & & \\ & \frac{1}{h^2} - \frac{2\xi_i\omega_i}{2h} & \omega_i^2 - \frac{2}{h^2} & \frac{1}{h^2} + \frac{2\xi_i\omega_i}{2h} & & & & \\ & & \frac{1}{h^2} - \frac{2\xi_i\omega_i}{2h} & \ddots & \ddots & & & \\ & & & \ddots & \omega_i^2 - \frac{2}{h^2} & \frac{1}{h^2} + \frac{2\xi_i\omega_i}{2h} & & \\ & & & & \frac{1}{h^2} - \frac{2\xi_i\omega_i}{2h} & \omega_i^2 - \frac{2}{h^2} & \frac{1}{h^2} + \frac{2\xi_i\omega_i}{2h} & \\ & & & & & \frac{1}{h^2} - \frac{2\xi_i\omega_i}{2h} & \omega_i^2 - \frac{2}{h^2} & \frac{1}{h^2} + \frac{2\xi_i\omega_i}{2h} \\ & & & & & & \frac{1}{h^2} - \frac{2\xi_i\omega_i}{2h} & \omega_i^2 - \frac{2}{h^2} \end{bmatrix}}_{K_{(n-2)\times(n-2)}} \underbrace{\begin{bmatrix} u_2 \\ u_3 \\ u_4 \\ \vdots \\ u_{n-4} \\ u_{n-3} \\ u_{n-2} \\ u_{n-1} \end{bmatrix}}_{U_{(n-2)}}$$

$$= \alpha \underbrace{\begin{bmatrix} \ddot{v}_{g_2} - (\frac{1}{h^2} - \frac{2\xi_i\omega_i}{2h})u_s \\ \ddot{v}_{g_3} \\ \ddot{v}_{g_4} \\ \vdots \\ \ddot{v}_{g_{n-4}} \\ \ddot{v}_{g_{n-3}} \\ \ddot{v}_{g_{n-2}} \\ \ddot{v}_{g_{n-1}} - (\frac{1}{h^2} + \frac{2\xi_i\omega_i}{2h})u_f \end{bmatrix}}_{P_{(n-2)}} \tag{6.51}$$

And finally the closed-form solution is obtained for the ith mode as

$$u_i = \sum_{j=1}^{n-2} \sum_{k=1}^{n-2} \left(\frac{\left(\sqrt{\frac{\frac{1}{h^2} - \frac{2\xi_i\omega_i}{2h}}{\frac{1}{h^2} + \frac{2\xi_i\omega_i}{2h}}}\right)^{i-j}}{\sum_{k=1}^{n} \mathrm{Sin}\left(\frac{k\pi}{n-1}\right)^2} \sin\left(\frac{ik\pi}{n-1}\right) \sin\left(\frac{jk\pi}{n-1}\right) \frac{1}{\omega_i} \ddot{v}_{g_{K+1}} \right) \tag{6.52}$$

where

$$\ddot{v}_{g_{k+1}} = \begin{cases} \ddot{v}_{g_2} - \left(\frac{1}{h^2} - \frac{2\xi_i\omega_i}{2h}\right)u_s & k = 1 \\ \ddot{v}_{g_{k+1}} & 1 < k < n-2 \\ \ddot{v}_{g_{n-1}} - \left(\frac{1}{h^2} + \frac{2\xi_i\omega_i}{2h}\right)u_f & k = n-2 \end{cases}$$

$$\ddot{v}_{g_{n-1}} - \left(\frac{1}{h^2} + \frac{2\xi_i\omega_i}{2h}\right)u_f = \frac{\ddot{v}_{g_1} - \left(\omega_i^2 - \frac{2}{h^2}\right)u_s - \left(\frac{2}{h^2}\right)a + 2hv_s\left(\frac{1}{h^2} - \frac{2\xi_i\omega_i}{2h}\right)}{\left(\frac{2}{h^2}\right)D_{1(n-2)}}$$

$$a = \underbrace{D_{11}\left[\ddot{v}_{g_2} - \left(\frac{1}{h^2} - \frac{2\xi_i\omega_i}{2h}\right)u_s\right]}_{a_1} + \underbrace{D_{12}\ddot{v}_{g_3}}_{a_2} + \underbrace{D_{13}\ddot{v}_{g_4}}_{a_3} + \cdots + \underbrace{D_{1(n-3)}\ddot{v}_{g_{(n-2)}}}_{a_{(n-3)}}$$

$$\omega_i = 2\cos\frac{(n-i+1)\pi}{2n+1}\sqrt{\frac{k}{m}}$$

It should be mentioned that $\ddot{v}_{g_{n-1}} - \left(\frac{1}{h^2} + \frac{2\xi_i\omega_i}{2h}\right)u_f$ and a are calculated only in the step $k = n-2$. Now, the solution of the structure is obtained as

$$y_i = \sum_{j=1}^{n} \varphi_{ij} u_j$$

$$\varphi_{ij} = \sin\left(\frac{(2i-1)j\pi}{2n+1}\right)\frac{2}{\sqrt{m(2n+1)}} \tag{6.53}$$

All that should be done for the modal analysis of a structure is presented in Eq. (6.52) and Eq. (6.53). As the degrees of freedoms and dimension of the corresponding matrices increase, the efficiency of the method is more apparent. This is because a usual solution, involved in eigensolution and matrix inversion with the computational complexities of $O(n^{2.373})$ and $O(n^3)$ respectively [5], is converted to the closed-form formula above with simple summation operations. It is achieved by converting an initial-value problem to a boundary-value one where the graph product rules can be applied to the obtained tri-diagonal and block tri-diagonal matrices.

The seismic analysis using the proposed modal solution is summarized as follows:

(1) Constructing dynamic matrix equation for a multi DOF system and changing the coordinate system to decompose the problem into n single DOF problem (i.e. modal analysis) through Eq. (6.36) to (6.38)
(2) Obtaining the eigenpairs using the closed-form relationships in Eq. (6.39) and (6.40)
(3) Finding the closed-form modal loading in Eq. (6.50) obtained through the eigenpairs and mathematical manipulations in Eq. (6.41) to (6.49)
(4) Defining the matrix equation in Eq. (6.51) associated with differential equation of ith mode (i.e. ith single DOF) using finite difference method
(5) Calculating the components of unknown vector U using Eq. (6.51) obtained through the efficient inverse of matrix K in Eq. (6.51)
(6) Finding the whole response under seismic loading through superposition of all modes in Eq. (6.53).

6.5 Numerical Example

Consider the schematic of a 10-story building shown in Fig. 6.2 and its plan provided in Fig. 6.3. Typical frame structures are shown in Figs. 6.4 and 6.5. The structural system is a dual one composed of moment resisting and braced frames. The height of all stories is equal to 3.5 m. The dead load and live load are considered as 200 kg/m^2 and 650 kg/m^2, respectively. The first 4 stories are chosen as type 1, the second 3 stories as type 2 and the third 3 stories as type 3. The stories of the same type are given same cross sections during the design process (e.g. the beams of the stories 5, 6 and 7 (defined as type 2) hold the same cross sections). Eight cross sections are chosen for columns as well as 8 for beams and 8 for braces. The cross sections and their dimensions are shown in Figs. 6.6, 6.7 and Table 6.1. The record of Imperial Valley ground motion shown in Fig. 6.8 is applied to the structure in the direction X. For optimizing the structural design, a genetic algorithm with 50 individuals and 60 generations are utilized.

The material properties of the frame are considered as:

$$E = 2e8\left(\frac{kN}{m^2}\right), \rho = 76.82\left(\frac{kN}{m^3}\right), \text{ and } \nu = 0.3$$

The following constraints are used in the genetic algorithm:

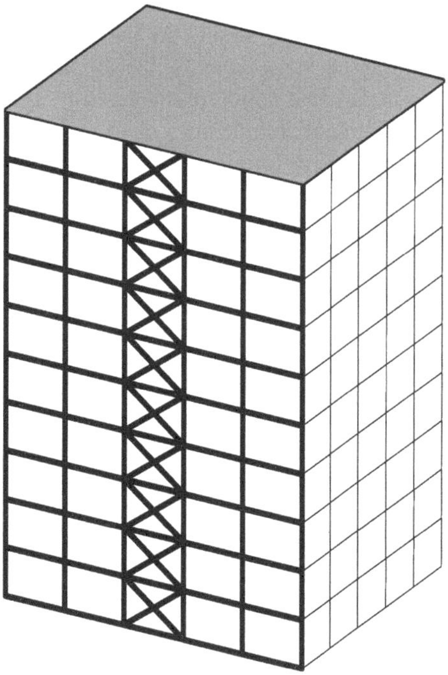

Fig. 6.2 Schematic of a 10-story building

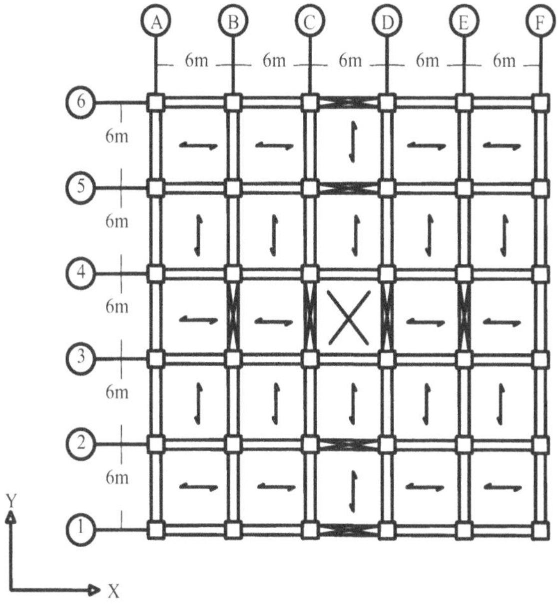

Fig. 6.3 The plan of the building

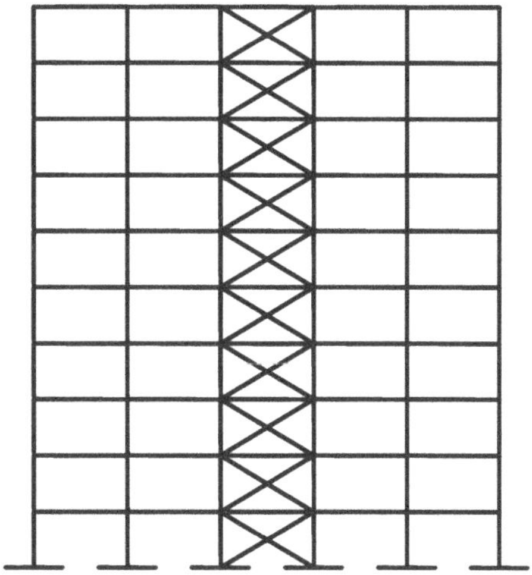

Fig. 6.4 The typical braced frame

(1) The constraints corresponding to the beams and columns include:

$$\text{if } \frac{p_r}{p_n \phi_c} < 0.2 \Rightarrow \frac{p_r}{2 p_n \phi_c} + \left(\frac{M_r}{\phi_b M_n} \right) \le 1$$

$$\text{if } \frac{p_r}{p_n \phi_c} \ge 0.2 \Rightarrow \frac{p_r}{p_n \phi_c} + \frac{8}{9} \left(\frac{M_r}{\phi_b M_n} \right) \le 1$$

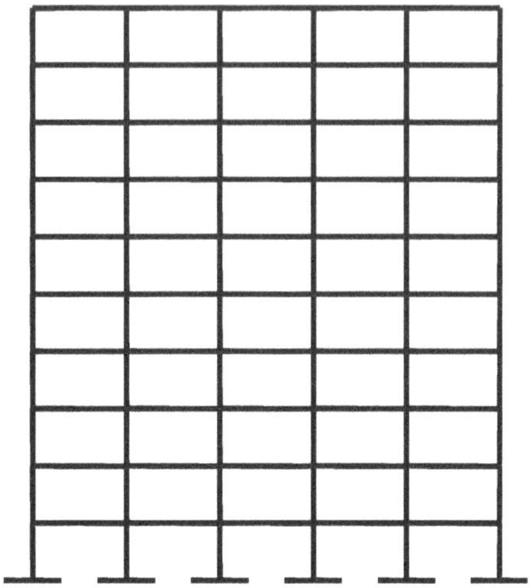

Fig. 6.5 The typical moment resisting frame

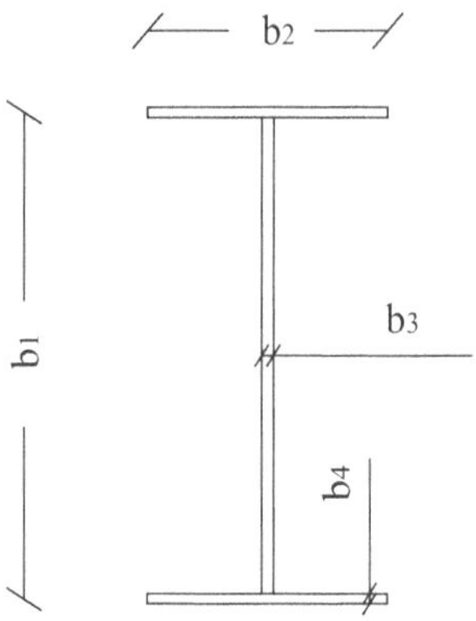

Fig. 6.6 The typical cross section of beams

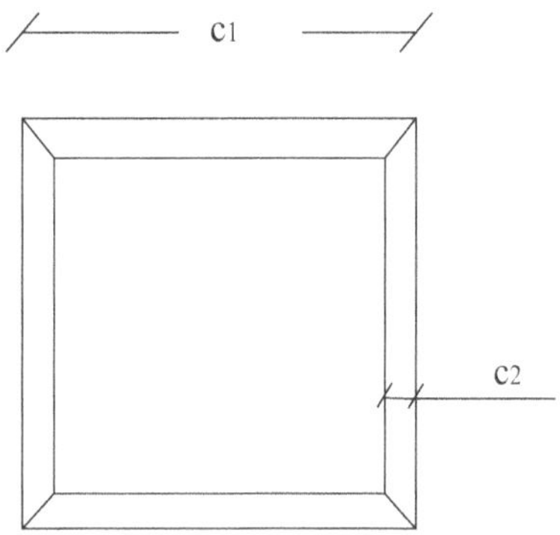

Fig. 6.7 The typical cross section of columns

Table 6.1 The cross sections used in the genetic algorithm

	Section 1	Section 2	Section 3	Section 4	Section 5	Section 6	Section 7	Section 8
Column	C_1	C_2	C_3	C_4	C_5	C_6	C_7	C_8
	$c_1 = 300$	$c_1 = 300$	$c_1 = 350$	$c_1 = 350$	$c_1 = 400$	$c_1 = 400$	$c_1 = 450$	$c_1 = 450$
	$c_2 = 10$	$c_2 = 15$	$c_2 = 15$	$c_2 = 20$	$c_2 = 20$	$c_2 = 25$	$c_2 = 25$	$c_2 = 30$
Beam	B_1	B_2	B_3	B_4	B_5	B_6	B_7	B_8
	$b_1 = 300$	$b_1 = 350$	$b_1 = 350$	$b_1 = 400$	$b_1 = 400$	$b_1 = 400$	$b_1 = 450$	$b_1 = 450$
	$b_2 = 150$	$b_2 = 150$	$b_2 = 150$	$b_2 = 150$	$b_2 = 200$	$b_2 = 200$	$b_2 = 250$	$b_2 = 250$
	$b_2 = 8$	$b_2 = 8$	$b_2 = 8$	$b_2 = 10$	$b_2 = 10$	$b_2 = 10$	$b_2 = 10$	$b_2 = 12$
	$b_2 = 12$	$b_2 = 15$	$b_2 = 20$	$b_2 = 20$	$b_2 = 20$	$b_2 = 25$	$b_2 = 25$	$b_2 = 30$
Brace	2UNP120	2UNP140	2UNP160	2UNP180	2UNP200	2UNP220	2UNP240	2UNP260

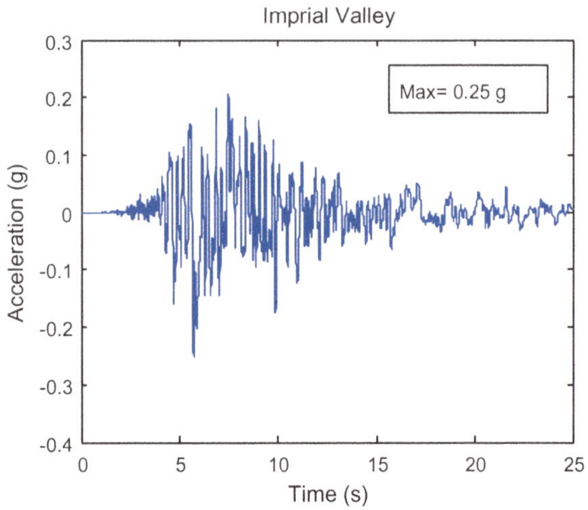

Fig. 6.8 The record of Imperial Valley ground motion

(2) The constraints corresponding to the bracing members include:

$$\lambda \leq 4.23\sqrt{\frac{E}{F_y}}$$

$$P_{cr} \leq \phi_c . P_{cn}$$

$$P_{tr} \leq \phi_t . P_{tn}$$

(3) The constraints of the drift of stories and the roof include:

$$\text{if } T \geq 0.7 \text{ sec} \Rightarrow \overline{\Delta_M} < 0.02\, h_s$$

$$\text{if } T < 0.7 \text{ sec} \Rightarrow \overline{\Delta_M} < 0.025\, h_s$$

See Kaveh et al. Ref. [6] for further details about the constraints.

Here, the adaptive coefficient is utilized for the penalty function introduced by [7]. The fitness of this coefficient was shown in (Goldberg [8]).

Because of 50 individuals and 60 generations, 3000 time-history analyses should be performed. Time-history analyses can be very laborious even for only one analysis. In Kaveh et al. [6] neural networks were used to avoid time-history analyses, however the method had its own problems like inaccurate answers in three-dimensional buildings. The importance of optimal analysis can be even seen in static analysis (Rahami et al. [9]). However, using the optimal seismic method, summarized in Eqs. (6.53) and

(6.53), the analyses are efficiently performed. Since in the record of the earthquake $\Delta t = 0.005$ sec, consequently the step size in the finite difference method is chosen as $h = 0.005$ sec. By considering the initial values equal to zero ($u_s = v_s = 0$) and $\xi = 3\%$ for all modes, all parameters in Eqs. (6.53) and (6.53) are defined. The relationships corresponding to ω_i and φ_{ij} in Eqs. (6.53) and (6.53) are valid for shear buildings with the same stiffness and mass in all stories as mentioned above. In the Kaveh Ref. [2], it was shown that the relationships are applicable to shear buildings in which the stories do not have the same stiffness and mass and the error was less than 5%. However, the structure in this example is not a shear frame and therefore the relationships corresponding to ω_i and φ_{ij} are not valid. Moreover, obtaining the eigenpairs for 3000 structures using a traditional method is very time-consuming. The weakness of the genetic algorithm is the numerous generations which should be solved. If we can take advantage of these generations in a way, the weakness is converted to a benefit. Here, an idea similar to Rayleigh–Ritz ratio is used in which instead of calculating the eigenpairs for all generations, the eigenpairs are only calculated in the first generation and the eigenpairs of the following generations are obtained via the previous generations. Suppose in the Nth generation for a structure we have

$$K_N \varphi_N = \omega_N^2 M_N \varphi_N$$

$$M_N^{-1} K_N \varphi_N = \omega_N^2 \varphi_N$$

$$A_N \varphi_N = \omega_N^2 \varphi_N$$

$$\varphi_N^t A_N \varphi_N = \varphi_N^t \omega_N^2 \varphi_N$$

$$\omega_N^2 = \frac{\varphi_N^t A_N \varphi_N}{\varphi_N^t \varphi_N}$$

Since little changes occur from a generation to another one, we can write

$$\omega_{N+1}^2 = \frac{\varphi_N^t A_{N+1} \varphi_N}{\varphi_N^t \varphi_N}$$

Using this equation, eigenvalues of the new generation are obtained via utilizing the eigenvectors of the previous generation. Then the new eigenvectors are found using the following equation:

$$A_{N+1} \varphi_{N+1} = \omega_{N+1}^2 \varphi_{N+1}$$

The error obtained from this method is less than 4 percent.

Table 6.2 Cross sections of the optimally designed structure

	Type 1	Type 2	Type 3
Column	C_5	C_3	C_1
Beam	B_4	B_3	B_3
Brace	2UNP220	2UNP200	2UNP180

Now, considering the first three modes and $\xi = 3\%$ for all modes, the optimal designed structure is obtained as shown in Table 6.2.

References

1. Kaveh, A., Rahami, H., Shojaei, I.: An efficient method for seismic analysis of structures. Eng. Comput. **32**(6), 1708–1721 (2015)
2. Kaveh, A.: Optimal analysis of structures by concepts of symmetry and regularity. Springer, GmbH, Wien-NewYork. https://doi.org/10.1007/978-3-7091-1565-7
3. Kaveh, A., Rahami, H.: Tri-diagonal and penta-diagonal block matrices for efficient eigensolutions of problems in structural mechanics. Acta Mech, http://www.springerlink.com/content/p62629m74120/?p=cc0dadc6e55a410daf169b171a7aca74&pi=0192:77–87
4. Chopra, A.K..: Dynamics of structures: theory and applications to earthquake engineering. 2nd edn, Prentice Hall, USA (2007)
5. Miller, W.: Computational complexity and numerical stability, SIAM. J. Comput. **4**, 97–107 (1975)
6. Kaveh, A., Shojaei, I., Golipour, Y., Rahami, H.: Seismic design of eccentric braced frames using multi-objective optimization. Struct. Eng. Mech. **45**, 211–232 (2013)
7. Barbosa, H.J.C., Lemonge, A.C.C.: A new adaptive penalty scheme for genetic algorithms. Inform. Sci. **156**, 215–251 (2003)
8. Goldberg, D.E.: Genetic Algorithms in Search, Optimization and Machine Learning. Addison-Wesley, Reading, MA (1989)
9. Rahami, H., Kaveh, A., Shojaei, I., Gholipour, Y.: Analysis of irregular structures composed of regular and irregular parts using graph product. J. Comput. Civil Eng. **28**(4), https://doi.org/10.1061/(asce)cp.1943-5487.0000375 (2014)

Chapter 7
Swift Analysis of Linear and Non-linear Structures and Applications Using Reanalysis

7.1 Introduction

In this chapter efficient methods are presented for reducing the computational complexity of analysis of structures in the process of size and geometry optimization. These methods result in simultaneous swift analysis and optimal design of structures [1]. Analysis of structures in the optimal design via genetic algorithm (GA), includes the solution of the relationship $\boldsymbol{F} = \boldsymbol{K}\Delta$ for each structure in each generation of the genetic algorithm. Inverting the stiffness matrix for thousands structures in the optimization process is time-consuming and expensive. Here, using a closed-form matrix solution, compatible with the features of genetic algorithms, computational complexity of the problem is reduced. A swift analysis for optimal size is performed using a modified solution for structures with changed members and the optimal analysis of optimal geometry is done via a modified solution for structures with changed supports and nodes. The methods are further generalized to non-linear analyses within optimization procedures and to mechanical/biomechanical systems other than structures [2].

7.2 Modification of the Solution of a Structure

In this section, we first review the efficient solution of structures with some changed members and in the Sect. 7.3 we utilize the obtained relationships for optimal analyses of structures designed using genetic algorithms. In addition to modified members, in many cases the modification of supports or nodes in a structure should be done. In this section, efficient solution of structures with such changes is presented and in the following sections the application of the method in optimal analysis is indicated.

A. Kaveh et al., *Swift Analysis of Civil Engineering Structures Using Graph Theory Methods*, Studies in Systems, Decision and Control 290,
https://doi.org/10.1007/978-3-030-45549-1_7

7.2.1 Modification with Some Changed Members

Consider the structure shown in Fig. 7.1. Suppose the structure was analyzed previously and now some members (like the typical member mn) should be rehabilitated. Let the rehabilitation includes increasing or decreasing the stiffness of the members or adding or removal of the members between two nodes.

The solution of the originally analyzed structure has the form shown in Eq. (7.1). The blocks corresponding to the members intended to become rehabilitated, are put in a separated partition

$$\begin{bmatrix} \Delta_m \\ \Delta_n \\ \vdots \\ \cdots \\ \Delta_k \\ \Delta_s \\ \vdots \end{bmatrix} = \left[\begin{array}{ccc:ccc} \boldsymbol{D}_{mm} & \boldsymbol{D}_{mn} & & \boldsymbol{D}_{mk} & \boldsymbol{D}_{ms} & \\ & \boldsymbol{D}_{nn} & & \boldsymbol{D}_{nk} & \boldsymbol{D}_{ns} & \\ & & \ddots & & & \ddots \\ \cdots & \cdots & \cdots & \cdots & \cdots & \cdots \\ & & & \boldsymbol{D}_{kk} & \boldsymbol{D}_{ks} & \\ & & & & \boldsymbol{D}_{ss} & \\ Sym & & & & & \ddots \end{array}\right] \begin{bmatrix} \boldsymbol{F}_m \\ \boldsymbol{F}_n \\ \vdots \\ \cdots \\ \boldsymbol{F}_k \\ \boldsymbol{F}_s \\ \vdots \end{bmatrix} \tag{7.1}$$

Or in a more compact form it is written as

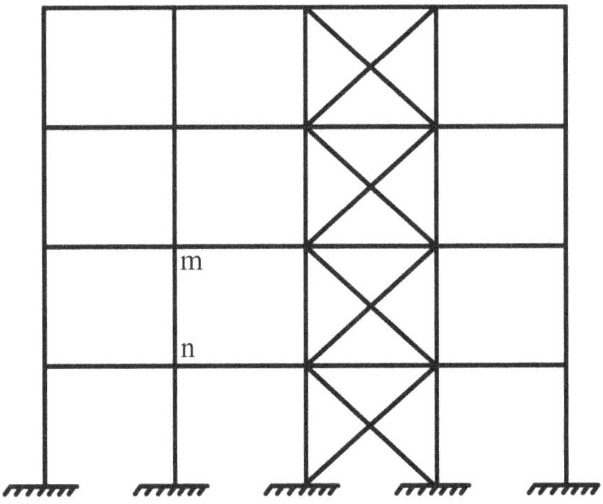

Fig. 7.1 A structure with some changed members

$$\begin{bmatrix} \Delta_I \\ \cdots \\ \Delta_{II} \end{bmatrix} = \begin{bmatrix} \boldsymbol{D}_{I,I} & \vdots & \boldsymbol{D}_{I,II} \\ \cdots & \vdots & \cdots \\ \boldsymbol{D}_{II,I} & \vdots & \boldsymbol{D}_{II,II} \end{bmatrix} \begin{bmatrix} \boldsymbol{F}_I \\ \cdots \\ \boldsymbol{F}_{II} \end{bmatrix} \tag{7.2}$$

Now, we form the stiffness matrix of the changed structure. It will have the following pattern:

$$\begin{bmatrix} \boldsymbol{F}_m \\ \boldsymbol{F}_n \\ \vdots \\ \cdots \\ \boldsymbol{F}_k \\ \boldsymbol{F}_s \\ \vdots \end{bmatrix} = \begin{bmatrix} \boldsymbol{K}_{mm} + \boldsymbol{K}'_{mm} & \boldsymbol{K}_{mn} + \boldsymbol{K}'_{mn} & & \vdots & \boldsymbol{K}_{mk} & \boldsymbol{K}_{ms} & \\ & \boldsymbol{K}_{nn} + \boldsymbol{K}'_{nn} & & \vdots & \boldsymbol{K}_{nk} & \boldsymbol{K}_{ms} & \\ & & \ddots & \vdots & & & \ddots \\ \cdots & \cdots & \cdots & \vdots & \cdots & \cdots & \cdots \\ & & & \vdots & \boldsymbol{K}_{kk} & \boldsymbol{K}_{ks} & \\ & & & \vdots & & \boldsymbol{K}_{ss} & \\ Sym & & & \vdots & & & \ddots \end{bmatrix} \begin{bmatrix} \Delta'_m \\ \Delta'_n \\ \vdots \\ \cdots \\ \boldsymbol{F}'_k \\ \boldsymbol{F}'_s \\ \vdots \end{bmatrix} \tag{7.3}$$

where the blocks $\boldsymbol{K}$ belong to the original structure and the blocks $\boldsymbol{K}'$ are the rehabilitated ones due to applying the changes. In a more compact form:

$$\begin{bmatrix} \boldsymbol{F}_I \\ \cdots \\ \boldsymbol{F}_{II} \end{bmatrix} = \begin{bmatrix} \boldsymbol{K}_{I,I} + \boldsymbol{K}'_{I,I} & \vdots & \boldsymbol{K}_{I,II} \\ \cdots & \vdots & \cdots \\ \boldsymbol{K}_{II,I} & \vdots & \boldsymbol{K}_{II,II} \end{bmatrix} \begin{bmatrix} \Delta'_I \\ \cdots \\ \Delta'_{II} \end{bmatrix} \tag{7.4}$$

where $\boldsymbol{K}'_{I,I}$ represents the amount of changes in the stiffness matrix caused by the rehabilitation of the members. Solving Eq. (7.4) results in:

$$\begin{bmatrix} \Delta'_I \\ \cdots \\ \Delta'_{II} \end{bmatrix} = \begin{bmatrix} \boldsymbol{D}'_{I,I} & \vdots & \boldsymbol{D}'_{I,II} \\ \cdots & \vdots & \cdots \\ \boldsymbol{D}'_{II,I} & \vdots & \boldsymbol{D}'_{II,II} \end{bmatrix} \begin{bmatrix} \boldsymbol{F}_I \\ \cdots \\ \boldsymbol{F}_{II} \end{bmatrix} \tag{7.5}$$

In the present approach we aim to obtain the solution of the modified structure using the inverse of the stiffness matrix of original structure. In other words, we want to refuse the direct inversion of the stiffness matrix in Eq. (7.3). Let us expand Eq. (7.4) as

$$\begin{bmatrix} \boldsymbol{F}_I \\ \cdots \\ \boldsymbol{F}_{II} \end{bmatrix} - \begin{bmatrix} \boldsymbol{K}'_{I,I}\Delta'_I \\ \cdots \\ \boldsymbol{0} \end{bmatrix} = \begin{bmatrix} \boldsymbol{K}_{I,I} & \vdots & \boldsymbol{K}_{I,II} \\ \cdots & \vdots & \cdots \\ \boldsymbol{K}_{II,I} & \vdots & \boldsymbol{K}_{II,II} \end{bmatrix} \begin{bmatrix} \Delta'_I \\ \cdots \\ \Delta'_{II} \end{bmatrix} \tag{7.6}$$

Now, multiplying the two sides of Eq. (7.6) by the inverse of the stiffness matrix of the original structure leads to:

$$\begin{bmatrix} \Delta'_I \\ \cdots \\ \Delta'_{II} \end{bmatrix} = \begin{bmatrix} \boldsymbol{D}_{I,I} & \vdots & \boldsymbol{D}_{I,II} \\ \cdots & \vdots & \cdots \\ \boldsymbol{D}_{II,I} & \vdots & \boldsymbol{D}_{II,II} \end{bmatrix} \begin{bmatrix} \boldsymbol{F}_I - \boldsymbol{K}'_{I,I}\Delta'_I \\ \cdots \\ \boldsymbol{F}_{II} \end{bmatrix} \tag{7.7}$$

Then

$$\begin{aligned} \Delta'_I &= \left[\boldsymbol{I} + \boldsymbol{D}_{I,I}\boldsymbol{K}'_{I,I}\right]^{-1}\left[\boldsymbol{D}_{I,I}\boldsymbol{F}_I + \boldsymbol{D}_{I,II}\boldsymbol{F}_{II}\right] \\ \Delta'_{II} &= \boldsymbol{D}_{II,I}\left\{\boldsymbol{F}_I - \boldsymbol{K}'_{I,I}\left[\boldsymbol{I} + \boldsymbol{D}_{I,I}\boldsymbol{K}'_{I,I}\right]^{-1}\left[\boldsymbol{D}_{I,I}\boldsymbol{F}_I + \boldsymbol{D}_{I,II}\boldsymbol{F}_{I,I}\right]\right\} + \boldsymbol{D}_{II,II}\boldsymbol{F}_{II} \end{aligned} \tag{7.8}$$

The blocks $\boldsymbol{D}$ were previously calculated through analyzing the original structure. Considering Eq. (7.2), we will have

$$\begin{aligned} \Delta_I &= \boldsymbol{D}_{I,I}\boldsymbol{F}_I + \boldsymbol{D}_{I,II}\boldsymbol{F}_{II} \\ \Delta_{II} &= \boldsymbol{D}_{II,I}\boldsymbol{F}_I + \boldsymbol{D}_{II,II}\boldsymbol{F}_{II} \end{aligned} \tag{7.9}$$

Regarding the following definition

$$\left[\boldsymbol{I} + \boldsymbol{D}_{I,I}\boldsymbol{K}'_{I,I}\right]^{-1} = \overline{\boldsymbol{D}}_{I,I} \tag{7.10}$$

Equation (7.8) is written as

$$\begin{aligned} \Delta'_I &= \overline{\boldsymbol{D}}_{I,I}\Delta_I \\ \Delta'_{II} &= -\boldsymbol{D}_{II,I}\boldsymbol{K}'_{I,I}\Delta'_I + \Delta_{II} \end{aligned} \tag{7.11}$$

and finally using Eqs. (7.9) and (7.11) we will have

$$\begin{bmatrix} \Delta'_I \\ \cdots \\ \Delta'_{II} \end{bmatrix} = \begin{bmatrix} \overline{D}_{I,I} D_{I,I} & \vdots & \overline{D}_{I,I} D_{I,II} \\ \cdots & \vdots & \cdots \\ -D_{II,I} K'_{I,I} \overline{D}_{I,I} D_{I,I} + D_{II,I} & \vdots & -D_{II,I} K'_{I,I} \overline{D}_{I,I} D_{I,II} + D_{II,II} \end{bmatrix} \begin{bmatrix} F_I \\ \cdots \\ F_{II} \end{bmatrix} \tag{7.12}$$

and

$$D_{modified} = \begin{bmatrix} \overline{D}_{I,I} D_{I,I} & \vdots & \overline{D}_{I,I} D_{I,II} \\ \cdots & \vdots & \cdots \\ -D_{II,I} K'_{I,I} \overline{D}_{I,I} D_{I,I} + D_{II,I} & \vdots & -D_{II,I} K'_{I,I} \overline{D}_{I,I} D_{I,II} + D_{II,II} \end{bmatrix} \tag{7.13}$$

As such, the exact solution of the modified structure is obtained. The whole process comprises inverting the small matrix $\overline{D}_{I,I}$, of the dimension equal to that of $K'_{I,I}$, and some other simple matrix operations. Δ_I and Δ_{II} are calculated through the results of the analyzed original structure. Consequently, the required time and effort for the solution of the modified structure depends on the number of rehabilitated members that define the dimension of $\overline{D}_{I,I}$. The obtained equations are independent of the type of structure and, therefore, can be applied to any structure. In the present method in addition to displacements, the inverse of the stiffness matrix $(D_{modified})$ is obtained as well. The matrix $D_{modified}$ can be used in potential future rehabilitations where this matrix would be the available solution (inverse) of the new original structure.

As mentioned, the efficiency of the method depends on the number of rehabilitated members affecting the dimension of the matrix $\overline{D}_{I,I}$. However, the efficiency is always guaranteed as in the worst case scenario, where all members should be modified, the dimension of the matrix $\overline{D}_{I,I}$ is equal to that of the stiffness matrix of the original structure. In such a case, computational complexity of the solution is equal to the one in a conventional solution where the stiffness matrix of the original structure should be inverted.

7.2.2 Modification with Some Changed Supports and Nodes

Consider the structure shown in Fig. 7.2 with a changed support shown in Fig. 7.3.

For the original structure, we will have

$$\begin{bmatrix} \Delta_I \\ \cdots \\ \Delta_{II} \end{bmatrix} = \begin{bmatrix} D_{I,I} & \vdots & D_{I,II} \\ \cdots & \vdots & \cdots \\ D_{II,I} & \vdots & D_{II,II} \end{bmatrix} \begin{bmatrix} F_I \\ \cdots \\ F_{II} \end{bmatrix} \tag{7.14}$$

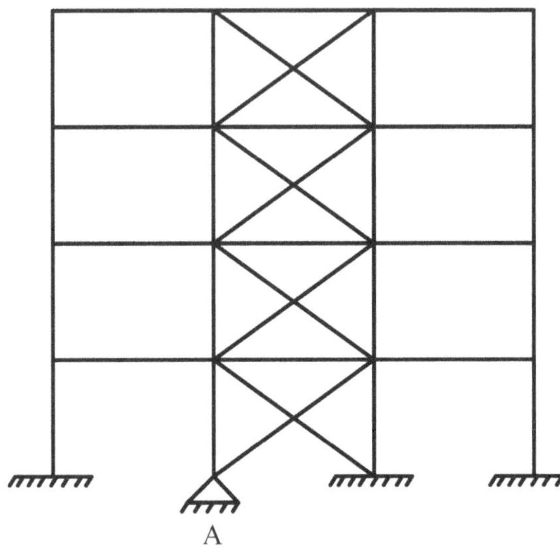

Fig. 7.2 The original structure

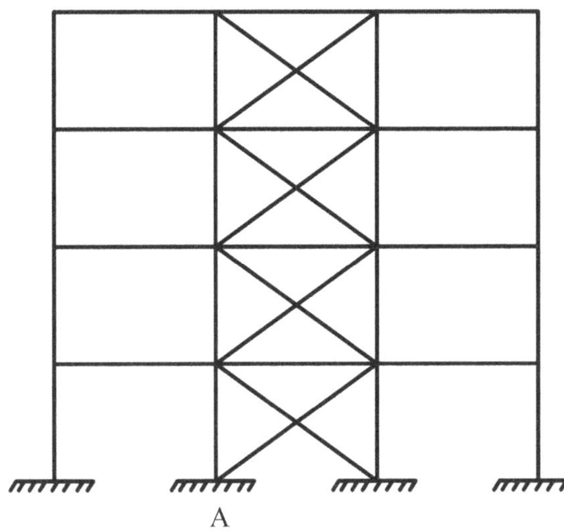

Fig. 7.3 The structure with changed support

And for the rehabilitated structure, we have

$$\begin{bmatrix} \Delta'_I \\ \cdots \\ 0 \end{bmatrix} = \begin{bmatrix} \boldsymbol{D}_{I,I} & \vdots & \boldsymbol{D}_{I,II} \\ \cdots & \vdots & \cdots \\ \boldsymbol{D}_{II,I} & \vdots & \boldsymbol{D}_{II,II} \end{bmatrix} \begin{bmatrix} \boldsymbol{F}_I \\ \cdots \\ \boldsymbol{F}'_{II} \end{bmatrix} \tag{7.15}$$

Expanding the equation results in

$$\begin{aligned} \Delta_I' &= \boldsymbol{D}_{I,I}\boldsymbol{F}_I + \boldsymbol{D}_{I,II}\boldsymbol{F}_{II}' \\ \boldsymbol{F}_{II}' &= -\boldsymbol{D}_{II,II}^{-1}\boldsymbol{D}_{II,I}\boldsymbol{F}_I \end{aligned} \tag{7.16}$$

Therefore,

$$\Delta_I' = [\boldsymbol{D}_{I,I} - \boldsymbol{D}_{I,II}\,\boldsymbol{D}_{II,II}^{-1}\,\boldsymbol{D}_{II,I}]\boldsymbol{F}_I \tag{7.17}$$

And finally

$$\boldsymbol{D}_{modified} = \boldsymbol{D}_{I,I} - \boldsymbol{D}_{I,II}\,\boldsymbol{D}_{II,II}^{-1}\,\boldsymbol{D}_{II,I} \tag{7.18}$$

The equation shows for obtaining the results of the modified structure using the original structure only inverting the small matrix $\boldsymbol{D}_{II,II}$ is required.

In the state where the structure shown in Fig. 7.3 is the original structure and the one in Fig. 7.2 is the modified structure, for the original structure it can be written

$$\Delta_I = \boldsymbol{D}_{I,I}\boldsymbol{F}_I \tag{7.19}$$

And for the changed structure

$$\begin{bmatrix} \Delta_I' \\ \cdots \\ \Delta_{II}' \end{bmatrix} = \begin{bmatrix} \boldsymbol{D}_{I,I}' & \vdots & \boldsymbol{D}_{I,II}' \\ \cdots & \vdots & \cdots \\ \boldsymbol{D}_{II,I}' & \vdots & \boldsymbol{D}_{II,II}' \end{bmatrix} \begin{bmatrix} \boldsymbol{F}_I \\ \cdots \\ \boldsymbol{F}_{II} \end{bmatrix} \tag{7.20}$$

And

$$\begin{bmatrix} \boldsymbol{F}_I \\ \cdots \\ \boldsymbol{F}_{II} \end{bmatrix} = \begin{bmatrix} \boldsymbol{K}_{I,I} & \vdots & \boldsymbol{K}_{I,II} \\ \cdots & \vdots & \cdots \\ \boldsymbol{K}_{II,I} & \vdots & \boldsymbol{K}_{II,II} \end{bmatrix} \begin{bmatrix} \Delta_I' \\ \cdots \\ \Delta_{II}' \end{bmatrix} \tag{7.21}$$

Since

$$\begin{bmatrix} \boldsymbol{D}_{I,I}' & \vdots & \boldsymbol{D}_{I,II}' \\ \cdots & \vdots & \cdots \\ \boldsymbol{D}_{II,I}' & \vdots & \boldsymbol{D}_{II,II}' \end{bmatrix} \begin{bmatrix} \boldsymbol{K}_{I,I} & \vdots & \boldsymbol{K}_{I,II} \\ \cdots & \vdots & \cdots \\ \boldsymbol{K}_{II,I} & \vdots & \boldsymbol{K}_{II,II} \end{bmatrix} = \boldsymbol{I} \tag{7.22}$$

Expanding Eq. (7.22) and using Eqs. (7.20) and (7.21) lead to

$$\begin{bmatrix} \Delta'_I \\ \cdots \\ \Delta'_{II} \end{bmatrix} = \begin{bmatrix} \boldsymbol{D}_{I,I} + \boldsymbol{D}_{I,I}\boldsymbol{D}_{I,I}\boldsymbol{K}_{I,II}[\boldsymbol{K}_{II,II} - \boldsymbol{K}_{II,I}\boldsymbol{D}_{I,I}\boldsymbol{K}_{I,II}]^{-1}\boldsymbol{K}_{II,I} & \vdots & -\boldsymbol{D}_{I,I}\boldsymbol{K}_{I,II}[\boldsymbol{K}_{II,II} - \boldsymbol{K}_{II,I}\boldsymbol{D}_{I,I}\boldsymbol{K}_{I,II}]^{-1} \\ \cdots & \vdots & \cdots \\ [\boldsymbol{K}_{II,II} - \boldsymbol{K}_{II,I}\boldsymbol{D}_{I,I}\boldsymbol{K}_{I,II}]^{-1}\boldsymbol{K}_{II,I}\boldsymbol{D}_{I,I} & \vdots & [\boldsymbol{K}_{II,II} - \boldsymbol{K}_{II,I}\boldsymbol{D}_{I,I}\boldsymbol{K}_{I,II}]^{-1} \end{bmatrix} \tag{7.23}$$

And finally

$$\boldsymbol{D}_{modified} = \begin{bmatrix} \boldsymbol{D}_{I,I} + \boldsymbol{D}_{I,I}\boldsymbol{D}_{I,I}\boldsymbol{K}_{I,II}[\boldsymbol{K}_{II,II} - \boldsymbol{K}_{II,I}\boldsymbol{D}_{I,I}\boldsymbol{K}_{I,II}]^{-1}\boldsymbol{K}_{II,I} & \vdots & -\boldsymbol{D}_{I,I}\boldsymbol{K}_{I,II}[\boldsymbol{K}_{II,II} - \boldsymbol{K}_{II,I}\boldsymbol{D}_{I,I}\boldsymbol{K}_{I,II}]^{-1} \\ \cdots & \vdots & \cdots \\ [\boldsymbol{K}_{II,II} - \boldsymbol{K}_{II,I}\boldsymbol{D}_{I,I}\boldsymbol{K}_{I,II}]^{-1}\boldsymbol{K}_{II,I}\boldsymbol{D}_{I,I} & \vdots & [\boldsymbol{K}_{II,II} - \boldsymbol{K}_{II,I}\boldsymbol{D}_{I,I}\boldsymbol{K}_{I,II}]^{-1} \end{bmatrix} \tag{7.24}$$

In this equation the matrix $[\boldsymbol{K}_{II,II} - \boldsymbol{K}_{II,I}\boldsymbol{D}_{I,I}\boldsymbol{K}_{I,II}]$ should be inverted. The dimension of this matrix is equal to the number of degrees of freedom which are added to the structure due to the changed supports. Therefore, it is a small matrix. Again, in the worst state the method is as difficult as a common method that shows the method is always efficient.

In addition to supports, the obtained equations can be applied to each arbitrary node of the structure other than the supports. It means the method is applicable in the structures where the degrees of freedom of some nodes change. But since the main application of the method is in the changed supports, the formulation was presented for them.

Up to now, structures with changed members and structures with changed degrees of freedom were studied. The last sort of modification which complete the present discussion is related to the structures in which the location of nodes changes. Consider the structure shown in Fig. 7.4.

The location of the node m is modified and we aim to solve this structure via the solution of the original unchanged structure. The formulation of this sort of modification is somehow similar to the formulation of rehabilitated members. The difference is that the modification of a specific node like m affects the nodes n, p and q which are connected to it but a changed member affects only the nodes at its two ends. This comparison shows the solution of structures with modified nodes is more difficult than the ones with modified members. However, in the worst state, the present methods in this paper are as complex as the usual methods and it is desirable. Using an approach similar to the approach of structures with changed members the matrix $D_{modified}$ is obtained as

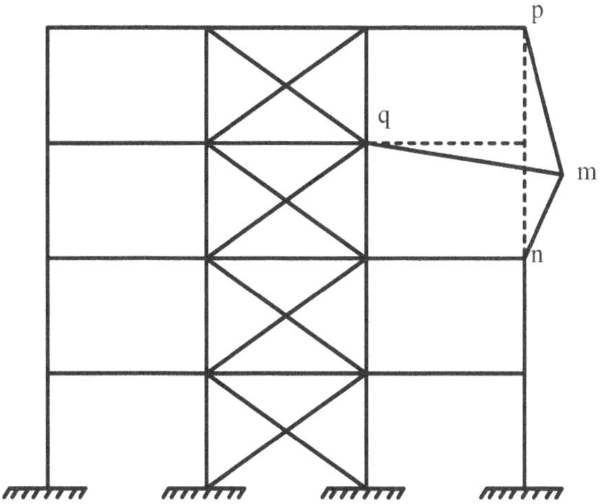

Fig. 7.4 A structure with some changed nodes

$$\boldsymbol{D}_{modified} = \begin{bmatrix} \overline{\boldsymbol{D}}_{I,I}\boldsymbol{D}_{I,I} & \vdots & \overline{\boldsymbol{D}}_{I,I}\boldsymbol{D}_{I,II} \\ \cdots & \vdots & \cdots \\ -\boldsymbol{D}_{II,I}\boldsymbol{K}'_{I,I}\overline{\boldsymbol{D}}_{I,I}\boldsymbol{D}_{I,I} + \boldsymbol{D}_{II,I} & \vdots & -\boldsymbol{D}_{II,I}\boldsymbol{K}'_{I,I}\overline{\boldsymbol{D}}_{I,I}\boldsymbol{D}_{I,II} + \boldsymbol{D}_{II,II} \end{bmatrix} \tag{7.25}$$

This equation seems to be similar to Eq. (7.13), however, as it was mentioned, in this equation the blocks with the index I,I include the degrees of freedom of the changed nodes and the nodes connected to the changed nodes. But in Eq. (7.13) blocks with the index I,I contain the degrees of freedom of the nodes ate the two ends of the modified members.

7.3 Optimal Analysis of Genetic Algorithms via the Application of Modified Solutions

In this section first the size optimization is performed followed by geometric optimization.

7.3.1 Optimal Size Analysis

One of the disadvantages of structural design problems using genetic algorithms includes excessive and time-consuming computations; especially in the case of calculating the governed equation. This issue has been rarely discussed by researchers in the literature. In this paper an analytical optimal approach is applied to minimize the amount of computations in the optimal design of structures. Usually, the most complex stage in a genetic algorithm is to be the solution of the governed equation. In a structural problem, it includes the solution of the equation $\boldsymbol{F} = \boldsymbol{K}\Delta$ and finding the results as $\Delta = \boldsymbol{K}^{-1}\boldsymbol{F} = \boldsymbol{D}\boldsymbol{F}$. The solution contains inverting the matrix $\boldsymbol{K}$ which can be very large for a real-life structure. It gets worse when it comes to an iterative method where numerous steps should be done. As an example, in a genetic algorithm with a population of 40 individuals and a generation number equal to 30, 1200 stiffness matrices should be inverted which is very laborious. Now, the obtained formulation in the previous section is applied into a genetic algorithm to solve the problem quickly.

Generally, the basic operations in a genetic algorithm include reproduction, crossover and mutation. There are several methods for performing each operation in the literature. In the reproduction operation better fit individuals are chosen and copied. In the crossover operation, individuals are placed in a matting pool to mate. In this operation, in general, two strings of the mating pool are selected and some parts of them are exchanged and two offsprings are produced. In a single-point crossover it is done by selecting a random point of the parents' strings and swapping the bits in the right side of the point. In this stage, the solution based on the current method in this paper is applied. Consider the truss with 10 members (design variables) shown in Fig. 7.5.

Suppose a string of 50-bits length is used as a chromosome for each solution (a 5-bits length for each design variable). Consider the two strings below with a random point in the 12th bit

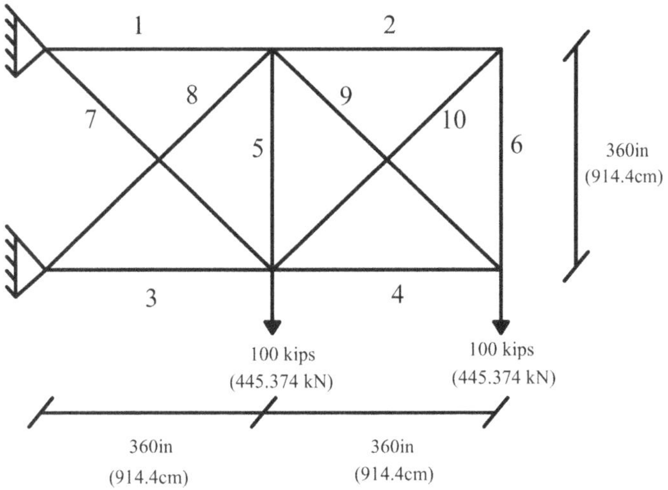

Fig. 7.5 A truss with 10 members

Parent 1: 01001 10110 10⋮ 010 11101 10010 01001 11101 01000 11100 10010

Parent 2: 10001 11000 00⋮ 111 10110 00110 11010 10011 11010 00110 10100

Now, the offsprings will be

Offspring 1: 01001 10110 10111 10110 00110 11010 10011 11010 00110 10100

Offspring 2: 10001 11000 00010 11101 10010 01001 11101 01000 11100 10010

It can be observed that the variables numbers 4–10 (green ones) are equal for the parent 2 and offspring 1, as they are equal for the parent 1 and offspring 2. If it is applied on the corresponding structures, we will have Figs. 7.6 and 7.7.

And also Figs. 7.8 and 7.9.

According to the previous section, it is seen that the truss in Fig. 7.6 is an original structure for the truss in Fig. 7.7 as the rehabilitated one where the members shown in red are changed. Such a comparison is also true for the trusses in Figs. 7.8 and 7.9. Consequently, the solution of offsprings can be quickly done using these two previously-mentioned Eqs. (7.12) and (7.13).

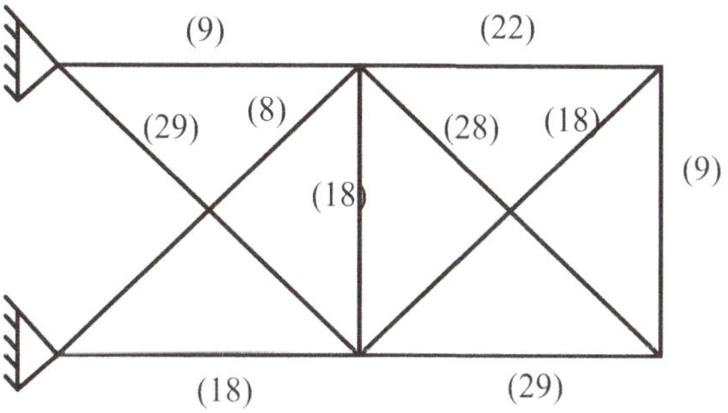

Fig. 7.6 The truss corresponding to parent 1

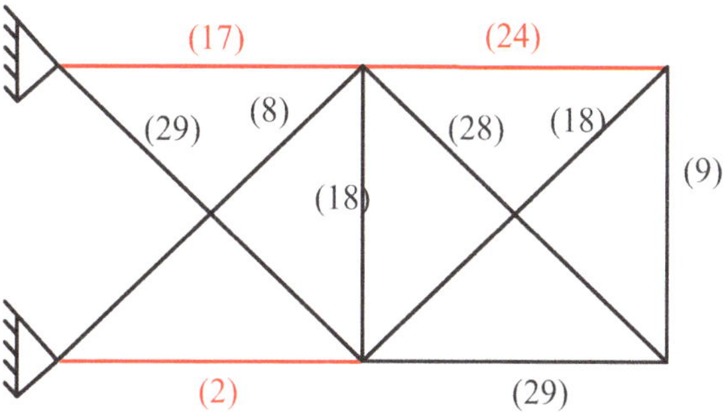

Fig. 7.7 The truss corresponding to offspring 2

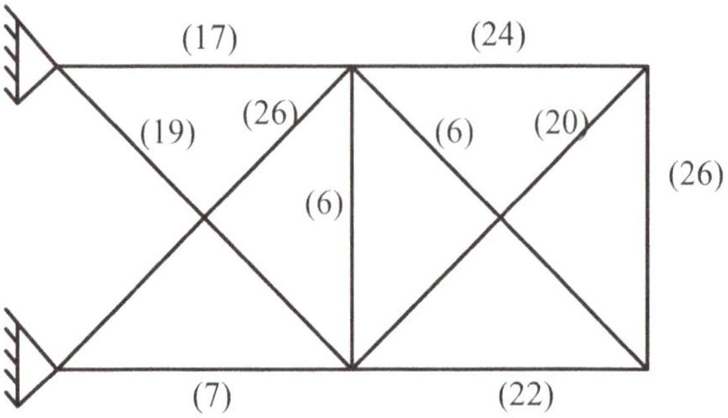

Fig. 7.8 The truss corresponding to parent 2

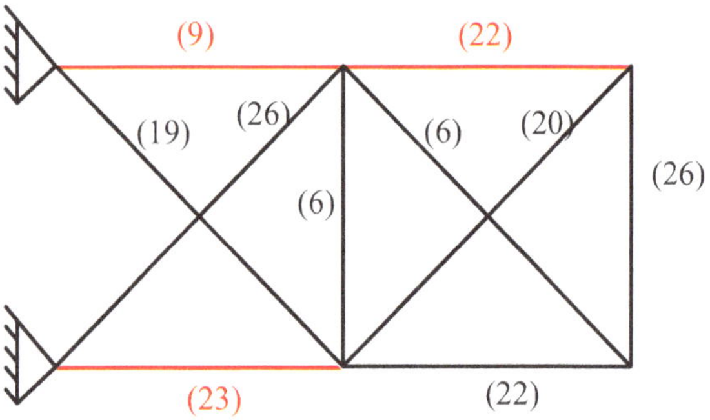

Fig. 7.9 The truss corresponding to offspring

$$\begin{bmatrix} \Delta'_I \\ \cdots \\ \Delta'_{II} \end{bmatrix} = \begin{bmatrix} \overline{D}_{I,I} D_{I,I} & \vdots & \overline{D}_{I,I} D_{I,II} \\ \cdots & \vdots & \cdots \\ -D_{II,I} K'_{I,I} \overline{D}_{I,I} D_{I,I} + D_{II,I} & \vdots & -D_{II,I} K'_{I,I} \overline{D}_{I,I} D_{I,II} + D_{II,II} \end{bmatrix} \begin{bmatrix} F_I \\ \cdots \\ F_{II} \end{bmatrix}$$

$$D_{modified} = \begin{bmatrix} \overline{D}_{I,I} D_{I,I} & \vdots & \overline{D}_{I,I} D_{I,II} \\ \cdots & \vdots & \cdots \\ -D_{II,I} K'_{I,I} \overline{D}_{I,I} D_{I,I} + D_{II,I} & \vdots & -D_{II,I} K'_{I,I} \overline{D}_{I,I} D_{I,II} + D_{II,II} \end{bmatrix}$$

The matrix $D_{modified}$ will be used as the available solution for the next generation where the offsprings in this step will play the role of parents in the next generation. This process will be continued up to the final generation.

In the present method, the number of rehabilitated members depends on the location of the applied random point. The amount of changes increases when the point moves toward the middle of the string. Then it again decreases when the point passes the middle of the string and approaches the end of the strings. As an example, when the random point is put in location of the bits number 1–5 or the bits number 46–50, in both state just one member changes. The difference is that in the first state (1–5) the [parent 1, offspring 2] and [parent 2, offspring 1] are the structural pairs, while in the second state (46–50] the [parent 1, offspring 1] and [parent 2, offspring 2] are the structural pairs. Consequently, the worst state happens when the random point is located in the middle of the string where 50% of the members change and the best state occur in the two ends with no changes. It means 25% changes happen, averagely, in this method and only the effect of these changes should be considered. It is obvious that for the first generation a usual method should be utilized since there are no parents for them.

For making the enough diversity in the population the mutation operation is usually used. The Mutation operation with a small probability is applied so that a design variable changes. Therefore, a member may change in a mutation operation. This small probable change can be easily considered too as it was mentioned.

It was seen 25% changes occur averagely in the present method. However, it does not mean 75% of computations are saved. Consider the 15-members trusses in Figs. 7.10 and 7.11.

In each generation less than 4 members change averagely ($0.25 \times 15 = 3.75$). According to Eq. (7.13) these 4 changed members define the dimension of the matrix $\overline{\boldsymbol{D}}_{I,I}$ which should be inverted. In a more exact concept, the affected nodes by the 4 members are the real changes and define the dimension of the matrix $\overline{\boldsymbol{D}}_{I,I}$. Each member affects the two points at its ends i.e. 4 members affects 8 nodes. However, some members hold shared nodes that decrease the number of affected nodes. The truss in Fig. 7.10 holds modified members with no shared nodes. All nodes in the truss are influenced and therefore the present method is changed to usual method since the dimension of the matrix $\overline{\boldsymbol{D}}_{I,I}$ is equal the dimension of the main stiffness matrix. On the other hand, in the structure shown in Fig. 7.11, since the modified members have shared nodes, the efficiency of the method is revealed. In this truss, dimension of the matrix $\overline{\boldsymbol{D}}_{I,I}$ is equal to 4, while the main stiffness matrix is of

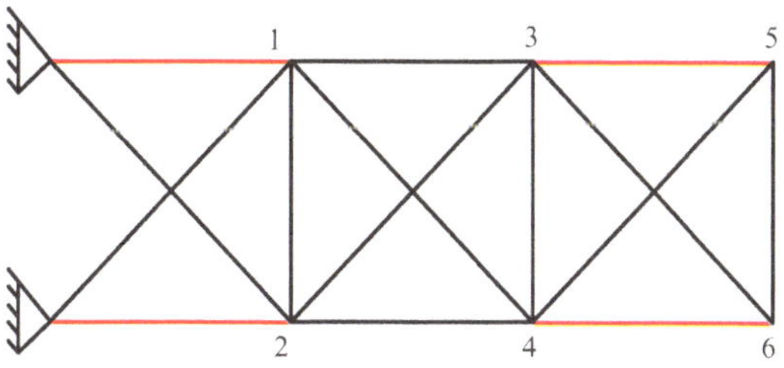

Fig. 7.10 A truss with 4 changed members holding no shared nodes

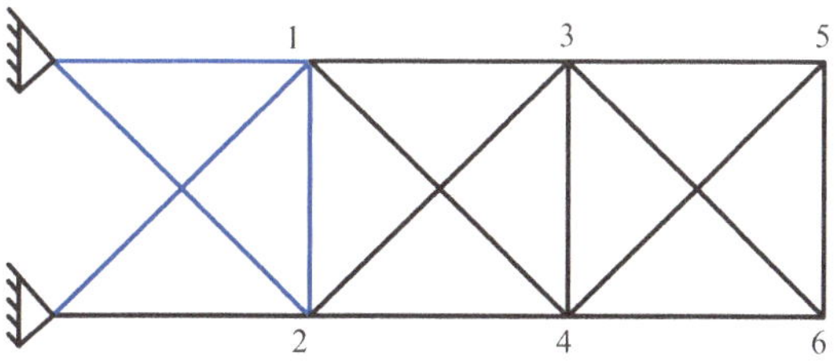

Fig. 7.11 The truss with 4 changed members holding some shared nodes

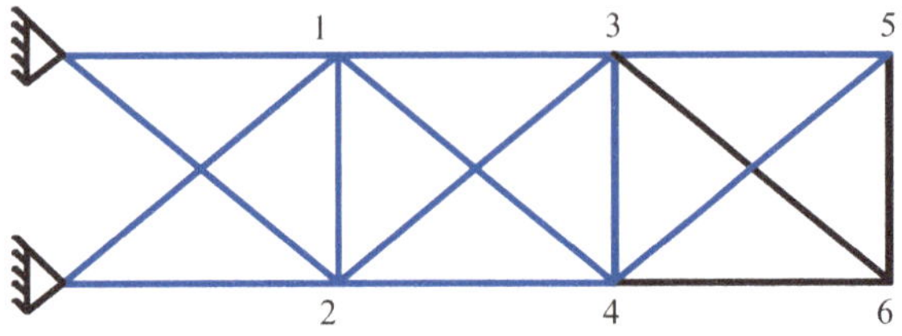

Fig. 7.12 The truss with 12 changed members

dimension 12. Thus, the size of the matrix, which should be inverted, is decreased by 67%. The importance of shared nodes is so that it can overcome the importance of the number of changed members. For instance, consider the truss in Fig. 7.12 with 12 changed members

Although in this truss, 12 members change, however, because of shared nodes a matrix of dimension 10 should be inverted instead of a matrix of dimension 12. It means, a structure with 12 changed members (Although such a state will not happen since it was mentioned that in the worst state only 50% of the members would change) may be solved faster than a structure with 4 changed nodes. Since the obtained matrix $\overline{\boldsymbol{D}}_{I,I}$ is always smaller or equal (in the worst state) to the main stiffness matrix, the efficiency of the present method is ensured. However, we aim to optimize the method even more by presenting a way for selecting the members with the most shared nodes. In a chromosome we attribute a string with a defined length to each variable. Therefore, the location of each string depends on the number of its corresponding member in the structure. Thus, it is a good idea to number the members of a structure in the way so that the changed strings (members) in the crossover operation have the most shared nodes. Since crossover is performed in a random way, it is not defined in advance how many members will change i.e. they can be from 1 member to n/2 members depending on the location of the random point. Consequently, the algorithm of numbering the members should be a general algorithm suitable for every number of selected members imposed by crossover operation.

Previously, the Fiedler vector was utilized in the optimal partitioning of graphs. It also used in the optimal numbering of the nodes of a graph (structure) so that the difference of the numbers of the nodes at the two ends of a member is minimum with the final goal of obtaining banded and spares matrices. Here, the problem is different since firstly we want to order the members (for using in the chromosome) not the nodes and secondly our final goal is having the most shared nodes in a set of selected members. By converting the concept of our problem to the nodal ordering and using a virtual graph, the goal of selecting members with the most common nodes is gained. Consider a graph (structure) with arbitrary ordered members. For converting the problem, the members of this graph are considered as the nodes of a virtual graph so that the nodes of the virtual graph are connected if their corresponding members in the main graph have a shared node and otherwise are not connected. Now, in this virtual graph with numbered nodes the Fiedler method for the optimal ordering is used. The new gained numbering for the nodes are now utilized as the modified

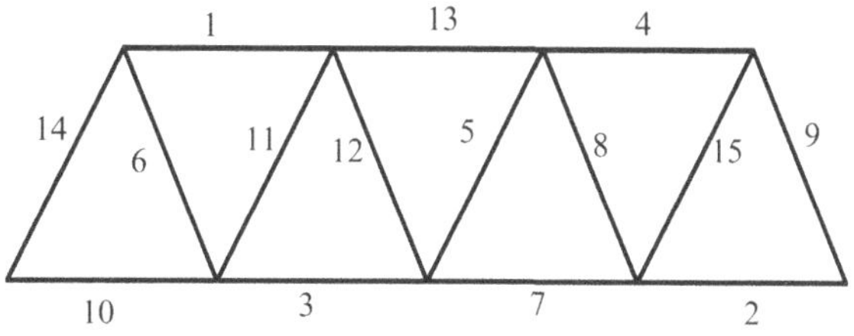

Fig. 7.13 A graph with an arbitrary numbering

numbers of the members in the main graph. Now, every set of selected members with this new numbering has the most shared nodes. The reason is that in the virtual graph the connected nodes will finally have the numbers with the minimum difference. In other words, the numbers of connected nodes are close to each other that means in the main graph the members with shared nodes hold close numbers. Now, when in the crossover operation it is imposed that for example 4 members should be changed, the members 1–4 are selected. Consider the graph shown in Fig. 7.13 and its corresponding virtual graph in Fig. 7.14.

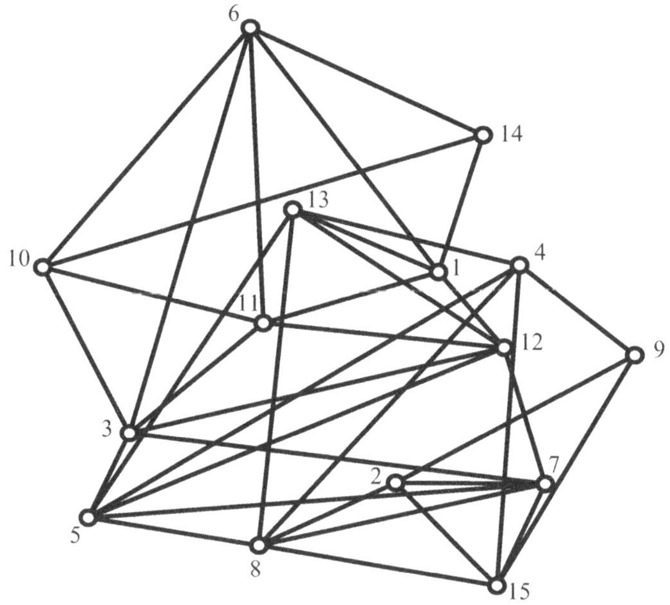

Fig. 7.14 The virtual graph

The Laplacian matrix of the virtual graph will be

$$
\boldsymbol{L} = \begin{bmatrix}
5 & 0 & 0 & 0 & 0 & -1 & 0 & 0 & 0 & 0 & -1 & -1 & -1 & -1 & 0 \\
 & 4 & 0 & 0 & 0 & 0 & -1 & -1 & -1 & 0 & 0 & 0 & 0 & 0 & -1 \\
 & & 6 & 0 & -1 & -1 & -1 & 0 & 0 & -1 & -1 & -1 & 0 & 0 & 0 \\
 & & & 5 & -1 & 0 & 0 & -1 & -1 & 0 & 0 & 0 & -1 & 0 & -1 \\
 & & & & 6 & 0 & -1 & -1 & 0 & 0 & 0 & -1 & -1 & 0 & 0 \\
 & & & & & 5 & 0 & 0 & 0 & -1 & -1 & 0 & 0 & -1 & 0 \\
 & & & & & & 6 & -1 & 0 & 0 & 0 & -1 & 0 & 0 & -1 \\
 & & & & & & & 6 & 0 & 0 & 0 & 0 & -1 & 0 & -1 \\
 & & & & & & & & 3 & 0 & 0 & 0 & 0 & 0 & -1 \\
 & & & Sym & & & & & & 4 & -1 & 0 & 0 & -1 & 0 \\
 & & & & & & & & & & 6 & -1 & -1 & 0 & 0 \\
 & & & & & & & & & & & 6 & -1 & 0 & 0 \\
 & & & & & & & & & & & & 6 & 0 & 0 \\
 & & & & & & & & & & & & & 3 & 0 \\
 & & & & & & & & & & & & & & 5
\end{bmatrix}
$$

The second eigenpairs of the Laplacian matrix are written as

$$\lambda_2 = 0.7981$$

$$
\begin{aligned}
\boldsymbol{V}_2 = [&0.2387, -0.3339, 0.1388, -0.2387, -0.0726, 0.3156, -0.1388, -0.2114, \\
&-0.4034, 0.3339, 0.2114, 0.0726, 0.0000, 0.4034, -0.3156]
\end{aligned}
$$

Now, sorting the vector results in

$$
\begin{aligned}
\boldsymbol{V}_{2sort} = [&-0.4034, -0.3339, -0.3156, -0.2387, -0.2114, -0.1388, -0.0726, \\
&0.0000, 0.0726, 0.1388, 0.2114, 0.2387, 0.3156, 0.3339, 0.4034]
\end{aligned}
$$

Now, using the sorted Fiedler vector ordering of the nodes in the virtual graph shown in Fig. 7.15 is performed and then the ordering of the members in the main graph shown in Fig. 7.16 is obtained.

Now, each set of members with the most shared nodes can be selected. For instance, for a crossover operation where the random point is placed in the 5th string, 5 members are changed; therefore the members 1–5 are selected. It can be seen that these 5 members affect 4 nodes, while in the initial ordering the members 1–5 affect 8 nodes. Considering the fact that the computational complexity of inverting a matrix using the fastest method is $O(n^{2.373})$, by decreasing the 8 nodes to the 4 nodes $4^{2.373} = 26.8342$ operations are saved. The key point is that firstly the modified numbering is performed only one time, and secondly it is applicable for every set of changed members no matter how many members are changed. Although the proper ordering in the truss of Fig. 7.16 is realizable by eye without using the present method, in the complex structures with no regularity the efficiency of the method becomes more obvious.

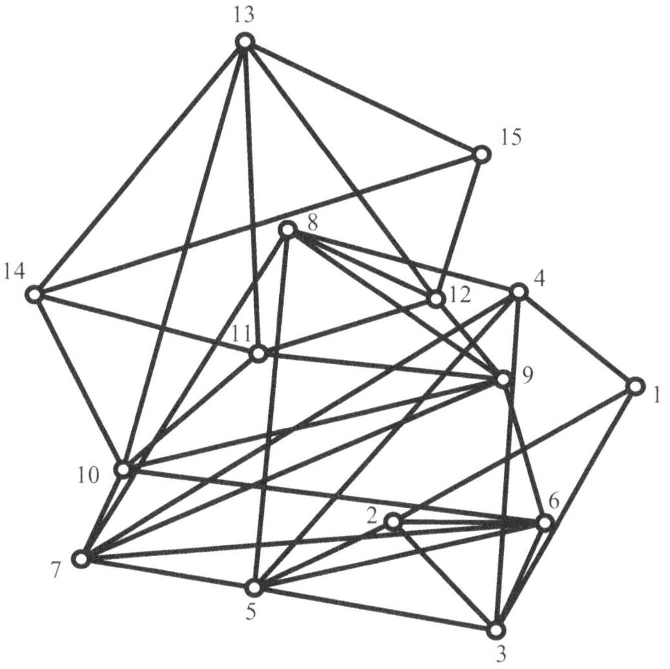

Fig. 7.15 The virtual graph ordered by the Fiedler vector

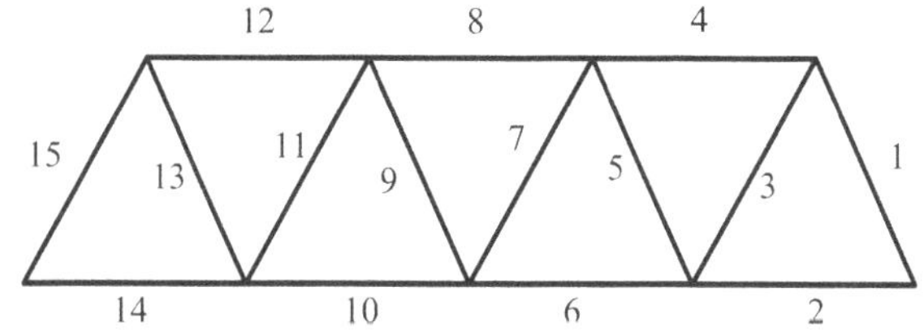

Fig. 7.16 The graph with modified numbering

7.3.2 Optimal Geometry Analysis

The application of member modification in the optimal analysis of minimizing size was shown. Now, the application of node modification in the optimal geometry design is presented. Consider the structure shown in Fig. 7.17. The location of the nodes 1, 3, 4 and 6 and the type of the supports can change.

Suppose a string of 52-bits length is used as a chromosome for each solution. The 52-bits length includes 6-bits lengths for the location of the changed nodes as their radiuses and 6-bits lengths as their angles and also one-bit lengths representing the

Fig. 7.17 A structure in which some nodes and supports can change

two possible forms of each support. Consider the two strings below with a random two-point crossover in the 11th bit and 50th bit

And then,

Parent 1: 011001 11010⁞ 0 110110 101101 011001 100010 001101 110100 0 1⁞ 1 0

Parent 2: 100110 11010⁞ 1 010011 101101 000110 001011 010011 010110 0 0⁞ 1 1

Offspring 1: 011001 110101 010011 101101 000110 001011 010011 010110 0 0 1 0

Offspring 2: 100110 110100 110110 101101 011001 100010 001101 110100 0 1 1 1

It can be seen the variables numbers 3–10 (green ones) are equal for the parent 2 and offspring 1, as they are equal for the parent 1 and offspring 2. For instance for the parent 1 shown in Fig. 7.18 and offspring 2 shown in Fig. 7.19, we will have.

According to the structures with modified nodes and supports, the structure in Fig. 7.18 is the original structure for the modified structure in Fig. 7.19 where the node and support shown in red are changed. Consequently, the solution of offsprings is quickly found using Eq. (7.25). Here we used a two-point crossover. When the length of the string between the two points enlarges, the number of changes increases up to the length equal to the half of the total length. After that, enlarging the length results in the decrease of the changes and in the length equal to the total length, there is no change. In the geometry optimization, when a node changes it influence the other nodes which are connected to it. Therefore, when a set of nodes change the

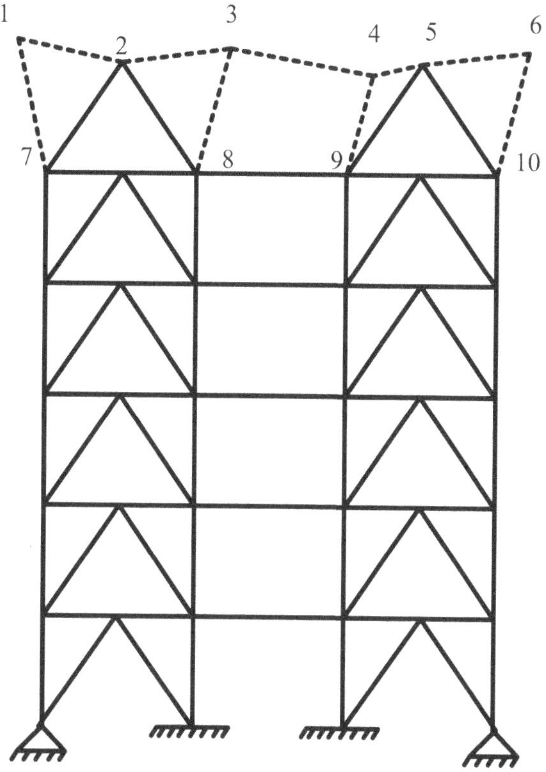

Fig. 7.18 The structure corresponding to parent 1

number of changes is equal to $m + n$, where m is the number of changed nods, n is the number of the influenced nodes which are connected to the set. Thus, appropriate selection of nodes can result in a more optimal analysis. In the case of changed nodes, similar to the changed members, finding a pattern of selecting is performed by the Fiedler vector. Consider the graph shown in Fig. 7.20 and suppose the 3 nodes 5, 6 and 11 should change.

By changing these nodes, the connected nodes i.e. 1, 4, 8, 3, 12 and 2 will change too. Now, the question is that which set of 3 nodes will have the less connected members. In partitioning graphs, the problem of dividing a graph into two parts with the minimum crossed members is solved using Fiedler vector. Consequently, the proper ordering is obtained using this vector.

$$\lambda_2 = 1.2767$$

$$\boldsymbol{V}_2 = [-0.2987, 0.0858, -0.0909, -0.0788, -0.1637, -0.2435, 0.4615, 0.0390, 0.0174, 0.6920, -0.1074, -0.3126]$$

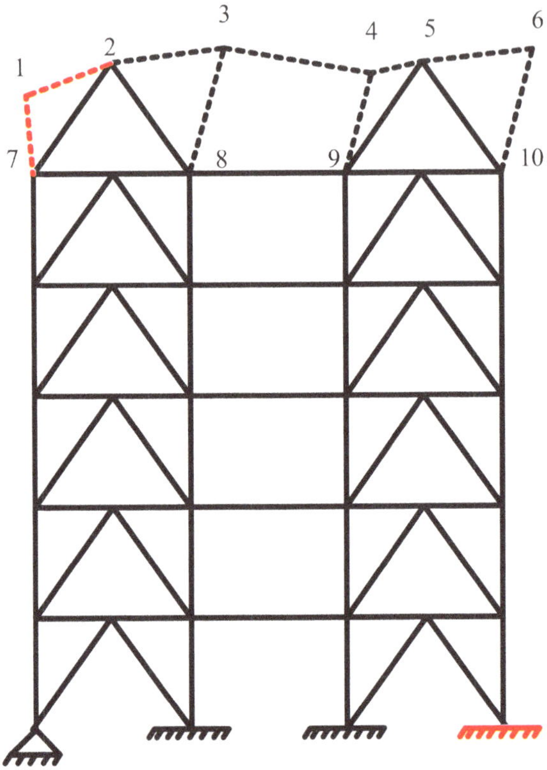

Fig. 7.19 The structure corresponding to offspring 2

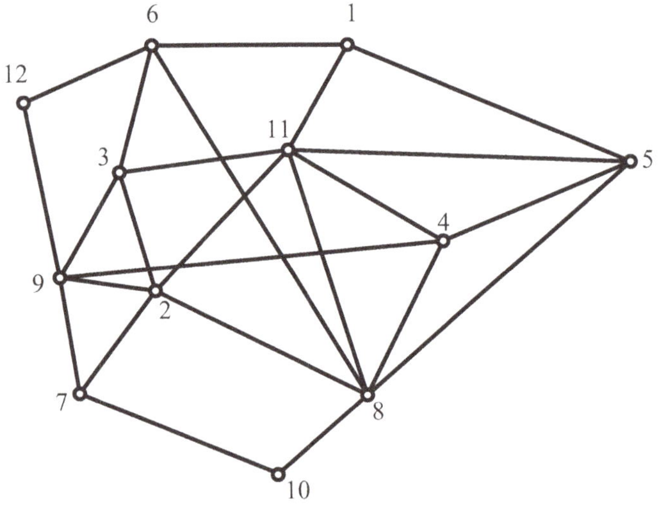

Fig. 7.20 The graph with some changed nodes

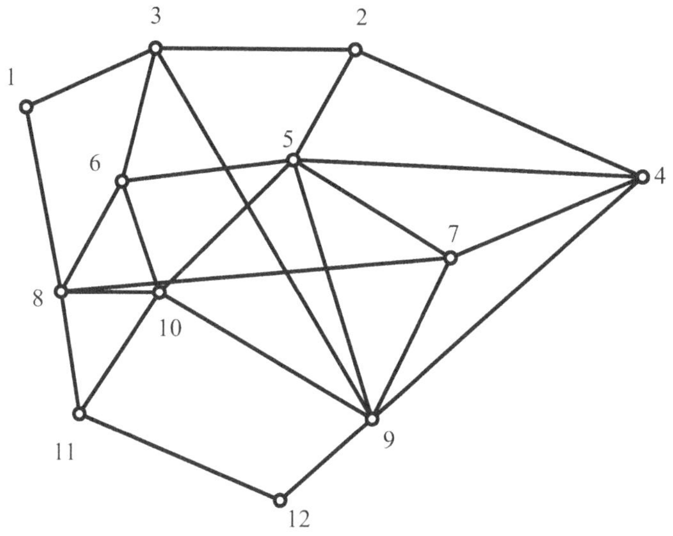

Fig. 7.21 The graph with proper ordering

Now, sorting the vector results in

$$V_{2sort} = [-0.3126, -0.2987, -0.2435, -0.1637, -0.1074, -0.0909, -0.0788, 0.0174, 0.0390, 0.0858, 0.4615, 0.6920]$$

Now, according to Fig. 7.21, we select the 3 nodes 1, 2 and 3 which less influence other nodes of the structure. Each other set of nodes can be selected using this ordering. Again, in the present method in the worst state the method is as efficient as a normal solution and it ensured the effectiveness of the solution. In the optimal geometry, in contrast with the optimal size, usually some specific nodes are moveable not all nodes and it causes less changes happen that is desirable.

The optimal analysis of structures using modified solutions was presented for optimal size and optimal geometry, independently. In an optimal design where both size and geometry are applied simultaneously, the suggested optimal analyses can be combined and used together. The only point is that in this case all changed degrees of freedom should be placed in the block with the index I,I. Most developed methods of optimal analysis focus on parallel solutions and multi-population methods. One of the advantages of the present method is that since the method is applied directly on the governed equation, it can be simply entered in the algorithms of parallel and multi-population methods to quickly solve their governed equation.

7.4 A Numerical Example

Consider the planar EBF bare-frame structure with the loading and dimensions shown in Fig. 7.22. The optimal design of the structure using an optimal analysis is aimed. Each story is considered as an independent type so that all beams in a story are of the same sections as well as the columns and brace members in the story. The lengths of link beams i.e. x_1 to x_5 are in the interval [30, 61 cm] and the used sections are shown in Table 7.1.

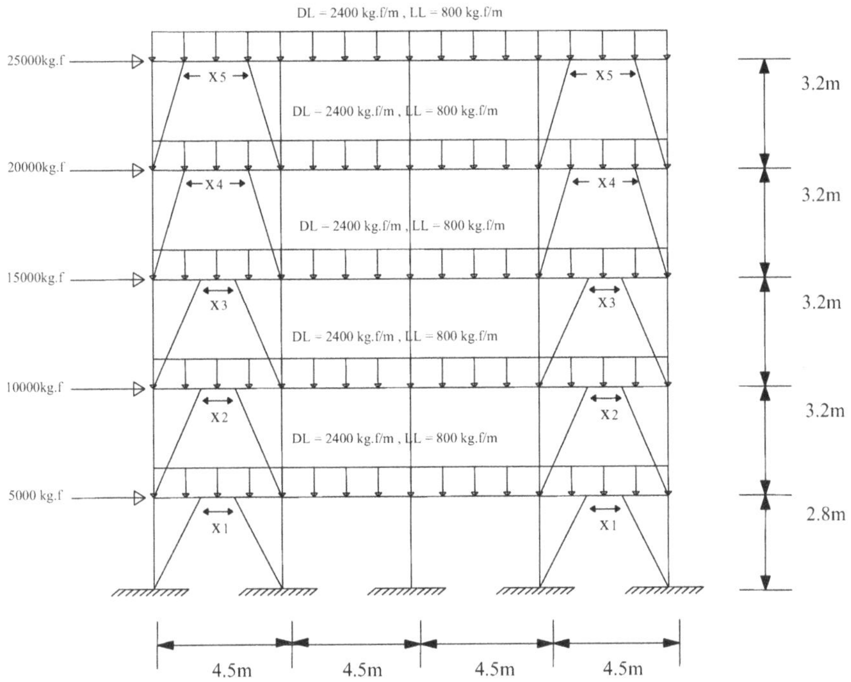

Fig. 7.22 A planar EBF bare-frame structure

Table 7.1 The used sections in the optimization algorithm

Member type	Section
Column	IPB14, IPB16, IPB18, IPB20 IPB22, IPB24, IPB27, IPB30
Beam	IPE16, IPE18, IPE20, IPE22 IPE24, IPE27, IPE30, IPE33
Brace	2L8, 2L10, 2L12, 2L14, 2UNP8 2UNP10, 2UNP12, 2UNP14

The material properties of the structure are considered as:

$$E = 2e8\left(\frac{kN}{m^2}\right),\ \rho = 76.82\left(\frac{kN}{m^3}\right),\ \nu = 0.3 \tag{7.26}$$

The constraints used in the genetic algorithm include:

1. The constraints corresponding to the beams, excentric beams and columns

$$if\,\frac{P_r}{P_n\Phi_c} < 0.2 \Rightarrow \frac{P_r}{2P_n\Phi_c} + \left(\frac{M_r}{\Phi_b M_n}\right) \leq 1 \tag{7.27}$$

$$if\,\frac{P_r}{P_n\Phi_c} \geq 0.2 \Rightarrow \frac{P_r}{P_n\Phi_c} + \frac{8}{9}\left(\frac{M_r}{\Phi_b M_n}\right) \leq 1 \tag{7.28}$$

2. The constraints corresponding to the bracing members

$$\lambda \leq 4.23\sqrt{\frac{E}{F_y}} \tag{7.29}$$

$$P_{cr} \leq \boldsymbol{\Phi}_c.P_{cn} \tag{7.30}$$

$$P_{tr} \leq \boldsymbol{\Phi}_t.P_{tn} \tag{7.31}$$

3. The constraints of the drift of stories and the roof

$$if\ T < 0.7\,\mathrm{s} \Rightarrow \overline{\Delta}_M < 0.025\,h_s \tag{7.32}$$

$$if\ T \geq 0.7\ \mathrm{s} \Rightarrow \overline{\Delta}_M < 0.02h_s \tag{7.33}$$

4. The constraints of link beams:

$$if\ e \leq 1.6\frac{M_P}{V_P} \Rightarrow \gamma_P = 0.08\,rad \tag{7.34}$$

$$if\ e \geq 2.6\frac{M_P}{V_P} \Rightarrow \gamma_P = 0.02\,rad \tag{7.35}$$

$$1.6\frac{M_P}{V_P} \leq e \leq 2.6\frac{M_P}{V_P} \tag{7.36}$$

$$V_u \leq \Phi_v . V_n \tag{7.37}$$

Further detailed information about the constraints can be found in [3].

This example includes both size and geometry optimization. The geometry optimization is subject to the length of link beams that can change in the mentioned interval. Since our sections are discrete and we have 8 sections for each size variable, strings of length 3 are used ($2^3 = 8$). Moreover, strings of length 5 are used for the variables x_1 to x_5 ($2^5 = 32 = 61–30 + 1$). There are 20 variables in this structure including beam section, column section, brace section and link beam length in each story ($4 \times 5 = 20$). Therefore, we will have a chromosome of length 70 ($15 \times 3 + 5 \times 5 = 70$).

Now, the order of variables in the chromosome should be obtained using Fiedler vector. Since there are two types of variables including size (members section) and geometry (link beams length), Fiedler vector is constructed for both size and geometry and then the ordering results of the two vectors are sorted simultaneously. Let us name the size variables as

y_1 to y_5: beams in the stories 1–5.
z_1 to z_5: columns in the stories 1–5.
k_1 to k_5: braces in the stories 1–5.

Now, as it is seen in Table 7.2, using the virtual graph the ordering for size variables, the real number of influenced nodes (N_1) and the absolute number of influenced nodes (N_2) are obtained.

It is seen that the N_1s are the real values and N_2s are absolute values. For instance, y_1 influences 9 nodes in reality and therefore its N_1 is equal to 9, but since all these 9 nodes are already influenced by k_1 and z_1, y_1 influences no new node anymore and its N_2 is equal to zero. Now, according to Table 7.3, using the Fiedler vector the ordering for geometry variables (x_1 to x_5) is obtained.

Now, we should consider both orderings together. k_1 holds the smallest N_2 in both orderings and then it is the first string. After k_1, the strings z_1 or x_1 can be placed. z_1 holds $N_2 = 5$, while x_1 holds $N_2 = 4$ (it should be considered that although $N_2 = 8$ for x_1 in the shown ordering, however, when we select k_1 as the first string, N_2 for x_1 should be modified because k_1 is now affecting x_1). Now, after selecting k_1 and x_1, we should select z_1 or x_2. Considering the effects of k_1 and x_1, $N_2 = 1$ for z_1 and $N_2 = 8$ for x_2 are obtained. After selecting z_1 as the third string, we should compare y_1 and x_2 as the forth string. By following this pattern the order of strings are simply obtained as

$$k_1\ x_1\ z_1\ y_1\ k_2\ x_2\ z_2\ y_2\ k_3\ x_3\ z_3\ y_3\ k_4\ x_4\ z_4\ y_4\ k_5\ x_5\ z_5\ y_5$$

Now, using the chromosomes with this ordering the optimization process can be started. An initial population including 50 individuals is created and the first generation is solved using a usual method. Subsequently, the other generations are quickly solved using their previous generation. For instance, in the second generation for two individuals (parents) with a random crossover point in 59th bit results in

Table 7.2 The ordering of size variables obtained by Fiedler vector

Ordering	k_1	z_1	y_1	k_2	z_2	y_2	k_3	z_3	y_3	k_4	z_4	y_4	k_5	z_5	y_5
N_1	4	5	9	8	10	9	8	10	9	8	10	9	8	10	9
N_2	4	5	0	4	5	0	4	5	0	4	5	0	4	5	0

Table 7.3 The ordering of geometry variables obtained by Fiedler vector

Ordering	x_1	x_2	x_3	x_4	x_5
N_1	8	12	12	12	12
N_2	8	8	8	8	8

Parent1:
11001001011010011011100100011100101110010100011001010100101⋮ 01110001011
Parent2:
10011101001101110100111010010011001010101000111001001110011⋮ 01011101100
Offspr1: 1100100101101001101110010001110010111001010001100101010010101011101100
Offspr2: 1001110100110111010011101001001100101010100011100100111001101110001011

The corresponding structures are shown in Figs. 7.23, 7.24, 7.25 and 7.26. The solution of the structures corresponding to the parents is available and therefore the solution of offsprings is quickly found using Eq. (7.12)

$$\begin{bmatrix} \Delta'_I \\ \cdots \\ \Delta'_{II} \end{bmatrix} = \left[\begin{array}{c:c} \overline{\boldsymbol{D}}_{I,I}\boldsymbol{D}_{I,I} & \overline{\boldsymbol{D}}_{I,I}\boldsymbol{D}_{I,II} \\ \cdots & \cdots \\ -\boldsymbol{D}_{II,I}\boldsymbol{K}'_{I,I}\overline{\boldsymbol{D}}_{I,I}\boldsymbol{D}_{I,I} + \boldsymbol{D}_{II,I} & -\boldsymbol{D}_{II,I}\boldsymbol{K}'_{I,I}\overline{\boldsymbol{D}}_{I,I}\boldsymbol{D}_{I,II} + \boldsymbol{D}_{II,II} \end{array} \right] \begin{bmatrix} \boldsymbol{F}_I \\ \cdots \\ \boldsymbol{F}_{II} \end{bmatrix}$$

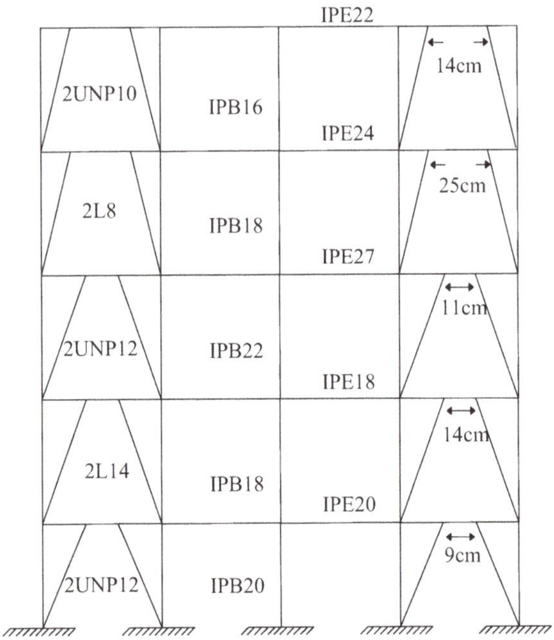

Fig. 7.23 The structure corresponding to parent 1

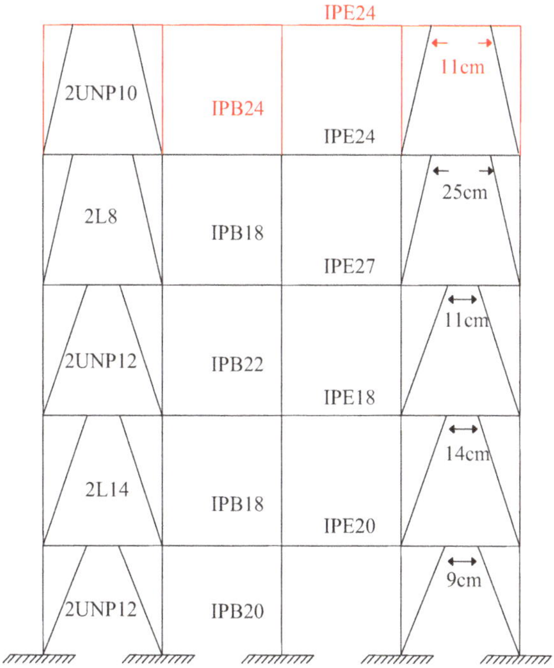

Fig. 7.24 The structure corresponding to offspring 1

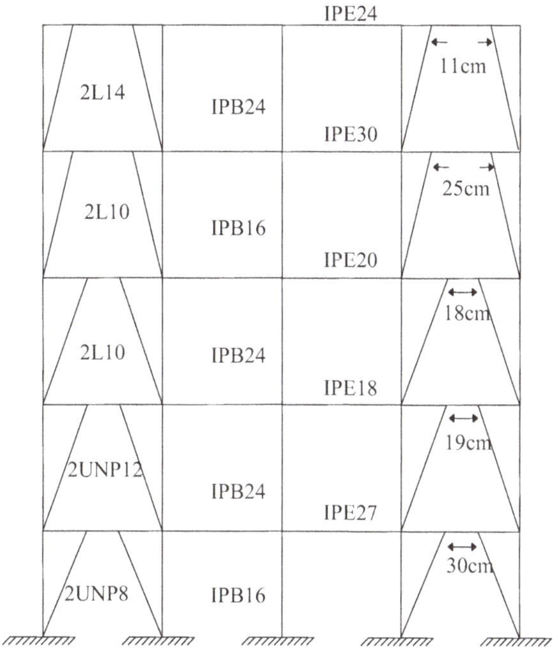

Fig. 7.25 The structure corresponding to parent 2

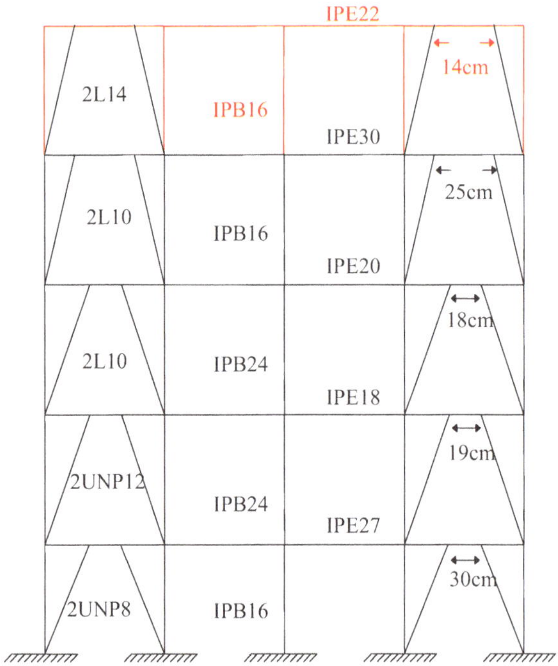

Fig. 7.26 The structure corresponding to offspring 2

The stiffness matrix of the structure is of dimension 135 (45 $\times$ 3). For the mentioned parents and offsprings, instead of inverting matrices of dimension 135, matrices of dimension 42 (14 $\times$ 3) should be solved that means reducing the computations by 68.89%. It can be seen the efficient ordering could decrease the number of affected nodes.

After 60 generations the optimal structure is obtained as shown in Table 7.4.

In comparison with a common analysis, the present analysis could reduce the computations by more than 60%. This way simultaneous optimal analysis and optimal design are obtained that ensure both quickness of the analysis and light weighting of the design.

7.5 Application: Efficient Linear Analysis in Optimal Design/Posture with GA

One critical issue in structural/mechanical design using GA is extreme and time-wasting computations. The most laborious phase in a GA is usually the solution of the governed equation. In structural/mechanical designs, solving the relationship

Table 7.4 The obtained sections of the optimal design

y_1	y_2	y_3	y_4	y_5	z_1	z_2	z_3	z_4	z_5
IPE22	IPE20	IPE20	IPE20	IPE18	IPB22	IPB20	IPB20	IPB16	IPB14
k_1	k_2	k_3	k_4	k_5	x_1	x_2	x_3	x_4	x_5
2UNP12	2L14	2L14	2UNP8	2UNP8	49	35	43	57	37

$F = K\Delta$ and finding the results as $\Delta = K^{-1}F = DF$ is the main part of calculations. To design real-life systems such as huge buildings not only the stiffness matrix K is very large to be inverted but also the procedure should be repeated in each generation of a heuristic method like GA. In this section, efficient formulations are developed to decrease the computational complexity of mechanical/biomechanical problems in optimal design/postures by GA. Effective methods of solving the equation $F = K\Delta$ were previously presented using finite element formulation in Ref. [4] and group theoretical methods [5]. Formulations for the swift analysis of arbitrary structural/mechanical systems with some changed members, supports and nodes were developed by our research group [1]. In the mentioned study, we assumed that an analyzed system is available (Eqs. 7.1 and 7.2) (i.e. the inverse of its stiffness matrix exists) and changes are applied to some members, supports or nodes (Eqs. 7.3 and 7.4) (i.e. rehabilitation). It was aimed to obtain a swift solution for the modified system using its available initial solution before the applied changes. An excellent application of the method is in the GAs wherein for each generation the solution of the previous generation is available (i.e. existing solution) and changes occur in the current generation [i.e. in some members (size) or in some nodes (geometry)] for obtaining the optimal design/posture. Although GAs are powerful tools for optimal design, numerous time-consuming analyses are required within the procedure. Alternatively, finding efficient solutions in GAs leads to the valuable simultaneous efficient analysis and optimal design. Efficient solution of mechanical/biomechanical models includes inverting the stiffness matrix in a swift way. Four inverted matrices $D_{modified}$ were previously formed for member, support and nodal modifications.

Now, consider a GA with a population of 40 individuals and a generation number equal to 30. In this optimization procedure 1200 stiffness matrices should entirely be inverted that is very arduous. Instead, the equations above are inserted into the GA to solve the problem. Normally, main operations in a GA consist of reproduction, crossover and mutation. There are different ways for performing any of these operations. In the reproduction operation better fit individuals are picked and duplicated. In the crossover operation, individuals are positioned in a matting pool to mate. In this operation, usually, two strings of the mating pool are selected and some parts of them are swapped to produce two new offsprings. In a single-point crossover it is done by selecting a random point of the parents' strings and exchanging the bits in the right side of the point. This is where the solution based on the modified formulations above is applied. Consider the biomechanical model of spine with 17 flexible discs shown in Fig. 7.27. Rotations of discs are considered as variables in the GA.

Suppose a string of 85-bits length is utilized as a chromosome for a set of rotational variables in the posture optimization (i.e. a gene of 5-bits length for each disc). Consider the following strings with a random point in the 12th bit.

Parent 1: 01001 10110 10⋮ 010 11101 10010….. 01001 11101 01000 11100 10010

Parent 2: 10001 11000 00⋮ 111 10110 00110….. 11010 10011 11010 00110 10100

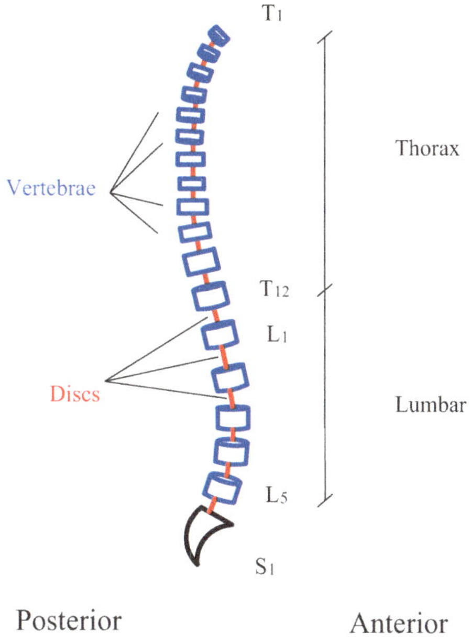

Fig. 7.27 The biomechanical model of spine

Therefore, the offsprings will be

Offspring 1: 10001 11000 00010 11101 10010….. 01001 11101 01000 11100 10010

Offspring 2: 01001 10110 10111 10110 00110….. 11010 10011 11010 00110 10100

It is observed the variable numbers 4–17 (black ones) are equal for parent 1 and offspring 1 and for parent 2 and offspring 2. Applying this concept to the corresponding postures leads to Figs. 7.28 and 7.29.

The postures in Fig. 7.28 are original postures (previous generation) for the postures in Fig. 7.29 (current generation) as the modified new ones. Consequently, regarding Eq. (7.5), obtained for the solution of structures with some changed members, the solution of offsprings (postures with some changed rotations) is quickly found:

$$\begin{bmatrix} \Delta'_I \\ \cdots \\ \Delta'_{II} \end{bmatrix} = \begin{bmatrix} \overline{D}_{I,I} D_{I,I} & \vdots & \overline{D}_{I,I} D_{I,I} \\ \cdots & \vdots & \cdots \\ -D_{II,I} K'_{I,I} \overline{D}_{I,I} D_{I,I} + D_{II,I} & \vdots & -D_{II,I} K'_{I,I} \overline{D}_{I,I} D_{I,II} + D_{II,II} \end{bmatrix} \begin{bmatrix} F_I \\ \cdots \\ F_{II} \end{bmatrix} \tag{7.38}$$

and

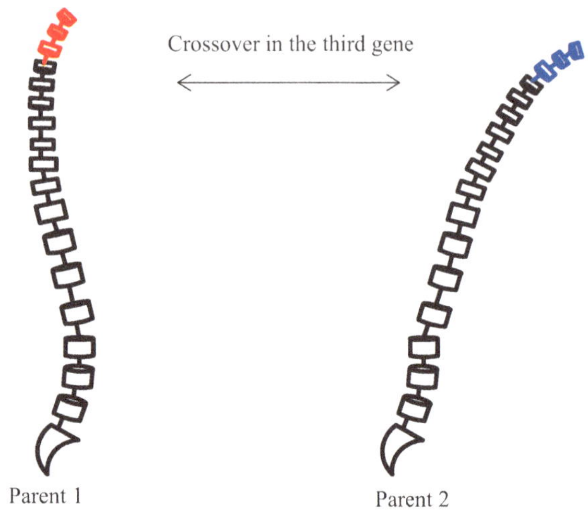

Fig. 7.28 The spinal postures corresponding to parents

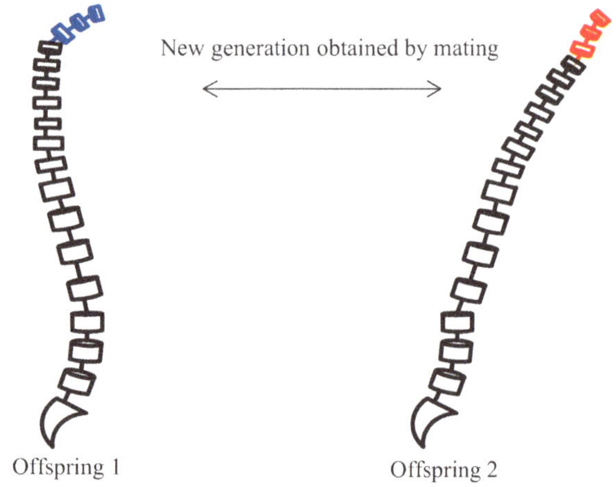

Fig. 7.29 The spinal postures corresponding to offsprings

$$D_{modified} = \begin{bmatrix} \overline{D}_{I,I} D_{I,I} & \vdots & \overline{D}_{I,I} D_{I,II} \\ \cdots & \vdots & \cdots \\ -D_{II,I} K'_{I,I} \overline{D}_{I,I} D_{I,I} + D_{II,I} & \vdots & -D_{II,I} K'_{I,I} \overline{D}_{I,I} D_{I,II} + D_{II,II} \end{bmatrix} \tag{7.39}$$

The matrix D_{modified} will itself be used as the available solution for the next generation where the offsprins in this step will play the role of parents in the subsequent generation. This process will be continued up to the final generation. More details about this section and efficiency of the method are discussed in [1]. Same but much simpler procedure can be completed during the mutation operation wherein an arbitrary bit in a chromosome is changed from its original state. Mutation operation is associated with the change in only one rotational variable above (i.e. only one gene). Therefore, the changed gene will be corresponding to the block K' in Eqs. (7.3) and (7.4). Due to the small change of the modified offspring relative to its corresponding parent during the mutation operations, the dimension of the block K' will be small such that the matrix D_{modified} is calculated using less computations.

Optimal design of a mechanical system is associated with finding the optimal size and geometry of a structure such to minimize the material cost. Optimal posture of a biomechanical system is associated with obtaining a set of optimal kinematics which results in minimizing a cost function (i.e. sum of muscle forces, sum of muscles stress, squared sum of muscles stress and so on). However, the procedure which is completed to define this optimal design/posture in the mechanical/biomechanical system is another aspect of the problem. During optimization procedure numerous iterations are completed to find the final design/posture. Each iteration of optimization algorithm itself involves an iterative non-linear analysis of the system that is very laborious. Therefore, while it is aimed to find the optimal design/posture, reducing the computational complexity of the solution (i.e. optimal analysis) is of interest. In the following sections, efficient non-linear analyses (i.e. optimal analyses) are presented within the optimal design/posture of mechanical/biomechanical systems.

7.6 Efficient Non-linear (Geometric and Material) Solution of Structural/Mechanical Systems Used in GAs for Optimal Design

In this section after reviewing some concepts of non-linear analysis, optimal design of structures is performed considering material and geometric nonlinearity.

7.6.1 Nonlinear Analysis of Structural/Mechanical Systems

Newton-Raphson method is the usual method for non-linear analysis of a structural/mechanical system. The method starts with a linear solution

$$K_0\Delta_0 = F_0 \tag{7.40}$$

where K_0, Δ_0 and F_0 are the initial stiffness matrix, initial displacement vector and external force vector.

The final goal of a non-linear structural/mechanical analysis is finding a displacement vector Δ such that its corresponding internal nodal forces F_I vector is in equilibrium with the external loads vector F_0

$$g(\Delta) = F_0 - F_I = 0 \tag{7.41}$$

Using the first two terms of the Taylor series about the vector of the initial displacements Δ_0 results in

$$g(\Delta) = g(\Delta_0) + \left.\frac{\partial g}{\partial \Delta}\right|_{\Delta_0} (\Delta - \Delta_0) \tag{7.42}$$

Substituting Eq. (7.41) into Eq. (7.42) leads to

$$\left.\frac{\partial F_I}{\partial \Delta}\right|_{\Delta_0} (\Delta - \Delta_0) = F_0 - F_I \tag{7.43}$$

The tangent stiffness matrix can be obtained as

$$K_t = \left.\frac{\partial F_I}{\partial \Delta}\right|_{\Delta_0} \tag{7.44}$$

The terms of the tangent stiffness matrix K_t include

$$K_t = K_0 + K_g + K_m \tag{7.45}$$

where $K_g = K_g(\Delta)$ and $K_m = K_m(\Delta)$ are the geometric stiffness matrix and material stiffness matrix, respectively. The increment nodal displacement is obtained as

$$\delta\Delta = K_t^{-1}\delta F \tag{7.46}$$

Now, the start point is updated

$$\Delta = \Delta_0 + \delta\Delta \tag{7.47}$$

where Δ_0 is obtained using the initial stiffness matrix K_0.

The calculations are repeated until the stop conditions $\|\delta F\| \leq \varepsilon_F$ and $\|\delta\Delta\| \leq \varepsilon_\Delta$ are satisfied.

Now, this iterative analysis should be inserted into a GA for the optimal design. In each generation of the GA iterative non-linear analyses should be performed that is laborious and time-consuming. The difficulty with the non-linear analysis comes from: (1) calculation inverse of the matrix K_t in each iteration (2) the number of iteration to reach to the stop criteria. In the following sections efficient analyses for reducing the computational complexity of the problem is presented. The optimal solutions are presented for both optimal size design and optimal geometry design.

7.6.2 Efficient Analysis in the Optimal Size and Geometry Design (Non-linear Material)

Consider the optimal size and geometry design of a structure using non-linear analysis. Suppose the non-linear behavior is only due to the non-linear material. Now, one is in the first generation of the GA and wants to start the non-linear analysis for the initial set of members and node locations (size and geometry). In each iteration the matrix K_t should be inverted. The matrix can be written as

$$K_{t_{i+1}} = K_{t_i} + K_m \tag{7.48}$$

where K_{t_i} and $K_{t_{i+1}}$ are the tangent stiffness matrices in the ith and i + 1th iteration respectively and K_m is the material stiffness matrix. From one iteration to another one the stiffness of some members change due to material nonlinearity, therefore

$$K_{t_{i+1}} = \begin{bmatrix} \left(K_{t_i}\right)_{I,I} + K_m & \vdots & \left(K_{t_i}\right)_{I,II} \\ \cdots & \vdots & \cdots \\ \left(K_{t_i}\right)_{II,I} & \vdots & \left(K_{t_i}\right)_{II,II} \end{bmatrix} \tag{7.49}$$

where the block I,I is corresponding to the nodes which are affected by K_m.

The inverse of the stiffness matrix $K_{t_{i+1}}$ is simply calculated as

$$D_{i+1} = \begin{bmatrix} (\overline{D}_i)_{I,I}(D_i)_{I,I} & \vdots & (\overline{D}_i)_{I,I}(D_i)_{I,II} \\ \cdots & \vdots & \cdots \\ -(D_i)_{II,I}K_m(\overline{D}_i)_{I,I}(D_i)_{I,I} + (D_i)_{II,I} & \vdots & -(D_i)_{II,I}K_m(\overline{D}_i)_{I,I}(D_i)_{I,II} + (D_i)_{II,II} \end{bmatrix} \tag{7.50}$$

where D_i is the inverse of the matrix K_{t_i} that is available from ith iteration and the matrix $(\overline{D}_i)_{I,I}$ is defined as

$$\left[I + (D_i)_{I,I}K_m\right]^{-1} = (\overline{D}_i)_{I,I} \tag{7.51}$$

The computational complexity of inverting a matrix of the dimension n is $O(n^{2.373})$. Here, the whole process comprises inverting the small matrix $\left[I + (D_i)_{I,I}K_m\right]$, of the dimension equal to that of small matrix K_m (K_m is small since material nonlinearity extends gradually from one iteration to another one). Using the present equations the inverse of the matrix K_t is simply obtained in each iteration up to the point the stop criteria is achieved (the end of first generation).

Analyses of the first generation of the GA got done. Now, offsprings are produced by parents similar to what is shown in Figs. 7.28 and 7.29. In a traditional method of analysis, the new generation (offspring) is considered as an independent structure and is analyzed from the first step. However, since only the size of some members change from parents to offsprings, one wants to get advantages of the available solution of parents (previous generation) in the solution of offsprings (new generation).

Suppose in the jth iteration of the first generation the stop criteria got satisfied and the corresponding displacement vector is Δ_j. Now, the second generation can be solved for this displacement vector (start point) and find the internal nodal forces vector. Using Δ_j as the start point of the second generation not only make a swift solution for the tangent stiffness matrix of the second generation (see below) but also decrease the number of iterations significantly (see Fig. 7.30).

When using the vector Δ_j as the start point, the members of the structure in the jth iteration of the first generation and the corresponding members in the first iteration of the second generation, except the members with changed size and members connected to changed-location nodes (size and geometry modifications applied by GA), will have the same position on the σ-ε curve and therefore the same stiffness matrices. This way the stiffness matrices of the two structures are similar except in the changed-location nodes and the nodes affected by the changed-size members

$$(K_{t_1})_2 = \left(K_{t_j}\right)_1 + K_{s,g} \tag{7.52}$$

where $\left(K_{t_j}\right)_1$ is the tangent stiffness matrix of the jth (i.e. last) iteration of the first generation, $(K_{t_1})_2$ is the tangent stiffness matrix of the first iteration of the second generation and $K_{s,g}$ is the stiffness matrix because of changed size and geometry applied by GA.

$$(K_{t_1})_2 = \begin{bmatrix} \left(\left(K_{t_j}\right)_1\right)_{I,I} + K_{s,g} & \vdots & \left(\left(K_{t_j}\right)_1\right)_{I,II} \\ \cdots & \vdots & \cdots \\ \left(\left(K_{t_j}\right)_1\right)_{II,I} & \vdots & \left(\left(K_{t_j}\right)_1\right)_{II,II} \end{bmatrix} \tag{7.53}$$

where the block I,I is corresponding to the nodes which are affected by $K_{s,g}$.

The inverse of the stiffness matrix $(K_{t_1})_2$ is calculated as

$$(D_1)_2 = \begin{bmatrix} ((\bar{D}_j)_1)_{I,I}((D_j)_1)_{I,I} & \vdots & ((\bar{D}_j)_1)_{I,I}((D_j)_1) \\ \cdots & \vdots & \cdots \\ -((D_j)_1)_{II,I}K_s((\bar{D}_j)_1)_{I,I}((D_j)_1)_{I,I} + ((D_j)_1)_{II,I} & \vdots & -((D_j)_1)_{II,I}K_s((\bar{D}_j)_1)_{I,I}((D_j)_1) \end{bmatrix} \tag{7.54}$$

where $(D_j)_1$ is the inverse of the matrix $\left(K_{t_j}\right)_1$ which is available from jth iteration of the first generation and the matrix $((\overline{D}_j)_1)_{I,I}$ is defined as

$$\left[I + ((D_j)_1)_{I,I} K_{s,g}\right]^{-1} = ((\overline{D}_j)_1)_{I,I} \tag{7.55}$$

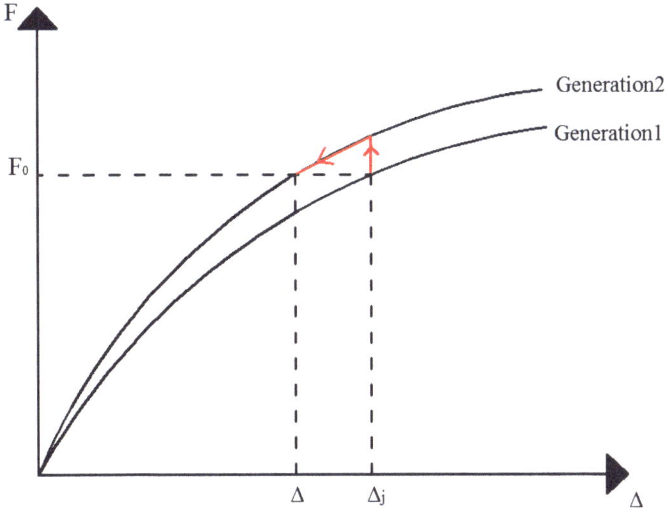

Fig. 7.30 The illustrative interpretation of jumping from one generation to another one

Therefore, the inverse of the stiffness matrix $(K_{t_1})_2$ is obtained by inverting the small matrix $\left[I + \left((D_j)_1\right)_{I,I} K_{s,g}\right]$, of the dimension equal to that of the small matrix $K_{s,g}$ (see below).

Briefly, the first iteration of the second generation is solved for Δ_j and the corresponding internal nodal forces vector is found. This is like jumping from the F-Δ curve of the first generation to that of the second generation vertically at the point Δ_j. Then, the tangent stiffness matrix of the first iteration of the second generation is obtained and inverted using Eqs. (7.53) and (7.54), respectively. Now, moving toward the point of solution is started iteratively (i.e. the point $\varDelta$ corresponding to the external load F_0). Similar to the previous generation, from one iteration to another one, the tangent stiffness matrix is swiftly inverted using Eq. (7.50). In comparison with a traditional method, one is now much closer to the point of solution $\varDelta$ since utilizing Δ_j as the start point.

The illustrative interpretation of the descriptions above is shown in Fig. 7.30.

The efficiency of the optimal solution depends on three factors [1, 6] the size of matrix K_m in Eq. (7.51) (1) the size of matrix $K_{s,g}$ in Eq. (7.55) and (2) the similarity of the curves of two following generations (more similar the curves are, less iterations are needed). Previously, it was shown that in the GA from one generation to another one about 75% of a structure remains unchanged, averagely (Ref. [1]). This assures that the second and third factors work efficiently. Moreover, since non-linearity extends gradually in the elements of a structure, the first factor works efficiently as well. The method can always be used with no concern because even in the worst condition its efficiently is similar to that of a traditional method.

7.6.3 Efficient Analysis in the Optimal Size and Geometry Design (Non-linear Geometry)

The optimal size and geometry design of a structure using an optimal non-linear analysis is presented. Suppose the non-linear behavior is only due to the non-linear geometry. Suppose one is in the first generation of the GA and want to start the non-linear analysis for the initial set of members and node locations (size and geometry). The tangent matrix K_t is inverted in each iteration. The matrix can be written as

$$K_{t_{i+1}} = K_{t_i} + K_g \tag{7.56}$$

where K_{t_i} and $K_{t_{i+1}}$ are the tangent stiffness matrices in the ith and i + $1th$ iteration respectively and K_g is the geometric stiffness matrix. However, from one iteration to another one, the size of matrices K_{t_i} and K_g are the same because unlike the material non-linearity which happens gradually, geometric non-linearity happens entirely in each iteration. Therefore, in the first generation up to the point Δ_j a usual method is utilized.

Now, solving the second generation for Δ_j and finding the internal nodal forces vector are aimed. When using the vector Δ_j as the start point, the members of the structure in the jth iteration of the first generation and the corresponding members in the first iteration of the second generation, except the members with changed size and members connected to the changed-location nodes, will have the same stiffness matrices. This way the stiffness matrices of the two structures are similar except in the changed-location nodes and the nodes affected by the changed members

$$(K_{t_1})_2 = \left(K_{t_j}\right)_1 + K_{s,g} \tag{7.57}$$

where $\left(K_{t_j}\right)_1$ is the tangent stiffness matrix of the jth iteration of the first generation, $(K_{t_1})_2$ is the tangent stiffness matrix of the first iteration of the second generation and $K_{s,g}$ is the stiffness matrix because of changed size and geometry.

$$(K_{t_1})_2 = \begin{bmatrix} \left(\left(K_{t_j}\right)_1\right)_{I,I} + K_{s,g} & \vdots & \left(\left(K_{t_j}\right)_1\right)_{I,II} \\ \cdots & \vdots & \cdots \\ \left(\left(K_{t_j}\right)_1\right)_{II,I} & \vdots & \left(\left(K_{t_j}\right)_1\right)_{II,II} \end{bmatrix} \tag{7.58}$$

where the block I,I is corresponding to the nodes which are affected by $K_{s,g}$.

Similar to the calculations shown in Eqs. (7.54) and (7.55), the stiffness matrix $(K_{t_1})_2$ is inverted. Now, moving toward the point of solution is started iteratively. Again from one iteration to iteration the tangent stiffness matrix is inverted using a usual method because geometric non-linearity happens entirely in each iteration. In comparison with a traditional method, one is now much closer to the point of solution Δ since utilizing Δ_j as the start point (see Fig. 7.30).

The efficiency of the optimal solution depends on two factors: (1) the size of matrix $K_{s,g}$ in Eq. (7.58) and (2) the similarity of the curves of two following generations (more similar the curves are, less iteration is needed). As mentioned, in the GA from one generation to another one about 75% of a structure remains unchanged, averagely. This assures that the two factors work efficiently [1].

7.7 Application: Efficient Analysis of a Biomechanical Model of the Lumbar Spine for Predicting the Muscle Forces (Non-linear Material and Geometry)

Unlike the structural/mechanical problems where the external loads are usually available and displacements are found, in biomechanical problems of human musculoskeletal system, measuring the kinematics data is much simpler than finding the applied loads. In such problems the applied loads include both the external loads and the internal tissue responses. However, direct measurement of these internal tissue responses (i.e. muscle forces, ligament forces and so on) is almost impossible such that alternative inverse solutions (i.e. measuring the kinematics data and estimating the internal tissue responses using those kinematics measures and external loads) are required. For instance, in problems related to the biomechanics of the lumbar spine (i.e., the lower portion of vertebral column starting from the T12 vertebra and ending at the S1), kinematics of the spine can be obtained using motion capture system and by tracking markers attached to different bony land marks on the lumbar spine. Though knowledge of mass distribution along the spine facilitates estimation

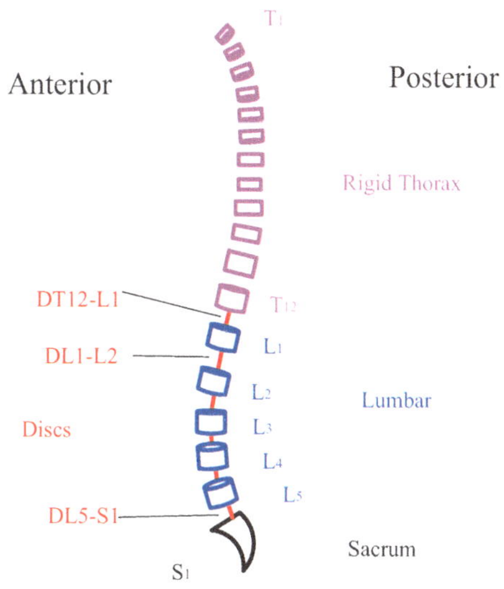

Fig. 7.31 A schematic model of the spine

of external loads (e.g., gravity and inertia) that act on the spine, measurement of internal muscle responses is not possible. Existing mechanical models of the lumbar spine predict such muscle responses by solving an inverse dynamic problem and obtaining moments at one or multiple spinal levels to be balanced by muscles attached to those levels. The generated equilibrium problems for estimation of muscle forces are however redundant; there are more unknown muscle responses as compared to equilibrium equations. Different methods have been adopted to resolve such redundancy problem including optimization method wherein from all possible solution a set of muscle forces that optimize a given cost function is selected. One of the cost functions that has been shown to predict muscle forces consistent with electromyography measurement of muscle activity is the minimization of sum of squared muscle stress.

Here, a previously developed and validated non-linear model of the lumbar spine [7] is used to show the time-efficiency of the algorithm in this paper. This model, shown in Figs. 7.31 and 7.32, will be used to estimate muscle forces and spinal loads under different conditions and using three methods: (1) non-linear inverse- dynamic

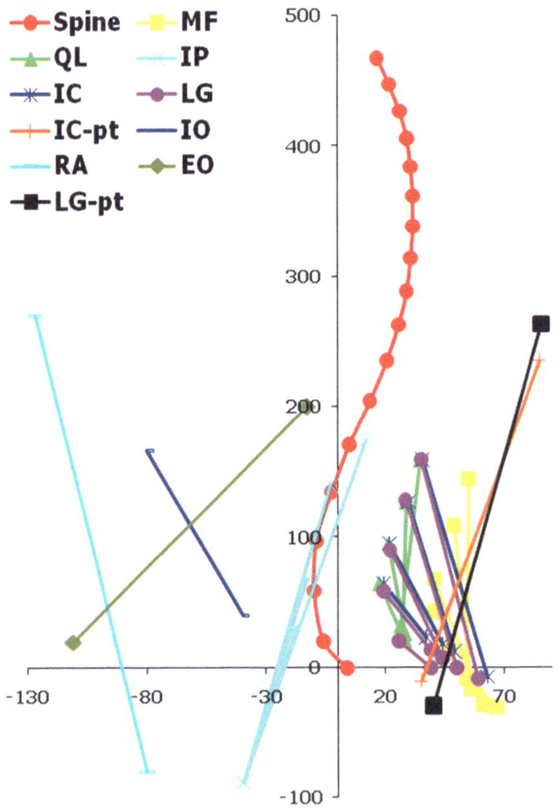

Fig. 7.32 The lateral view of the spine and the attached muscles

analysis with assumption of constant segmental rotations (Method-1) [7, 8] non-linear inverse dynamic analysis with segmental rotations obtained using the optimal design (i.e. posture) (Method-2) and (3) modified non-linear analysis, as presented in this paper, with segmental rotations obtained using the optimal design (i.e. posture) (Method-3). Available Kinematics data for these simulations included rotation of the T12 and the S1 spinal vertebrae in the sagittal plane (i.e., the plane that divide the body to right and left parts) and were obtained from an earlier study [8]. Vertebral rotations in the lumbar region in the Method-1 are estimated as constant percentages of total lumbar rotation (i.e., T12 minus S1 rotations). These percentages have been obtained from earlier studies of segmental range of motion and include 8% for the T12-L1, 13% L1-L2, 16% for L2-L3, 23% for L3-L4, 26% for L4-L5 and 14% for L5-S1 [8]. In the Method-2 and the Method-3 however vertebral rotations in the lumbar region were considered as unknown (dependent on input muscle forces) in the optimization problem (see Eq. 7.30) but were bounded within their reported physiological range of motion. The Method-2 and the Method-3 are expected to predict similar results but different execution times because the analyses in Method-2 are performed using a usual algorithm while the analyses in Method-3 are done using the present optimal analysis in this paper. However, vertebral rotation and spinal loads will be different between the two last methods and those of the Method-1 and will be hence compared here. Description of using the nonlinear model of spine in the Method-1 can be found elsewhere [7], but its application within Method-2 and Method-3 are described here.

The models account for 56 muscles that are symmetrically distributed with respect to sagittal plane and will be used as input to the optimization procedure. These include 10 muscles in each level from the T12 to the L4 and 6 muscles in the level L5 that is 56 muscles totally. Because of the symmetry of the muscles and symmetry of the motion task (i.e. flexion in sagittal plane) 28 muscles are considered as optimization inputs. The optimization problem is formulated as

$$\left\{\begin{array}{c} \boldsymbol{F} = [F_1, F_2, F_3, \ldots, F_{28} \\ 0 \le F_i \le (\sigma_{\max})_m \times PCSA_i \quad i = 1:28 \\ \text{Minmize}\left(\sum_{i=1}^{n=28}\left(\frac{F_i}{PCSA_i}\right)^2\left(1+\alpha\sum_{i=1}^{nc=14}\max[0, g_m]\right)\right) \\ \text{Subject to} \\ \theta_{T_{12}} = \theta_1 \\ \theta_{S_1} = \theta_2 \\ -9.6^\circ \le \theta_{T_{12}} - \theta_{L_1} \le 6^\circ \\ -9.6^\circ \le \theta_{L_1} - \theta_{L_2} \le 6^\circ \\ -12^\circ \le \theta_{L_2} - \theta_{L_3} \le 3.6^\circ \\ -14.4^\circ \le \theta_{L_3} - \theta_{L_4} \le 1.2^\circ \\ -15.6^\circ \le \theta_{L_4} - \theta_{L_5} \le 2.4^\circ \\ -10.8^\circ \le \theta_{L_5} - \theta_{S_1} \le 6^\circ \\ \left|\sigma_{D_j}\right| \le (\sigma_{\max})_d \quad j = 1:6 \end{array}\right. \tag{7.59}$$

where F_i and $PCSA_i$ respectively denote the force and the physiological cross section area of ith lower back muscle, $(\sigma_{max})_m$ is the maximum allowable stress in the muscle (i.e., 1.0 MPa), *nc* is the number of constraints (i.e. 14), *gm* is the violations of the optimization constraints, α is a large number [3], θ_{T12} and θ_{S1} are respectively the rotation of T12 and S1 vertebrae and are inserted in the constraints by the user for a given flexion angle, θ_1 and θ_2 are calculated rotations of T12 and S1 vertebrae in any generation for the corresponding set of muscle forces, θ_{L1} to θ_{L5} are respectively vertebral rotations of L1 to L5 in any generation for the corresponding set of muscle forces and $(\sigma_{max})_d$ is the maximum allowable stress in the disc. The rotational inequality constraints denote sagittal plane range of motion of lumbar motion segments with negative sign denoting flexion.

To solve the defined optimization problem in Eq. (7.30), a heuristic method is employed wherein a GA with 100 generations and 30 individuals in each generation is utilized. Therefore, 3000 non-linear analysis should be performed totally that is very laborious and time-consuming. In this optimization procedure from one generation to another one, external loads (i.e. muscle forces) as the variables of the problem will change. However, as mentioned before, about 75% of the loads remain unchanged in two subsequent generations. Here, the non-linear behavior is due to both non-linear material and geometry. Suppose one is in the first generation of the GA and want to start the non-linear analysis for the initial set of muscle forces. The tangent matrix K_T is inverted in each iteration. The matrix can be written as

$$K_{t_{i+1}} = K_{t_i} + K_m + K_g \tag{7.60}$$

where K_{t_i} and $K_{t_{i+1}}$ are the tangent stiffness matrices in the ith and i + 1th iteration respectively, K_m is the material stiffness matrix and K_g is the geometric stiffness matrix. Although from one iteration to another one the size of matrix K_m is smaller than that of matrices K_{t_i} and K_g, since matrices K_{t_i} and K_g are of the same dimension, a usual method is utilized in the first generation up to the point Δ_j.

Since about 0.75 of the loads remain unchanged in the second generation, instead of solving the new generation from the first step, one starts to solve it for Δ_j which is much closer the target displacement Δ of the second generation. Solving the second generation for Δ_j and finding the corresponding internal nodal forces vector means jumping from the curve of generation 1 to the curve of generation 2 vertically. When using the vector Δ_j as the start point, the members of the structure in the *j*th iteration of the first generation and the corresponding members in the first iteration of the second generation will have the same position on the σ-ε curve as well as same geometry. Consequently, the stiffness matrices of the two structures are similar

$$(K_{t_1})_2 = \left(K_{t_j}\right)_1 \tag{7.61}$$

where $\left(K_{t_j}\right)_1$ is the tangent stiffness matrix of the *j*th iteration of the first generation, $(K_{t_1})_2$ is the tangent stiffness matrix of the first iteration of the second generation.

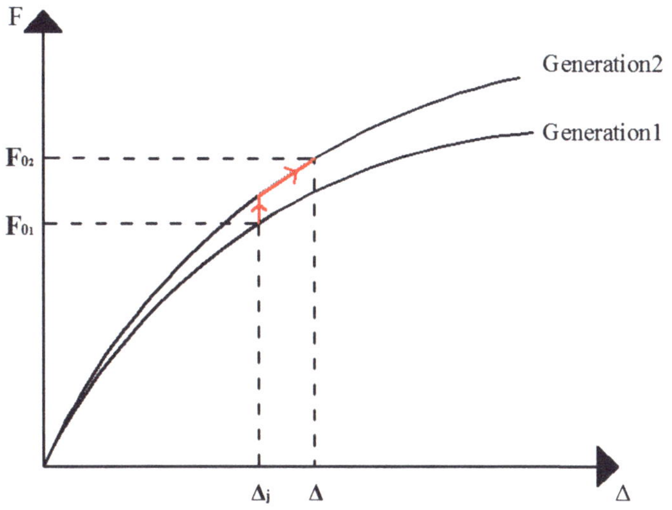

Fig. 7.33 The illustrative interpretation of jumping from one generation to another one

The inverse of the stiffness matrix $\left(K_{t_j}\right)_1$ is available from jth iteration of the first generation that means the inverse of matrix $(K_{t_1})_2$ is available as well. Now, moving toward the point of solution is started iteratively. From one iteration to iteration the tangent stiffness matrix is inverted using a usual method. As shown in Fig. 7.33, one is now much closer to the target point Δ in comparison with a usual method.

Unlike Fig. 7.30, where the external loads are the same but the size of members and nodes location change in two generations, here the external loads (i.e. muscle forces) change from one generation to another one (i.e. the different between structural (mechanical) and biomechanical approaches).

Using the three methods explained above, muscle forces and spinal loads were calculated for different lumbar rotations and external loading (with and without holding a 4.55 kg weight in hand) conditions. Tables 7.5 and 7.6 contain compression and shear forces at the lower most level of lumbar spine as calculated by the Method-1 and the Method-2 (or Method-3) for all condition studied here. For each condition,

Table 7.5 The value of cost function and compression and shear forces of the L5-S1 level (no external load)

Lumbar rotation (degree)		25	30	35	40	45	50	55
Cost function*e-12	Method-1	0.93	1.29	1.66	2.49	2.73	3.25	3.52
	Method-2	0.16	0.319	0.50	0.638	0.75	1.04	1.68
Compression (N)	Method-1	1127	1400	1579	1848	1978	2282	2473
	Method-2	935	1177	1343	1565	1716	2028	2400
Shear (N)	Method-1	378	424	440	447	437	400	362
	Method-2	402	449	468	480	477	450	385

Table 7.6 The value of cost function and compression and shear forces of the L5-S1 level (4.55 kg external load)

Lumbar rotation (degree)		25	30	35	40	45	50	55
Cost function*e-12	Method-1	1.20	1.66	2.18	2.93	3.59	4.17	4.49
	Method-2	0.26	0.48	0.67	0.94	1.46	1.74	2.22
Compression (N)	Method-1	1287	1618	1840	2136	2452	2778	2952
	Method-2	1059	1371	1570	1849	2260	2581	2882
Shear (N)	Method-1	430	491	513	520	504	458	420
	Method-2	454	517	546	564	549	511	445

Table 7.7 The average value for the execution time, number of iterations, and cost function during different generations of GA

Total analyses (Generations*Individuals)		500	1000	1500	2000	2500	3000
Execution time (h)	Method-2	2.14	4.19	6.45	8.62	10.65	12.50
	Method-3	1.26	2.42	3.67	4.97	6.21	7.25
Number of iterations*e-2	Method-2	5000	10,000	15,000	20,000	25,000	30,000
	Method-3	2831	5527	8257	12,016	14,283	17,531
Cost function*e-12		4.9	3.1	2.03	1.35	1.13	1.11

the predicted muscle forces were used to calculate the value of cost function considered in the Method-2 (or Method-3) and is also reported in Tables 7.5 and 7.6. As mentioned above, the only difference between Method-2 and Method-3 is related to the execution time. The time-efficiency of the Method-3 (the present algorithm) versus the Method-2 (the usual algorithm) was reflected in the total time needed to complete the optimization procedure for each condition. In average, simulation durations for completing the optimization procedure for one condition were respectively 12.5 and 7.25 h. This suggests an average of 42% decrease in execution time of the Method-3. The average value (different lumbar rotations) for the execution time, number of iterations, and cost function during different generations of GA is presented in Table 7.7.

References

1. Rahami, H., Kaveh, A., Shojaei, I.: Swift analysis for size and geometry optimization of structures. Adv. Struct. Eng. **18**(3), 365–380 (2015)
2. Shojaei, I., Kaveh, A., Rahami, H., Bazrgari, B.: Efficient non-linear analysis and optimal design of mechanical and biomechanical systems. Adv. Biomech. Appl. **2**(1), 11–27 (2015)
3. Kaveh, A., Shojaei, I., Golipour, Y., Rahami, H.: Seismic design of eccentric braced frames using multi-objective optimization. Struct. Eng. Mech. **45**(2), 211–232 (2013)
4. Bathe, K.J.: Finite Element Procedures. Englewood Cliffs, Prentice-Hall, NJ, USA (1996)

5. Zingoni, A.: Group-theoretic insights on the vibration of symmetric structures in engineering. Phil. Trans. R Soc. A **13**(372), 20120037 (2014)
6. Shojaei, I., Kaveh, A., Rahami, H.: Analysis of structures convertible to repeated structures using graph products. Comput. Struct. **125**, 153–163 (2013)
7. Bazrgari, B., Shirazi-Adl, A., Arjmand, N.: Analysis of squat and stoop dynamic liftings: muscle forces and internal spinal loads. Eur. Spine J. **16**(5), 687–699 (2007)
8. Arjmand, N., Shirazi-Adl, A.: Biomechanics of changes in lumbar posture in static lifting. Spine **30**(23), 2637–2648 (2005)

Chapter 8
Global Near-Regular Mechanical Systems

8.1 Introduction

In Chaps. 2–5 we extensively studied analysis and eigensolution of regular and near-regular skeletal and continuous mechanical systems. These mechanical systems involved locally concentrated irregularities where only a small number of systems' degrees of freedom were affected by the irregularities. On the other hand, there is another group of near-regular mechanical systems that have global deviations from their corresponding regular system (hereafter called global near-regular mechanical systems). In such systems a large number of degrees of freedom are affected by irregularities. Despite extensive research on mechanical systems with local irregularities, the literature on global near-regular mechanical systems is scant. In this chapter, methods are developed for static and dynamic analysis/reanalysis as well as eigensolution (vibration analysis) of such systems using Kronecker products and matrix manipulations [1].

8.2 Static Analysis and Reanalysis

In this section, the formulation for the static analysis/reanalysis of global (hereafter, the word "global" is removed for simplicity) near-regular systems is presented. The types of analysis and reanalysis include:

1. Analysis of near-regular structures: the original structure is a near-regular structure and we assume it is a modified (in terms of stiffness) version of a corresponding regular structure. In other words, the analysis of a near-regular structure is converted to the reanalysis of a regular structure.
2. Reanalysis of near-regular structures: a corresponding regular structure is considered as the initial structure and the altered stiffness associated with irregularity,

A. Kaveh et al., *Swift Analysis of Civil Engineering Structures Using Graph Theory Methods*, Studies in Systems, Decision and Control 290,
https://doi.org/10.1007/978-3-030-45549-1_8

design procedure, and/or non-linear behavior is added to the stiffness matrix of the initial regular structure.

Consider the matrix equation of a near-regular system as

$$\boldsymbol{kd} = \boldsymbol{F} \tag{8.1}$$

where $\boldsymbol{K}$ is the stiffness matrix, and $\boldsymbol{d}$ and $\boldsymbol{F}$ are displacement and force vectors, respectively.

The stiffness matrix is written as the summation of the stiffness matrix corresponding to initial regular structure ($\boldsymbol{K}_{\text{ini}}$) and an altered stiffness ($\Delta\boldsymbol{K}$) as

$$(\boldsymbol{K}_{ini} + \Delta\boldsymbol{K})\boldsymbol{d} = \boldsymbol{F} \tag{8.2}$$

Left multiplication of the two sides by $\boldsymbol{K}_{ini}^{-1}$ leads to

$$\left(\boldsymbol{I} + \boldsymbol{K}_{ini}^{-1}\Delta\boldsymbol{K}\right)\boldsymbol{d} = \boldsymbol{K}_{ini}^{-1}\boldsymbol{F} \tag{8.3}$$

By defining $\boldsymbol{G} = \boldsymbol{K}_{ini}^{-1}\Delta\boldsymbol{K}$ and $\boldsymbol{d}_{ini} = \boldsymbol{K}_{ini}^{-1}\boldsymbol{F}$, we will have

$$\boldsymbol{d} = (\boldsymbol{I} + \boldsymbol{G})^{-1}\boldsymbol{d}_{ini} \tag{8.4}$$

Now, consider the series [2, 3]

$$\sum_{r=0}^{n} \boldsymbol{G}^{r} = \boldsymbol{I} + \boldsymbol{G} + \boldsymbol{G}^{2} + \cdots \tag{8.5}$$

The series converges if and only if $\boldsymbol{G}^{r} \to 0$ [3]. We can write

$$(\boldsymbol{I} - \boldsymbol{G})\left(\boldsymbol{I} + \boldsymbol{G} + \boldsymbol{G}^{2} + \cdots + \boldsymbol{G}^{r}\right) = \boldsymbol{I} - \boldsymbol{G}^{r+1} \tag{8.6}$$

and

$$\left(\boldsymbol{I} + \boldsymbol{G} + \boldsymbol{G}^{2} + \cdots + \boldsymbol{G}^{r}\right) = (\boldsymbol{I} - \boldsymbol{G})^{-1} - (\boldsymbol{I} - \boldsymbol{G})^{-1}\boldsymbol{G}^{r+1}) \tag{8.7}$$

The second term on the right side approaches zero, therefore

$$\boldsymbol{I} + \boldsymbol{G} + \boldsymbol{G}^{2} + \cdots \to (\boldsymbol{I} - \boldsymbol{G})^{-1} \tag{8.8}$$

and

$$\boldsymbol{I} - \boldsymbol{G} + \boldsymbol{G}^{2} + \cdots \to (\boldsymbol{I} + \boldsymbol{G})^{-1}$$

Rewriting Eq. (8.4) results in

$$\boldsymbol{d} = \left(\boldsymbol{I} - \boldsymbol{G} + \boldsymbol{G}^2 - \cdots\right)\boldsymbol{d}_{ini} \tag{8.9}$$

And an efficient recurrence equation is obtained as

$$\boldsymbol{d}_n = -\boldsymbol{G}\boldsymbol{d}_{n-1}, \quad n = 2, 3, \ldots, m \tag{8.10}$$

where $\boldsymbol{d}_1 = \boldsymbol{d}_{ini}$.

For an arbitrary structure (not necessarily a regular or near-regular structure) with given $\boldsymbol{LU}$ decomposition ($\boldsymbol{LL}^T$ decomposition for symmetric positive definite matrices) of initial stiffness matrix, $\boldsymbol{K}_{ini} = \boldsymbol{LL}^T$, Eq. (8.10) is solved as follows:

$$\begin{aligned}
\boldsymbol{K}_{ini}\boldsymbol{d}_n &= -\Delta\boldsymbol{K}\boldsymbol{d}_{n-1}\\
\boldsymbol{LL}^T\boldsymbol{d}_n &= -\Delta\boldsymbol{K}\boldsymbol{d}_{n-1}\\
\boldsymbol{L}s &= -\Delta\boldsymbol{K}\boldsymbol{d}_{n-1}, \quad \boldsymbol{L}^T\boldsymbol{d}_n = s
\end{aligned} \tag{8.11}$$

For the known matrix $\Delta\boldsymbol{K}$ and vector $\boldsymbol{d}_{n-1}$, we first solve for the unknown vector s and then solve for the vector $\boldsymbol{d}_n$. Regardless of the required effort for the $\boldsymbol{LL}^T$ decomposition, the recurrence solution includes matrix by vector multiplications of the computational complexity $O(n^2)$. Since $\boldsymbol{LU}$ decomposition is of the order $O(n^3)$, the method is efficient only if there is a simple way to obtain the decomposition.

8.3 Decomposition of Regular Patterns Using Kronecker Products

For a regular structure/graph the full decomposition of the stiffness matrix/ Laplacian matrix can be found using Kronecker products. The two general groups of regular structures decomposable by Kronecker products include repetitive and circulant matrices with the following block patterns:

$$\begin{aligned}
\boldsymbol{M}_n &= \boldsymbol{F}_l(\boldsymbol{A}_m, \boldsymbol{B}_m, \boldsymbol{A}_m) = \boldsymbol{I}_l \otimes \boldsymbol{A}_m + \boldsymbol{T}_l \otimes \boldsymbol{B}_m\\
\boldsymbol{M}_n &= \boldsymbol{G}_l(\boldsymbol{A}_m, \boldsymbol{B}_m, \boldsymbol{A}_m) = (\boldsymbol{P}_1)_l \otimes \boldsymbol{A}_m + (\boldsymbol{P}_2)_l \otimes \boldsymbol{B}_m + (\boldsymbol{P}_3)_l \otimes \boldsymbol{B}_m
\end{aligned} \tag{8.12}$$

where $\boldsymbol{I}$ is an identity matrix, $\boldsymbol{T} = \boldsymbol{F}(0, 1, 0)$, and $\boldsymbol{P}_1$, $\boldsymbol{P}_2$ and $\boldsymbol{P}_3$ are permutation matrices.

Using the following lemma, matrix $\boldsymbol{M}$ can be decomposed:

Sufficient conditions for converting Hermitian matrices $\boldsymbol{A}_1$ and $\boldsymbol{A}_2$ into upper triangular ones using one orthogonal matrix are [4]:

$$\boldsymbol{A}_1\boldsymbol{A}_2 = \boldsymbol{A}_2\boldsymbol{A}_1 \tag{8.13}$$

Let the matrix $\boldsymbol{M}$ be the sum of two (it can also be generalized to n Kronecker products) Kronecker products as $\boldsymbol{M} = \boldsymbol{A}_1 \otimes \boldsymbol{B}_1 + \boldsymbol{A}_2 \otimes \boldsymbol{B}_2$. If $\boldsymbol{F}$ is the matrix that upper triangularizes the matrices $\boldsymbol{A}_1$ and $\boldsymbol{A}_2$, then $\boldsymbol{U} = \boldsymbol{F} \otimes \boldsymbol{I}$ block upper triangularizes the matrix $\boldsymbol{M}$, which means $\boldsymbol{U}^t\boldsymbol{M}\boldsymbol{U}$ is block upper diagonalized and thus one can write:

$$\lambda_{\boldsymbol{M}} = \bigcup_{i=1}^{l} eig(\boldsymbol{M}_i);\ \boldsymbol{M}_i = \lambda_i(\boldsymbol{A}_1)\boldsymbol{B}_1 + \lambda_i(\boldsymbol{A}_2)\boldsymbol{B}_2 \tag{8.14}$$

where matrices $\boldsymbol{A}_1$ and $\boldsymbol{A}_2$ are of dimension l, and matrices $\boldsymbol{B}_1$ and $\boldsymbol{B}_2$ are of dimension m.

Since $\boldsymbol{IT} = \boldsymbol{TI}$ in Eq. (8.12) and considering that $\lambda_{T_l} = 2\cos\frac{k\pi}{l+1}$ [5], we will have

$$\lambda_{\boldsymbol{M}} = \bigcup_{k=1}^{l} eig\left(2\cos\frac{k\pi}{l+1}\boldsymbol{B} + \boldsymbol{A}\right) \tag{8.15}$$

Permutation matrices also hold the following property:

$$\boldsymbol{P}_i\boldsymbol{P}_j = \boldsymbol{P}_j\boldsymbol{P}_i \tag{8.16}$$

The eigenpairs of the permutation matrix $\boldsymbol{P}_i$ are obtained as follows

$$\begin{aligned}\lambda = \left(e^{2\pi i/l}\right)^k = \omega^k;\quad i = \sqrt{-1};\quad k = 0 : l-1\\ \boldsymbol{\vartheta}_k = \frac{1}{\sqrt{l}}\left(1, \omega_k, \omega_k^2, \ldots, \omega_k^{l-1}\right)^T\end{aligned} \tag{8.17}$$

and

$$\lambda_{\boldsymbol{M}} = \bigcup_{k=1}^{l} eig\left(2\left(e^{2\pi i/l}\right)^{k-1}\boldsymbol{B} + \boldsymbol{A}\right) \tag{8.18}$$

The eigenvectors of matrix $\boldsymbol{M}$ are calculated through $\boldsymbol{u} \otimes \boldsymbol{v}$ [6], where $\boldsymbol{u}$ is the eigenvector of $\boldsymbol{A}_2$ ($\boldsymbol{T}$ in Eq. 8.12) and$\boldsymbol{v}$ is the eigenvector of $\boldsymbol{B}_1$. With the obtained eigenpairs, one can decompose $\boldsymbol{M}$ as follows:

$$\boldsymbol{M} = \boldsymbol{V}^T\boldsymbol{\lambda}\boldsymbol{V} \tag{8.19}$$

Now, by inserting Eq. (8.19) into Eq. (8.11) ($\boldsymbol{K}_{ini} = \boldsymbol{M}$), we will have

$$\boldsymbol{d}_n = -\boldsymbol{V}\boldsymbol{\lambda}^{-1}\boldsymbol{V}^T\Delta\boldsymbol{K}\boldsymbol{d}_{n-1},\quad n = 2, 3, \ldots, m \tag{8.20}$$

Using a proper order of multiplication (operating from right to left), Eq. (8.20) only includes matrix by vector multiplications of the computational complexity $O(n^2)$.

For large change in matrix $\Delta \boldsymbol{K}$ we may encounter slow convergence or divergence of the series. The approximation can be significantly improved if the terms of the series are used as the basis vectors in the functional extension (similar to the approach in finite element solution) of the displacement solution [2, 7]:

$$\boldsymbol{d} = x_1\boldsymbol{d}_1 + x_2\boldsymbol{d}_2 + \cdots + x_m\boldsymbol{d}_m = \boldsymbol{Dx} \tag{8.21}$$

where the number of basis vectors, m, is much smaller than the number of degrees of freedom, n. $\boldsymbol{D} = [\boldsymbol{d}_1, \boldsymbol{d}_2, \ldots, \boldsymbol{d}_m]$ is matrix of dimension nm and $\boldsymbol{x}^T = [x_1, x_2, \ldots, x_m]$.

By substituting Eq. (8.21) into Eq. (8.1) and left multiplying the two sides by $\boldsymbol{D}^T$, we will have

$$\boldsymbol{D}^T\boldsymbol{K}\boldsymbol{D}\boldsymbol{x} = \boldsymbol{D}^T\boldsymbol{F} \tag{8.22}$$

$$\boldsymbol{K}_r\boldsymbol{x} = \boldsymbol{F}_r; \quad \boldsymbol{K}_r = \boldsymbol{D}^T\boldsymbol{K}\,\boldsymbol{D} \text{ and } \boldsymbol{F}_r = \boldsymbol{D}^T\boldsymbol{F}$$

The unknown vector $\boldsymbol{x}$ is found using the reduced equation $\boldsymbol{K}_r\boldsymbol{x} = \boldsymbol{F}_r$ of dimension m. Substituting the vector $\boldsymbol{x}$ in Eq. (8.21) leads to finding the unknown vector $\boldsymbol{d}$. The components of matrix $\boldsymbol{D}$ are efficiently found using Eq. (8.20).

For the reanalysis of a near-regular structure the same formulation will be used but the matrix $\Delta\boldsymbol{K}$ in Eq. (8.2) will also include (in addition to the altered stiffness from regular to near-regular structure) the altered stiffness due to the change in the design.

Non-linear static analysis of a structure is inherently a reanalysis problem due to the iterative solution procedure. In the Newton-Raphson method for the kth iteration in ith step we can write

$${}^{i}\boldsymbol{K}^{(k-1)}\delta\boldsymbol{d}^{(k)} = \delta\boldsymbol{F}^{(k-1)} \tag{8.23}$$

where $\boldsymbol{K}^{(k-1)}$ is the current tangent stiffness matrix and $\delta\boldsymbol{F}^{(k-1)}$ is the imbalance load vector between the external load in the ith step, ${}^{i}\boldsymbol{F}_E$, and the internal load in the $(k-1)$ th iteration in the ith step, ${}^{i}\boldsymbol{F}_I^{(k-1)}$.

$$\delta\boldsymbol{F}^{(k-1)} = {}^{i}\boldsymbol{F}_E - {}^{i}\boldsymbol{F}_I^{(k-1)} \tag{8.24}$$

Therefore, the incremental displacement vector $\delta\boldsymbol{d}^{(k)}$ is calculated in each iteration. Consider the following simplified notation:

$$\boldsymbol{K}_T\delta\boldsymbol{d} = \boldsymbol{\delta} \tag{8.25}$$

The tangent stiffness matrix $\boldsymbol{K}_T$ can be considered as $\boldsymbol{K}_T = \boldsymbol{K}_{ini} + \Delta\boldsymbol{k}$, where $\Delta\boldsymbol{k}$ represents the alternations in stiffness due to non-linear behavior:

$$(\boldsymbol{K}_{ini} + \Delta \boldsymbol{k}_{NL})\delta \boldsymbol{d} = \delta \boldsymbol{F} \tag{8.26}$$

This equation is similar to Eq. (8.2) and can be solved using Eq. (8.20) to Eq. (8.22) and the developed graph product rules for full decomposition of the regular initial stiffness matrix $\boldsymbol{K}_{ini}$.

8.4 Dynamic Analysis and Reanalysis

Dynamic equation of motion can be written as

$$\boldsymbol{M}\ddot{\boldsymbol{d}} + \boldsymbol{c}\dot{\boldsymbol{d}} + \boldsymbol{k}\boldsymbol{d} = \boldsymbol{F} \tag{8.27}$$

For a near-regular structure the stiffness, mass and damping matrix can be expressed as follows

$$\boldsymbol{K} = \boldsymbol{K}_{ini} + \Delta \boldsymbol{K}, \quad \boldsymbol{M} = \boldsymbol{M}_{ini} + \Delta \boldsymbol{M}, \quad \boldsymbol{C} = \boldsymbol{C}_{ini} + \Delta \boldsymbol{C} \tag{8.28}$$

where $\boldsymbol{K}_{ini}$, M_{ini} and $\boldsymbol{C}_{ini}$ are the stiffness, mass and damping matrix of the corresponding regular initial structure, respectively.

Writing the equation of motion for the regular structure leads to

$$\boldsymbol{M}_{ini}\ddot{\boldsymbol{d}}_{ini} + \boldsymbol{C}_{ini}\dot{\boldsymbol{d}}_{ini} + \boldsymbol{K}_{ini}\boldsymbol{d}_{ini} = \boldsymbol{F}_{ini} \tag{8.29}$$

Using implicit integration Newmark method with the parameters α_0 and α_1, the effective initial stiffness matrix is formed as

$$\widehat{\boldsymbol{K}}_{ini} = \boldsymbol{K}_{ini} + \alpha_0 \boldsymbol{M}_{ini} + \alpha_1 \boldsymbol{C}_{ini} \tag{8.30}$$

Defining damping matrix $\boldsymbol{C}_{ini}$ as a linear combination of $\boldsymbol{K}_{ini}$ and $\boldsymbol{M}_{ini}$, $\boldsymbol{C}_{ini} = \boldsymbol{\alpha K}_{ini} + \beta M_{ini,}$ will hold the regular form shown in Eq. (8.12) and is factorized using the developed graph product relationships

$$\widehat{\boldsymbol{K}}_{ini} = \boldsymbol{V}^{T}\boldsymbol{\lambda}\boldsymbol{V} \tag{8.31}$$

The effective load at time $t + \Delta t$ is calculated as follows:

$$\begin{aligned}\widehat{\boldsymbol{F}}_{ini}^{t+\Delta t} &= \boldsymbol{F}_{ini}^{t+\Delta t} + \boldsymbol{M}_{ini}\left(\alpha_0 \boldsymbol{d}_{ini}^{t} + \alpha_2 \dot{\boldsymbol{d}}_{ini}^{t} + \alpha_3 \ddot{\boldsymbol{d}}_{ini}^{t}\right) \\ &+ \boldsymbol{C}_{ini}\left(\alpha_1 \boldsymbol{d}_{ini}^{t} + \alpha_4 \dot{\boldsymbol{d}}_{ini}^{t} + \alpha_5 \ddot{\boldsymbol{d}}_{ini}^{t}\right)\end{aligned} \tag{8.32}$$

And the effective equation of motion is expressed as

$$
\begin{aligned}
\widehat{K}_{ini}\,d_{ini}^{t+\Delta t} &= \widehat{F}_{ini}^{t+\Delta t} \\
V^{T}\lambda V d_{ini}^{t+\Delta t} &= \widehat{F}_{ini}^{t+\Delta t} \\
d_{ini}^{t+\Delta t} &= V\lambda^{-1}V^{T}\widehat{F}_{ini}^{t+\Delta t}
\end{aligned}
\tag{8.33}
$$

This equation is efficiently solved and then velocities and accelerations are obtained as

$$
\ddot{d}_{ini}^{t+\Delta t} = \alpha_0\left(d_{ini}^{t+\Delta t} - d_{ini}^{t}\right) - \alpha_2\dot{d}_{ini}^{t} - \alpha_3\ddot{d}_{ini}^{t}
$$

and

$$
\dot{d}_{ini}^{t+\Delta t} = \dot{d}_{ini}^{t} + \alpha_6\ddot{d}_{ini}^{t} + \alpha_7\ddot{d}_{ini}^{t+\Delta t}
\tag{8.34}
$$

Now, the original near-regular structure can be analyzed/reanalyzed using the changes shown in Eq. (8.28). In addition to the change of the stiffness matrix from a regular structure to a near-regular one, matrix ΔK can include the changes due to non-linear behavior and/or design procedure. Notice that a change in stiffness of members due to the change in design also leads to change in the mass of members and damping (it was mentioned damping is a function of mass and stiffness):

$$
\widehat{K} = \widehat{K}_{ini} + (\Delta K + \alpha_0\Delta M + \alpha_1\Delta C) = \widehat{K}_{ini} + \Delta\widehat{K}
\tag{8.35}
$$

Similar to Eq. (8.2) for static equations, we will have an equivalent system of equations

$$
\left(\widehat{K}_{ini} + \Delta\widehat{K}\right)d^{t+\Delta t} = \widehat{F}^{t+\Delta t}
\tag{8.36}
$$

This equation is efficiently solved using Eq. (8.20) to Eq. (8.22) considering that the full decomposition of matrix $\widehat{K}_{ini}$ is available using graph product rules.

Five steps of the proposed procedure for analysis of global near-regular mechanical systems are as follows:

1. Split the (general: Eq. (8.2), tangent: Eq. (8.25). effective Eq. (8.36)) stiffness matrix K into the initial regular stiffness K_{ini} and an altered stiffness ΔK.
2. Decompose the initial regular stiffness K_{ini} into the form $V^{T}\lambda V$ using graph product rules (Eqs. (8.14–8.19)).
3. Calculate the basis vectors using the recurrence equation $d_n = -V\lambda^{-1}V^{T}\Delta K d_{n-1}$ (Eq. (8.20)) of the computational complexity $O\left(n^2\right)$.
4. Approximate the displacement vector d using the functional expansions $d = x_1 d_1 + x_2 d_2 + \cdots + x_m d_m = Dx$ (Eq. (8.21)).

5. Solve the reduces equation $\boldsymbol{K}_r\boldsymbol{x} = \boldsymbol{F}_r$ (Eq. (8.22)) and substitute the vector $\boldsymbol{x}$ into the functional extension (Eq. (8.21)) to calculate the displacement vector $\boldsymbol{d}$.

8.5 A Comprehensive Example for Developing a Subject-Specific Finite Element Model of Human Spine

A non-linear finite element model of the spine was developed and used earlier for estimating muscle forces and spinal loads on the lower back (Fig. 8.1) [8–10]. The nonlinearity of model arise from (1) the nonlinear geometry of the model (i.e. force-deformation effect and large deformations), and (2) the nonlinear material properties of elements used to simulate the behavior of the passive spine components (i.e. ligaments and intervertebral discs). In the finite element model, as schematically shown in Fig. 8.1, the intervertebral discs are simulated by nonlinear flexible beam elements whereas vertebrae are represented by rigid elements.

The behavior of passive spine components have been obtained from earlier cadaver lumbar motion segments studies (Fig. 8.2) and used in the FE model [9, 10].

However, the passive behavior of lower back (i.e., the passive ligamentous spine along with the surrounding musculature) is subject-specific and changes with age, gender, spinal health and other personal features, and has a non-linear behavior. For instance, we have recently reported age-related changes in lower back stiffness obtained using passive flexion experiments (Figs. 8.3 and 8.4) [11].

Lower back moment (M) versus flexion angle (°) has been measured during passive flexion tests in upright standing posture such that the lower back passive stiffness can be estimated at each time increment [11]. Using these experimental measures

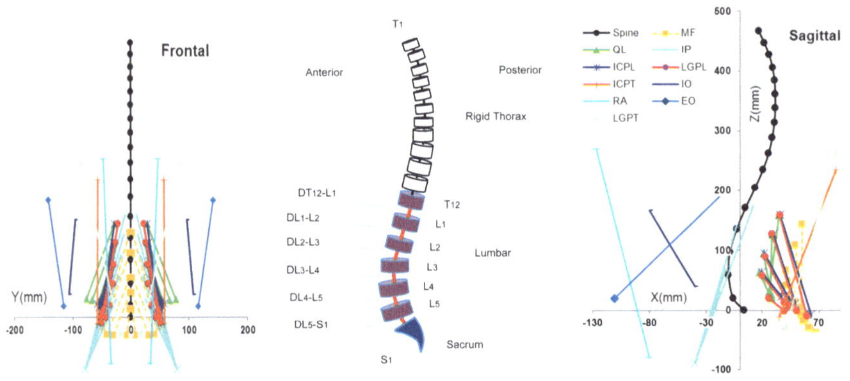

Fig. 8.1 Spinal column and the surrounding musculature

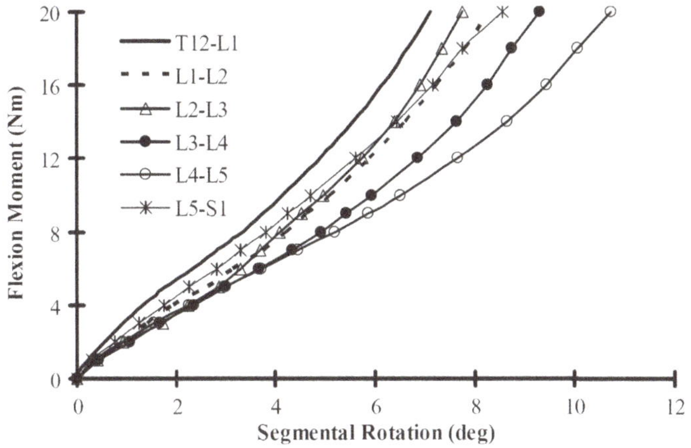

Fig. 8.2 Variation of flexion moment with segmental rotation for different lumbar motion segments under 1800 N axial compression

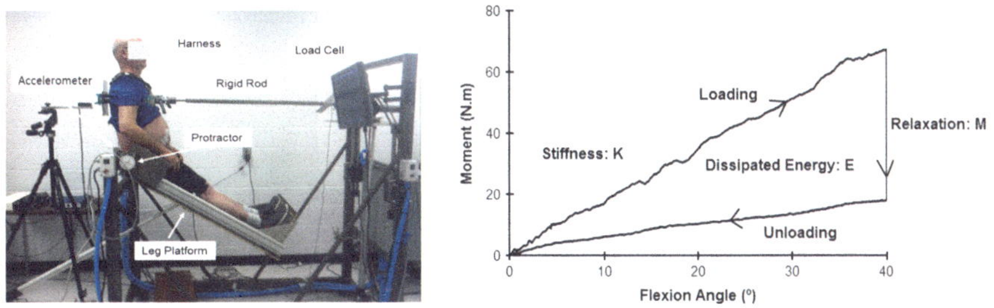

Fig. 8.3 Experimental setup for passive flexion test (left); Moment-Flexion curve of lower back (right) [11]

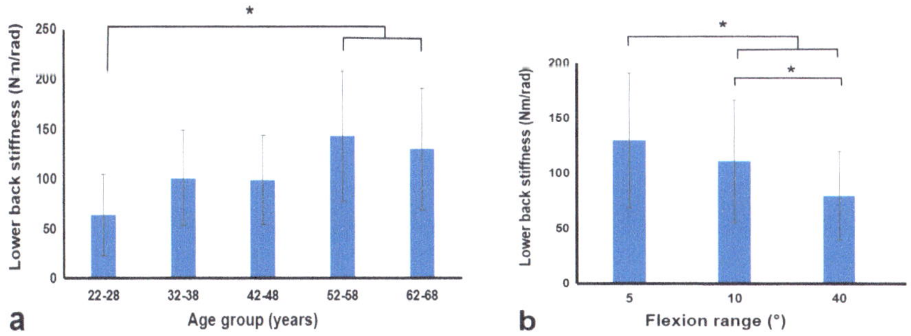

Fig. 8.4 Change in lower back stiffness with aging **a**; non-linear behavior of lower back stiffness **b** [11]. Error bars indicate standard deviations. Significant paired differences are indicated with brackets

applied to the model as well as optimization procedures, estimating age-specific passive stiffness of lumbar motion segments from lower back stiffness (obtained from passive flexion tests) is aimed.

To initially estimate the equivalent properties of beam elements from the passive behavior of cadaver lumbar motion segments (Fig. 8.2), we will have

$$\frac{1}{r} = k = \frac{\theta_2 - \theta_1}{l} = \frac{M}{EI}$$
$$M = \frac{EI}{l}(\theta_2 - \theta_1) = \frac{EI}{l}\theta$$

where $\frac{1}{r} = k$ is curvature, θ_2 and θ_1 are the rotations at two ends of the segment, and θ is the intervertebral disc rotation in Fig. 8.2. Therefore, EI/l will be the slope of moment rotation curves (Fig. 8.2) at each point. Given l from anthropometric measures, $\boldsymbol{EI}$ for each beam element is calculated. The instantaneous stiffness matrix for a beam element is constructed as follows:

$$\boldsymbol{K}_b = EI\begin{bmatrix} \frac{12}{l^3} & \frac{6}{l^2} & -\frac{12}{l^3} & \frac{6}{l^2} \\ \frac{6}{l^2} & \frac{4}{l} & -\frac{6}{l^2} & \frac{2}{l} \\ -\frac{12}{l^3} & -\frac{6}{l^2} & \frac{12}{l^3} & -\frac{6}{l^2} \\ \frac{6}{l^2} & \frac{2}{l} & -\frac{6}{l^2} & \frac{4}{l} \end{bmatrix}$$

By assembling the instantaneous stiffness matrix for the whole lower back in the global system, the following matrix pattern is obtained

$$\boldsymbol{K} = \begin{bmatrix} \boldsymbol{A}_1 & \boldsymbol{B}_1 & & & & \\ \boldsymbol{B}_1 & \boldsymbol{A}_2 & \boldsymbol{B}_2 & & & \\ & \boldsymbol{B}_2 & \boldsymbol{A}_3 & \boldsymbol{B}_3 & & \\ & & \boldsymbol{B}_3 & \boldsymbol{A}_4 & \boldsymbol{B}_4 & \\ & & & \boldsymbol{B}_4 & \boldsymbol{A}_5 & \boldsymbol{B}_5 \\ & & & & \boldsymbol{B}_5 & \boldsymbol{A}_6 \end{bmatrix}$$

where $\boldsymbol{A}_i$s and $\boldsymbol{B}_i$s are matrices of dimension 3.

In earlier studies this stiffness matrix was constructed using cadaver passive properties (Fig. 8.2), and used for individuals of all ages [9, 10]. We intend to estimate subject-specific, e.g., age-specific in the current study, passive properties utilizing the data obtained from individuals in different age and gender groups during in vivo passive flexion tests. The subject-specific (ss subscript notations below) equivalent beam element instantaneous stiffness matrix for a motion segment will have the form:

$$(\boldsymbol{K}_b)_{ss} = (EI)_{ss}\begin{bmatrix} \frac{12}{l_{ss}^3} & \frac{6}{l_{ss}^2} & -\frac{12}{l_{ss}^3} & \frac{6}{l_{ss}^2} \\ \frac{6}{l_{ss}^2} & \frac{4}{l_{ss}} & -\frac{6}{l_{ss}^2} & \frac{2}{l_{ss}} \\ -\frac{12}{l_{ss}^3} & -\frac{6}{l_{ss}^2} & \frac{12}{l_{ss}^3} & -\frac{6}{l_{ss}^2} \\ \frac{6}{l_{ss}^2} & \frac{2}{l_{ss}} & -\frac{6}{l_{ss}^2} & \frac{4}{l_{ss}} \end{bmatrix}$$

Assuming $((EI)_{ss})_i = \alpha_i (EI)_i$, where α_i is an unknown coefficient, i is the beam element number ($i = 1, 2,\ldots,6$), $(EI)_i$ is the instantaneous beam passive property in Fig. 8.2, and $(l_{ss})_i$ is obtained from anthropometric measures. The subject-specific global instantaneous stiffness matrix will have the following form:

$$K_{ss} = \begin{bmatrix} \hat{A}_1 & \hat{B}_1 & & & & \\ \hat{B}_2 & \hat{A}_2 & \hat{B}_2 & & & \\ & \hat{B}_2 & \hat{A}_3 & \hat{B}_3 & & \\ & & \hat{B}_3 & \hat{A}_4 & \hat{B}_4 & \\ & & & \hat{B}_4 & \hat{A}_5 & \hat{B}_5 \\ & & & & \hat{B}_5 & \hat{A}_6 \end{bmatrix} = f(\alpha_i, (EI)_i, (l_{ss})_i)$$

The only unknowns in the stiffness matrix are $\boldsymbol{\alpha}_i$ s. Consider the passive flexion test (Fig. 8.3) wherein the trunk at the T8 spinal level was fixed using a rigid rod and lower limbs were rotated around lumbo-pelvic joint. The instantaneous pelvic rotation and the developed force in the rod were measured using an accelerometer connected to the leg platform and a load cell on the connecting rod, respectively (Fig. 8.3). The translation of pelvis was constrained using straps. The flexion passive test was simulated using the model in Fig. 8.5. Since the test is performed slowly the governed equation will be a non-linear quasi-static equation:

$$\boldsymbol{K}_{ss}\delta\,\boldsymbol{d} = \delta\boldsymbol{F}$$

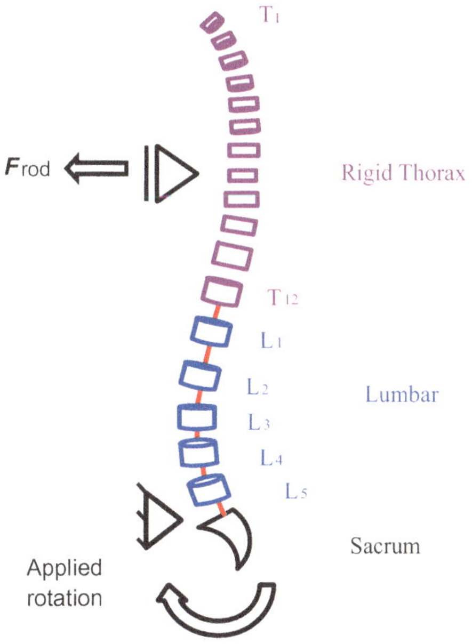

Fig. 8.5 Simulation of the passive flexion test

Separating the stiffness matrix into the earlier proposed form:

$$(\boldsymbol{K}_{ini} + \Delta\boldsymbol{K}_1 + \Delta\boldsymbol{K}_2)\delta\, \boldsymbol{d} = \delta\boldsymbol{F}$$

where $\boldsymbol{K}_{ini} = \boldsymbol{F}(\boldsymbol{A}, \boldsymbol{B}, \boldsymbol{A}) = \boldsymbol{I} \otimes \boldsymbol{A} + \boldsymbol{T} \otimes \boldsymbol{B}$, $\Delta\boldsymbol{K}_1 = \boldsymbol{K} - \boldsymbol{K}_{ini}$, and $\Delta\boldsymbol{K}_2 = \boldsymbol{K}_{ss} - \boldsymbol{K}$ which is a function of unknown values a_is. $\boldsymbol{K}_{\text{ini}}$ is a matrix with regular pattern corresponding to lower back column. The physical interpretation of such a regular matrix would be a lumbar column with similar passive tissues (equivalent beam elements) and vertebrae properties that are perfectly aligned in a vertical line. $\boldsymbol{A}$ and $\boldsymbol{B}$ can be considered as the average values of $\boldsymbol{A}_i$s and $\boldsymbol{B}_i$s in matrix $\boldsymbol{K}$. The force vector δF includes the body weights and muscle forces (assumed to be unchanged (passive test and controlled by EMG activity) and are obtained from upright standing posture) on each degree of freedom as well as the applied moment to the pelvis $M_P = F_{rod} \times h_{T8-S1}$ where F_{rod} and h_{T8-S1} are obtained from load cell and anthropometric measures, respectively. All components of the displacement vector $\delta\boldsymbol{d}$ are unknown except the rotation of pelvis (measured using an accelerometer connected to the leg platform (Fig. 8.3)). Let us consider $\Delta\boldsymbol{K} = \Delta\boldsymbol{K}_1 + \Delta\boldsymbol{K}_2$ and rewrite the governing equation as

$$(\boldsymbol{K}_{ini} + \Delta\boldsymbol{K})\delta\boldsymbol{d} = \delta\boldsymbol{F}$$

The left side of the equation includes the unknown a_is in $\Delta\boldsymbol{K}$, and the unknown components of $\delta\boldsymbol{d}$. Using optimization procedures one can obtain a set ofa_i values for which the calculated displacement vector $\delta\boldsymbol{d}$ meets the experimental requirements, i.e. the calculated component of vector $\delta\boldsymbol{d}$ corresponding to pelvis rotation should be equal to the one obtained from passive flexion test. Furthermore, the predicted segmental rotations at other degrees of freedom should satisfy the segmental range of flexions, as percentages of lumbar flexion (LF), obtained from imaging studies:

$$\begin{cases} \text{Var } \boldsymbol{\alpha} = [\alpha_1\ \alpha_2\ \alpha_3\ \alpha_4\ \alpha_5\ \alpha_6] \\ \quad \text{Minimize } |\delta\theta_{S1} - \delta\theta_e| \\ \delta\theta_{T12} \text{ to } \delta\theta_{S1} \text{ are functions of } \boldsymbol{\alpha} \\ \qquad \text{Subject to} \\ -0.08 \times LF \le \delta\theta_{T12} - \delta\theta_{L1} \le 0 \\ -0.13 \times LF \le \delta\theta_{L1} - \delta\theta_{L2} \le 0 \\ -0.16 \times LF \le \delta\theta_{L2} - \delta\theta_{L3} \le 0 \\ -0.23 \times LF \le \delta\theta_{L3} - \delta\theta_{L4} \le 0 \\ -0.26 \times LF \le \delta\theta_{L4} - \delta\theta_{L5} \le 0 \\ -0.14 \times LF \le \delta\theta_{L5} - \delta\theta_{S1} \le 0 \end{cases}$$

where $\delta\theta$ s (i.e., $\delta\theta_{T12}$ to $\delta\theta_{S1}$) are rotational components of $\delta\boldsymbol{d}$, and $\delta\theta_e$ is pelvis rotation from experimental test. This optimization procedure should be performed for each increment. Considering a genetic algorithm (GA) comprising of 20 individuals and 100 generations, and dividing the 40° total rotation of pelvis into 3° increments (~13 increments), the equation $(\boldsymbol{K}_{ini} + \Delta\boldsymbol{K})\delta\boldsymbol{d} = \delta\boldsymbol{F}$ should be solved $20 \times 10 \times 13 = 2600$ times. Each increment also includes several iterations in order to achieve internal forces $\boldsymbol{F}_I$ equal to external forces $\boldsymbol{F}_E$. Given such a calculation demand, efficient solutions are required as were developed using regularity concept and graph product rules. For 5 iterations in each increment, the proposed method will solve the problem in ~4.21 e^{+006} time unit versus ~75.82 e^{+006} time unit using a conventional method. The algorithm for this example is summarized as follows:

(1) Constructing the stiffness matrix $\boldsymbol{K}$ using the beam element stiffness matrices $\boldsymbol{K}_b$ for the initial (zero) segmental rotation in Fig. 8.2.
(2) Forming the regular stiffness matrix $\boldsymbol{K}_{ini}$ with $\boldsymbol{A}$ and $\boldsymbol{B}$ equal to the average values of $\boldsymbol{A}_i$s and $\boldsymbol{B}_i$s in matrix $\boldsymbol{K}$.
(3) Defining the subject specific beam element stiffness matrix $(\boldsymbol{K}_b)_{ss}$ and consequently the subject specific stiffness matrix $\boldsymbol{K}_{ss}$ regarding $((EI)_{ss})_i = a_i(EI)_i$.
(4) Writing the governing equation in the form $(\boldsymbol{K}_{ini} + \Delta\boldsymbol{K})\delta\boldsymbol{d} = \delta\boldsymbol{F}$ and defining the load increment $\delta\boldsymbol{F}$ from the body weights and muscle forces as well as the moment obtained during passive flexion test (Fig. 8.3).
(5) Decomposing the matrix $\boldsymbol{K}_{ini}$ using graph product rules $\boldsymbol{K}_{ini} = \boldsymbol{V}^T\boldsymbol{\lambda}\boldsymbol{V}$.
(6) Initialization of $\boldsymbol{\alpha}$ and starting the optimization procedure.
(7) Solving the governing equation within the iterations (related to both the optimization procedure and non-linear solution) using Eqs. (8.20) to (8.22).
(8) Continuation of the procedure until the termination criterion in the optimization formulation is satisfied.
(9) Using the obtained α_i s and $\delta\theta$ s (i.e., $\delta\theta_{T12}$ to $\delta\theta_{S1}$) define the first part of the moment-rotation curve for each segment similar to those in Fig. 8.3.
(10) Updating the stiffness matrix $\boldsymbol{K}$ using the updated beam element stiffness matrix $\boldsymbol{K}_b$ wherein $(EI)_i = a_i(EI)_i$.
(11) Returning to step 2 and continuing the procedure until the entire loading $\boldsymbol{F}$ is applied.

The final result of this procedure would be obtaining the subject specific passive behavior of lumbar motion segments using efficient analysis procedures.

8.6 Vibration Analysis of a Global Near-Regular System

Developing a simple algorithm for the vibration analysis of global near-regular mechanical systems from available vibration solution of their corresponding regular mechanical systems is aimed [12].

Consider the given eigenproblem for a global near-regular system

$$\boldsymbol{K}\boldsymbol{x} = \lambda\boldsymbol{M}\boldsymbol{x} \tag{8.37}$$

where $\boldsymbol{K}$ is the stiffness matrix, $\boldsymbol{x}$ and λ are the eigenpairs, and $\boldsymbol{M}$ is the mass matrix. Assume $\Delta\boldsymbol{K}$ and $\Delta\boldsymbol{M}$ are the deviations of the stiffness and mass matrices of the near-regular system from the stiffness ($\boldsymbol{K}_{reg}$) and mass ($\boldsymbol{M}_{reg}$) matrices of the regular system.

$$\boldsymbol{K} = \boldsymbol{K}_{reg} + \Delta\boldsymbol{K}; \quad \boldsymbol{M} = \boldsymbol{M}_{reg} + \Delta\boldsymbol{M} \tag{8.38}$$

and

$$\boldsymbol{K}_{reg}\boldsymbol{x}_{reg} = \lambda_{reg}\boldsymbol{M}_{reg}\boldsymbol{x}_{reg} \tag{8.39}$$

where $\boldsymbol{x}_{reg}$ and λ_{reg} are eigenpairs of the regular system.

Substituting Eq. (8.38) in (8.37)

$$\left(\boldsymbol{K}_{reg} + \Delta\boldsymbol{K}\right)\boldsymbol{x} = \lambda\boldsymbol{M}\boldsymbol{x} \tag{8.40}$$

Assume the eigenvector $\boldsymbol{x}$ can be approximated using a linear combination of m linearly independent basis vectors $\boldsymbol{y}_1, \boldsymbol{y}_2, \ldots, \boldsymbol{y}_m$.

$$\boldsymbol{x}_{(n\times 1)} = a_1\boldsymbol{y}_1 + a_2\boldsymbol{y}_2 + \cdots + a_m\boldsymbol{y}_m \text{ or } \boldsymbol{x}_{(n\times 1)} = \boldsymbol{Y}_{(n\times m)}\boldsymbol{A}_{(m\times 1)}$$

and

$$\boldsymbol{Y} = \left[\boldsymbol{y}_1, \boldsymbol{y}_2, \ldots, \boldsymbol{y}_m\right]_{(n\times m)}; \quad \boldsymbol{A}^T = [a_1, a_2, \ldots, a_m]_{(1\times m)} \tag{8.41}$$

where $\boldsymbol{A}$ is a vector of unknown coefficients. We set m to be much smaller than n (i.e., the number of degrees of freedom) and substitute Eq. (8.41) in Eq. (8.37):

$$\boldsymbol{K}\boldsymbol{x} = \lambda\boldsymbol{M}\boldsymbol{x} \rightarrow \boldsymbol{K}\boldsymbol{Y}\boldsymbol{A} = \lambda\boldsymbol{M}\boldsymbol{Y}\boldsymbol{A} \rightarrow \boldsymbol{Y}^T\boldsymbol{K}\boldsymbol{Y}\boldsymbol{A} = \lambda\boldsymbol{Y}^T\boldsymbol{M}\boldsymbol{Y}\boldsymbol{A}$$

$$\boldsymbol{K}_r = \boldsymbol{Y}^T\boldsymbol{K}\boldsymbol{Y} \text{ and } \boldsymbol{M}_r = \boldsymbol{Y}^T\boldsymbol{M}\boldsymbol{Y} \rightarrow \boldsymbol{K}_r\boldsymbol{A} = \lambda\boldsymbol{M}_r\boldsymbol{A} \tag{8.42}$$

where $\boldsymbol{K}_r$ and $\boldsymbol{M}_r$ are reduced stiffness and mass matrices of dimension m. Therefore, rather than solving a system of dimension n in Eq. (8.37), the smaller system of dimension m should be solved. Upon finding the vector $\boldsymbol{A}$ in Eq. (8.42), $\boldsymbol{x}$ is calculated from Eq. (8.41). However, the basis vectors are not determined yet.

Consider Eq. (8.40) and pre-multiply it by $\boldsymbol{K}_{reg}^{-1}$

$$\left(\boldsymbol{K}_{reg} + \Delta\boldsymbol{K}\right)\boldsymbol{x} = \lambda\boldsymbol{M}\boldsymbol{x} \xrightarrow{\boldsymbol{K}_{reg}^{-1}\times} \left(\boldsymbol{I} + \underbrace{\boldsymbol{K}_{reg}^{-1}\Delta\boldsymbol{K}}_{\boldsymbol{G}}\right)\boldsymbol{x} = \underbrace{\boldsymbol{K}_{reg}^{-1}\lambda\boldsymbol{M}\boldsymbol{x}}_{\boldsymbol{y}_1} \rightarrow (\boldsymbol{I} + \boldsymbol{G})\boldsymbol{x} = \boldsymbol{y}_1 \tag{8.43}$$

Therefore,

$$\boldsymbol{x} = (\boldsymbol{I} + \boldsymbol{G})^{-1}\boldsymbol{y}_1$$

But, we can write

$$(\boldsymbol{I} - \boldsymbol{G})\left(\boldsymbol{I} + \boldsymbol{G} + \boldsymbol{G}^2 + \cdots + \boldsymbol{G}^r\right) = \boldsymbol{I} - \boldsymbol{G}^{r+1}$$

$$\left(+\boldsymbol{G} + \boldsymbol{G}^2 + \cdots + \boldsymbol{G}^r\right) = (\boldsymbol{I} - \boldsymbol{G})^{-1} - \underbrace{(\boldsymbol{I} - \boldsymbol{G})^{-1}\boldsymbol{G}^{r+1}}_{0} \text{ since } \boldsymbol{G}^r \to 0$$

$$\boldsymbol{G} \to -\boldsymbol{G} : (\boldsymbol{I} + \boldsymbol{G})^{-1} = \left(\boldsymbol{I} - \boldsymbol{G} + \boldsymbol{G}^2 - \cdots\right)$$

Thus,

$$\boldsymbol{x} = (\boldsymbol{I} + \boldsymbol{G})^{-1}\boldsymbol{y}_1 = \left(\boldsymbol{I} - \boldsymbol{G} + \boldsymbol{G}^2 - \cdots\right)\boldsymbol{y}_1 \tag{8.44}$$

Now assuming $\lambda = \lambda_{reg}$ and $\boldsymbol{x} = \boldsymbol{x}_{reg}$

$$\boldsymbol{y}_1 = \boldsymbol{K}_{reg}^{-1}\lambda\boldsymbol{M}\boldsymbol{x} = \boldsymbol{K}_{reg}^{-1}\underbrace{\lambda_{reg}\boldsymbol{M}\boldsymbol{x}_{reg}}_{\boldsymbol{R}_{reg}} \to \boldsymbol{y}_1 = \boldsymbol{K}_{reg}^{-1}\boldsymbol{R}_{reg} \tag{8.45}$$

Finally,

$$\begin{cases} \boldsymbol{y}_i = -\boldsymbol{G}\boldsymbol{y}_{i-1}(i = 2 : m) \\ \boldsymbol{y}_1 = \boldsymbol{K}_{reg}^{-1}\boldsymbol{R}_{reg} \end{cases} \to \boldsymbol{x} = \sum_{i=1}^{m} \boldsymbol{y}_i;\ \lambda_j = \frac{\boldsymbol{x}_j^T\boldsymbol{K}\boldsymbol{x}_j}{\boldsymbol{x}_j^T\boldsymbol{M}\boldsymbol{x}_j} \tag{8.46}$$

For a regular mechanical system with given eigenpair matrices, Δ_{reg} and $\boldsymbol{X}_{reg}$, decomposition of stiffness matrix would be $\boldsymbol{K}_{reg} = \boldsymbol{X}_{reg}^T\Delta_{reg}\boldsymbol{X}_{reg}$ and $\boldsymbol{y}_1$ is solved as follows:

$$\boldsymbol{y}_1 = \left(\boldsymbol{X}_{reg}^T\Delta_{reg}\boldsymbol{X}_{reg}\right)^{-1}\boldsymbol{R}_{reg} = \boldsymbol{X}_{reg}^T\Delta_{reg}^{-1}\boldsymbol{X}_{reg}\lambda_{reg}\boldsymbol{M}\boldsymbol{x}_{reg} \tag{8.47}$$

With $\boldsymbol{x}$ and λ calculated from Eq. (8.46), we can update $\boldsymbol{y}_1$ (i.e., $\boldsymbol{R}_{reg}$) and recalculate $\boldsymbol{x}$ and λ to achieve the desired accuracy. Also, same procedures should be conducted to calculate other eigenvalues and eigenvectors of the mechanical system.

Regardless of the required effort for defining the eigenpair matrices Δ_{reg} and $\boldsymbol{X}_{reg}$, the solution of Eq. (8.46) includes matrix by vector multiplications of the computational complexity $O(n^2)$. However, the method is efficient only if there is a simple way to obtain Δ_{reg} and $\boldsymbol{X}_{reg}$. In Sect. 8.3, efficient methods to obtain these functions were introduced.

8.6.1 Example

Stiffness and mass matrices of a structure are given as

$$M = \begin{bmatrix} 1 & & & & \\ & 1 & & & \\ & & 1 & & \\ & & & 1 & \\ & & & & 1 \end{bmatrix} \quad K = \begin{bmatrix} 2.8 & -1 & & & \\ -1 & 2 & -1 & & \\ & -1 & 2 & -1 & \\ & & -1 & 2 & -1 \\ & & & -1 & 2.5 \end{bmatrix}$$

We can rewrite $\boldsymbol{K}$ as

$$\boldsymbol{K} = \boldsymbol{K}_{reg} + \Delta\boldsymbol{K}$$

where

$$\boldsymbol{K}_{reg} = \begin{bmatrix} 2 & -1 & & & \\ -1 & 2 & -1 & & \\ & -1 & 2 & -1 & \\ & & -1 & 2 & -1 \\ & & & -1 & 2 \end{bmatrix}; \quad \Delta\boldsymbol{K} = \begin{bmatrix} 0.8 & & & & \\ & 0 & & & \\ & & 0 & & \\ & & & 0 & \\ & & & & 0.5 \end{bmatrix}$$

The first eigenpair of $\boldsymbol{M}^{-1}\boldsymbol{K}_{\text{reg}}$ is

$$\lambda_{reg} = 0.2679 \text{ and } \boldsymbol{x}_{reg}^T = [1.0000,\ 1.7321,\ 2.0000,\ 1.7321,\ 1.0000]$$

	$m = 4$	$m = 6$	$m = 8$	$m = 10$	Exact solution
λ	0.3529	0.3510	0.3505	0.3503	0.3503
x	1.0000	1.0000	1.0000	1.0000	1.0000
	2.1431	2.2792	2.3589	2.4040	2.4497
	2.5811	2.7839	2.9032	2.9721	3.0413
	2.1712	2.3447	2.4475	2.5073	2.5675
	1.0477	1.1102	1.1487	1.1713	1.1944

References

1. Shojaei, I., Kaveh, A., Rahami, H., Shirazi, R., Bazrgari, B.: Analysis and reanalysis of mechanical systems: global near-regularity concept. Acta Mech. **228**(4), 1445–1456 (2017)
2. Kirsch, U.: Design-oriented analysis of structures-unified approach. J. Eng. Mech. **129**, 264–272 (2003)
3. Wilkinson, J.H., Wilkinson, J.H.: The algebraic eigenvalue problem. vol. 87, Clarendon Press Oxford (1965)
4. Kaveh, A., Shojaei, I., Rahami, H.: New developments in the optimal analysis of regular and near-regular structures: decomposition, graph products, force method. Acta Mech. **226**, 665–681 (2015)
5. Yueh, W.C.: Eigenvalues of several tridiagonal matrices. Appl. Math. E-Notes. **5**, 210–230 (2005)
6. Kaveh, A., Rahami, H.: An efficient analysis of repetitive structures generated by graph products. Int. J. Numer. Meth. Eng. **84**, 108–126 (2010)
7. Kirsch, U.: Combined approximations–a general reanalysis approach for structural optimization. Struct. Multidiscip. Optim. **20**, 97–106 (2000)
8. Shojaei, I., Arjmand, N., Bazrgari, B.: An optimization based method for prediction of lumbar spine segmental kinematics from the measurements of thorax and pelvic kinematics. Int. J. Numer. Meth. Biomed. Eng. P. 31 (2015)
9. Arjmand, N., Shirazi-Adl, A.: Sensitivity of kinematics-based model predictions to optimization criteria in static lifting tasks. Med. Eng. Phys. **28**, 504–514 (2006)
10. Bazrgari, B., Shirazi-Adl, A., Arjmand, N.: Analysis of squat and stoop dynamic liftings: muscle forces and internal spinal loads. Eur. Spine J. **16**, 687–699 (2007)
11. Shojaei, I., Allen-Bryant, K., Bazrgari, B.: Viscoelastic response of the human lower back to passive flexion: the effects of age. Ann. Biomed. Eng. pp. 1–10 (2016)
12. Shojaei, I., Rahami, H.: Vibration analysis of global near-regular mechanical systems. J. Algo. Comput. **49**(2), 113–118 (2017)

Chapter 9
Mappings for Transformation of Near-Regular Domains to Regular Domains

9.1 Introduction

In this chapter a numerical method is presented for efficient solution of many differential equations with arbitrary domains using geometrical transformation and Kronecker product rules [1]. Initially, a mesh free formulation for rectangular domains is developed and a full decomposition of matrix equations is achieved using Kronecker product rules. The solution of a governing equation on an arbitrary domain is sought through a geometrical transformation from the rectangular domain into the original domain using conformal mapping. Although such a transformation may change the governing equation into a more complicated differential equation, it can be shown that conformal mapping preserves the Laplace and Poisson's equations, which are broadly used in engineering problems. The numerical implication of the conformal mapping is the existence of a unique domain partitioning in the original domain that leads to matrix equations similar to those in rectangular domain. Such a unique domain partitioning, inspired by conformal mapping, reduces the computational complexity of the problem to that in a rectangular domain. The efficiency of the proposed method is examined using various engineering examples.

9.2 Preservation of Laplace and Poisson's Equations

Consider the following two-dimensional Laplace equation:

$$\nabla^2 \psi = \psi_{xx} + \psi_{yy} = 0 \tag{9.1}$$

And let ψ be C^2 in Ω.

A. Kaveh et al., *Swift Analysis of Civil Engineering Structures Using Graph Theory Methods*, Studies in Systems, Decision and Control 290,
https://doi.org/10.1007/978-3-030-45549-1_9

The geometrical shape of the domain Ω defines the degree of difficulty of solving the problem such that a standard method like separation of variables fails to solve the problem if the domain Ω is not a rectangle or circle. Using a change of variables from x, y to u, v

$$u = u(x, y), \quad v = v(x, y) \tag{9.2}$$

We try to obtain a new region, Ω' (u-v plane), which is hopefully less complicated than the original domain Ω. If so, the problem might be solved in Ω' and then the solution can be returned back to domain Ω. To have the inverse functions:

$$x = x(u, v), \quad y = y(u, v) \tag{9.3}$$

in Ω', u and v require to be C^1 in Ω and non-zero Jacobian in Ω to guarantee the existence of the desire inverse function;

$$\frac{\partial(u, v)}{\partial(x, y)} = \begin{vmatrix} u_x & u_y \\ v_x & v_y \end{vmatrix} = u_x v_y - u_y v_x \neq 0 \tag{9.4}$$

Denoting $\psi(x(u, v), y(u, v)) \equiv \Psi(u, v)$, restricting $u_x = v_y$ and $u_y = -v_x$, using the chain differentiation rules, and substituting Eq. (9.3) in Eq. (9.1) result in

$$(u_x^2 + u_y^2)(\Psi_{uu} + \Psi_{vv}) = 0 \tag{9.5}$$

Considering Eq. (9.4) and the restrictions $u_x = v_y$ and $u_y = -v_x$, we will have $(u_x^2 + u_y^2) \neq 0$ and

$$\nabla^2 \Psi = \Psi_{uu} + \Psi_{vv} = 0 \tag{9.6}$$

in Ω'.

The restrictions $u_x = v_y$ and $u_y = -v_x$ are the Cauchy-Riemann conditions which suggest it is helpful (though not necessary) to regard the x, y and u, v planes as complex z and w planes as follows:

$$z = x + iy, \quad w = f(z) = u(x, y) + iv(x, y) \tag{9.7}$$

Then

$$u_x v_y - u_y v_x = u_x^2 + v_x^2 = \left|f'(z)\right|^2 \tag{9.8}$$

This leads to the preservation of the Laplace equation theorem.

Theorem *Suppose $w = f(z) = u(x, y) + iv(x, y)$ is analytic in domain Ω such that $f'(z) \neq 0$. Consider Ω' as the image of Ω using a one-by-one mapping. If*

$\psi(x, y)$ is harmonic (i.e. satisfying Laplace equation) in Ω, then $\Psi(u, v)$ will also be harmonic in Ω'.

Although preservation of the differential equation is not a necessary condition, it makes the calculation simpler. Accordingly, for the Poisson's equation in Eq. (9.1) with $g(x, y)$ in the right side, Eq. (9.5) can be written as

$$(u_x^2 + u_y^2)(\Psi_{uu} + \Psi_{vv}) = \left|f'(z)\right|^2(\Psi_{uu} + \Psi_{vv}) = g(x, y) \tag{9.9}$$

Expressing $g(x, y)$ and $\left|f'(z)\right|^2$ in the u-v plane results in $g(x(u, v), y(u, v)) \equiv G(u, v)$ and $\left|f'(z)\right|^2 = F(u, v)$, therefore

$$(\Psi_{uu} + \Psi_{vv}) = \frac{G(u, v)}{F(u, v)} = R(u, v) \tag{9.10}$$

In this way the Poisson's equation in x-y plane (Eq. (9.1) with $g(x, y)$ in the right side) is converted to another Poisson's equation (Eq. 9.10).

The computational complexity of solving the problem in the new plane (i.e. u-v) depends on: (1) the form of the newly obtained differential equation, (2) the complexity of the domain's geometry (Ω') in u-v plane. The conversion from x-y plane to u-v plane does not change the Laplace equation. Furthermore, though the right side of Poisson's equation changes, the new PDE is still Poisson's equation. Accordingly, the computational complexity of solving the problem in the new plane will only depends on the complexity of the domain's geometry (Ω').

The main objective of this paper is to reduce the computational complexity of numerical solution of Laplace and Poisson's equations through geometrical transformations from an arbitrary domain to a rectangular domain. The efficient solution of the problem in the new domain is developed using Kronecker product rules within a mesh free formulation. Then, the solution of the main problem in the original domain is obtained using the solution in the rectangular domain. The matrix implication of the mapping and the necessity of such implication for numerical formulation is developed and discussed. In the next section efficient mesh free method using Kronecker product rules in the rectangular domain is developed.

9.3 Efficient Mesh Free Formulation of Laplace and Poisson's Equations on Rectangular Domain Using Kronecker Product Rules

In this section the first mesh method is investigated and then using this method the formulation is provided for Laplace and poisson equations on rectangular domain considering the rules of the Kronecker product.

9.3.1 Mesh Free Formulation and Computational Complexity

Consider the following differential equation with the given boundary condition on the domain Ω

$$\begin{cases} Du = f \text{ in } \Omega \\ Eu = g \text{ in } \Gamma_g \\ u = u_p \text{ in } \Gamma_u \end{cases} \tag{9.11}$$

where D and E are differential operators, and Γ_g and Γ_u are Neumann and Dirichlet boundaries, respectively.

The most common procedure for numerical solution of Eq. (9.11) is achieved through weighted residual method [2], wherein the function u_h approximates the solution of the differential equation

$$\int_{\Omega} w_1(Du_h - f)d\Omega + \int_{\Gamma_g} w_2(Eu_h - g)d\Gamma + \int_{\Gamma_u} w_3(u_h - u_p)d\Gamma = 0 \tag{9.12}$$

where w_1, w_2 and w_3 are weighting functions.

To preserve the local features of the problem, the function u is approximated over a set of n nodes $\{x_i, i = 1 : n,\ x_i \in \bar{\Omega}\}$ in the interpolation domain $\bar{\Omega}$ (i.e. a cloud defined by star point i) as follows (Figs. 9.1 and 9.2):

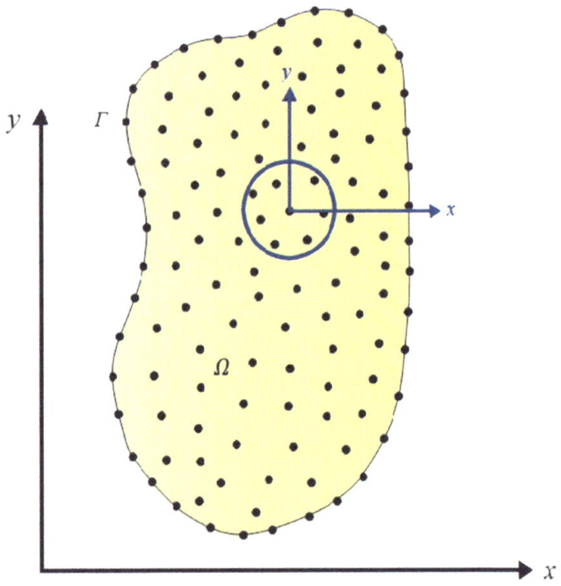

Fig. 9.1 The arbitrary domain Ω including a set of distributed points

Fig. 9.2 Local cloud i including the domain $\bar{\Omega}$

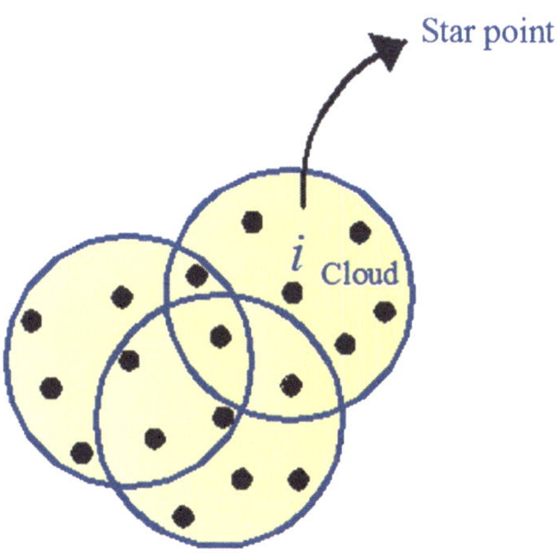

$$u(x) \cong u_h(x) = \sum_{i=1}^{k} p_i(x)a_i = p(x)^T a \quad x \in \bar{\Omega} \tag{9.13}$$

where the vector $p(x)^T$ includes a complete monomial basis of order k and $a = [a_1, a_2, a_3, \ldots, a_k]^T$.

To define the unknown vector $\boldsymbol{a}$ using the moving least square (MLS) approximation, the functional J is minimized

$$J = \sum_{j=1}^{n} \varphi(x - x_j)\left[u(x_j) - p^T(x_j)a\right]^2 \tag{9.14}$$

Minimizing the functional J with respect to the parameters a_i leads to

$$a = \left(P^T \varphi P\right)^{-1} P^T \varphi u = A^{-1} B u \tag{9.15}$$

where

$$P = \begin{bmatrix} p^T(x_1) \\ p^T(x_2) \\ \vdots \\ p^T(x_n) \end{bmatrix}, \quad \varphi = \begin{bmatrix} \varphi(x - x_1) & & & \\ & \varphi(x - x_2) & & \\ & & \ddots & \\ & & & \varphi(x - x_n) \end{bmatrix}, \quad u = \begin{bmatrix} u(x_1) \\ u(x_2) \\ \vdots \\ u(x_n) \end{bmatrix} \tag{9.16}$$

Substituting a from Eq. (9.15) into Eq. (9.13) results in

$$u_h(x) = N^T(x)u = \sum_{j=1}^{n} N_j(x)u_j \quad x \in \bar{\Omega} \tag{9.17}$$

where

$$N^T(x) = p^T(x)A^{-1}B$$

And

$$N_j(x) = \sum_{i=1}^{k} p_i(x)(A^{-1}B)_{ij} \tag{9.18}$$

The first order partial derivative of $u_h(\boldsymbol{x})$ with respect to x_l is obtained as

$$(u_h)_{,l}(x) = \sum_{j=1}^{n} N_{j,l}(x)u_j \quad x \in \bar{\Omega} \tag{9.19}$$

And

$$N_{j,l}(x) = \sum_{i=1}^{k} p_{i,l}(x)(A^{-1}B)_{ij} + p_i(x)(A^{-1}B_{,l} + \left(A_{,l}^{-1}B\right)_{ij} \tag{9.20}$$

where

$$A_{,l}^{-1} = (A^{-1})_{,l} = -A^{-1}A_{,l}A^{-1} \tag{9.21}$$

Repeating the same procedure, higher order derivatives are obtained.

To solve the problem using finite point method (i.e. point collocation), we set $w_1 = w_2 = w_3 = \delta$ (δ is Dirac delta function) and substitute $u_h(\boldsymbol{x})$ and its partial derivative in Eq. (9.12). This leads to a set of equations wherein the components of the coefficient or stiffness matrix M are found as follows:

$$M_{ij} = \left[D(N_j)\right]_i + \left[E(N_j)\right]_i \tag{9.22}$$

And the matrix equation is obtained

$$Mu = b \tag{9.23}$$

where u is the vector of unknown displacements and b is the vector of known forces at the points. The computational complexity of solving the matrix equation using LU decomposition is $O(N^{2.373})$ in the Optimized CW-like algorithms [3]. However, for the case that domain Ω is a rectangle, an efficient closed-form formulation can be obtained for finding the unknowns in Eq. (9.23) using Kronecker product rules.

9.3.2 Kronecker Products and Closed-Form Mesh Free Formulation on Rectangular Domains

Suppose matrix M of the dimension $l = n \times m$ has the following block form:

$$M = \begin{bmatrix} A_m & B_m & & & & \\ B_m & A_m & B_m & & & \\ & B_m & A_m & B_m & & \\ & & B_m & \ddots & \ddots & \\ & & & \ddots & A_m & B_m \\ & & & & B_m & A_m \end{bmatrix}_n = F_n(A_m, B_m, A_m) = I_n \otimes A_m + T_n \otimes B_m \tag{9.24}$$

where I is an identity matrix, $T = F_n(0, 1, 0)$ and $\otimes$ is the Kronecker product sign.

Using the following lemma, matrix M can be decomposed:

Sufficient conditions for converting Hermitian matrices A_1 and A_2 into upper triangular ones using one orthogonal matrix are [4, 5]:

$$A_1 A_2 = A_2 A_1 \quad \text{or } A_1^2 = A_2^2 \tag{9.25}$$

Let the matrix M be the sum of two Kronecker products as:

$$M = A_1 \otimes B_1 + A_2 \otimes B_2 \tag{9.26}$$

If P is the matrix that upper triangularize matrices A_1 and A_2, then $U = P \otimes I$ block upper triangularizes the matrix M which means $U^t M U$ is block upper diagonalized. Thus considering Eq. (9.26), one can write:

$$\lambda_M = \bigcup_{i=1}^{n} eig(M_i); \; M_i = \lambda_i(A_1)B_1 + \lambda_i(A_2)B_2 \tag{9.27}$$

where matrices A_1 and A_2 are of dimension l and matrices B_1 and B_2 are of dimension m.

The matrix in Eq. (9.24) holds for the condition in Eq. (9.25), that is $IT = TI$. In the mesh free analysis of Laplace equation on a rectangular domain with Dirichlet boundary condition, generating the matrix pattern in Eq. (9.24) is possible. To do so, the clouds in the domain should include the same number of points with the same symmetric distribution. These properties leads to generation of blocks $\boldsymbol{A}_m$ and $\boldsymbol{B}_m$. Furthermore, to generate the repetitive pattern of these two blocks (Eq. 9.24), a block by block numbering should be performed. Consider the rectangular domain Ω in Fig. 9.3. For the quadratic vector $\boldsymbol{p}(\boldsymbol{x}) = \left[1, x, y, x^2, xy, y^2\right]^T$, the number of points in a cloud (i.e. $n_p = 9$) should be greater than the order of vector $\boldsymbol{p}(\boldsymbol{x})$

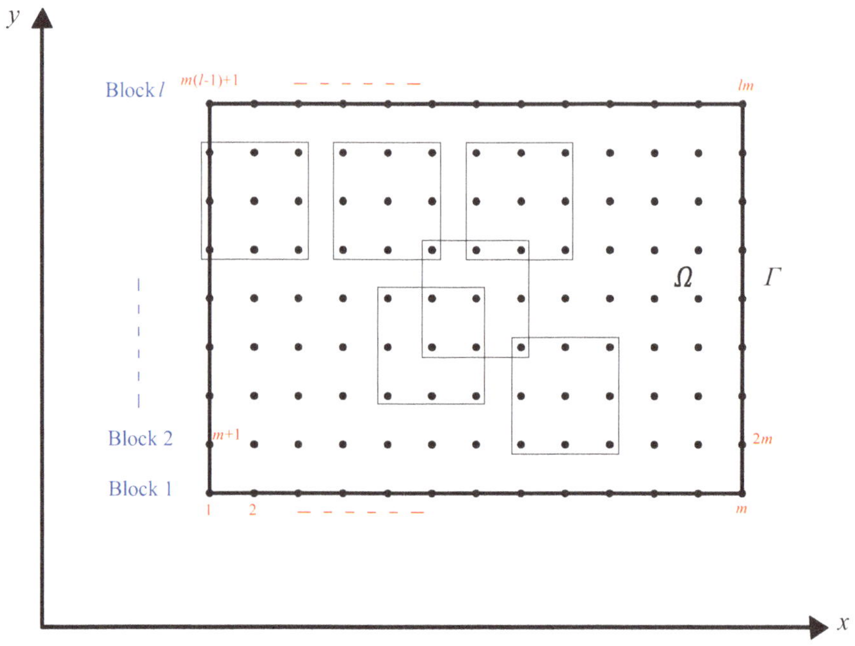

Fig. 9.3 The rectangular domain including horizontal and vertical blocks of points with similar 9-points clouds

(i.e. $k = 6$). The domain is localized using square clouds (Fig. 9.4), all points in the domain are numbered in a block by block order (Fig. 9.3), and a quartic spline weight function is employed.

Fig. 9.4 A square cloud including 9 points with the central point i as the star point

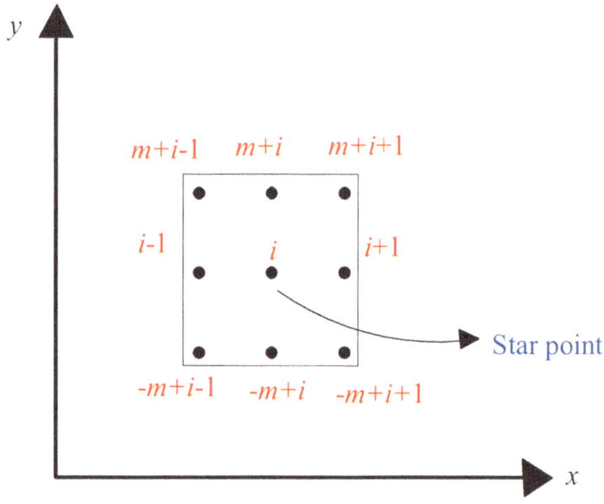

Accordingly, by applying the local displacement field of the cloud to the Laplace equation, the global coefficient matrix M with the following pattern is constructed

$$\begin{aligned} M &= A_1 \otimes B_1 + A_2 \otimes B_2 \\ M &= I_n \otimes A_m + T_n \otimes B_m \\ M &= I_n \otimes F_m(-53.33, 10.66, -53.33) + F_n(0, 1, 0) \otimes F_m(10.66, 2.66, 10.66) \end{aligned} \tag{9.28}$$

If λ and u are the eigenvalue and eigenvector of $A_2 = F_n(0, 1, 0)$, μ and v are eigenvalue and eigenvector of $B_1 = F_m(-53.33, 10.66, -53.33)$, and γ and v (eigenvectors of B_1 and B_2 are the same [6]) are eigenvalue and eigenvector of $B_2 = F_m((10.66, 2.66, 10.66)$ respectively, then it can be shown that $u \otimes v$ is the eigenvector of M:

Since $(A \otimes B)(C \otimes D) = AC \otimes BD$, one can write:

$$(A_1 \otimes B_1 + A_2 \otimes B_2)(u \otimes v) = (A_1 u) \otimes (B_1 v) + (A_2 u) \otimes (B_2 v) \tag{9.29}$$

From Eq. (9.28) we have $A_1 = I$

$$A_1 u = u; \ A_2 u = \lambda u; \ B_1 v = \mu v; \ B_2 v = \gamma v \tag{9.30}$$

And

$$(A_1 \otimes B_1 + A_2 \otimes B_2)(u \otimes v) = u \otimes \mu v + \lambda u \otimes \gamma v = (\mu + \lambda\gamma)(u \otimes v) \tag{9.31}$$

Thus, $\mu + \lambda\gamma$ and $u \otimes v$ are the eigenvalue and eigenvector of matrix M, respectively.

To solve the matrix equation $Mu = b$, obtained from mesh free solution (Eq. (9.23)) using the above defined eigenpairs, we can write:

$$Mu = b \Rightarrow \{\varphi\}_j^T M \{\varphi\}_j y_j = \lambda_j y_j = \{\varphi\}_j^T b$$

$$y_j = \frac{f_j}{\lambda_j} \Rightarrow \{u\}_l = \sum_{i=1}^{l} \{\varphi\}_i y_i = \sum_{i=1}^{l} \{\varphi\}_i \frac{f_i}{\lambda_i} = \sum_{i=1}^{l} \frac{\{\varphi\}_i \{\varphi\}_i^T}{\lambda_i} b \tag{9.32}$$

Since the eigenpairs of matrices of the form $F_n(a, b, c)$ are available, one can still simplify Eq. (9.32). Consider the Laplace equation on the rectangular domain discretized as follows Fig. 9.5:

Eigenpairs of the matrix $F_n(a, b, a)$ can be obtained using the following formulas [6]:

$$\lambda_i(n) = a + 2\sqrt{b^2} \cos\left(\frac{i\pi}{n+1}\right); \quad \varphi_i(n) = \sin\left(\frac{i(1{:}n)\pi}{n+1}\right), \ i = 1{:}n \tag{9.33}$$

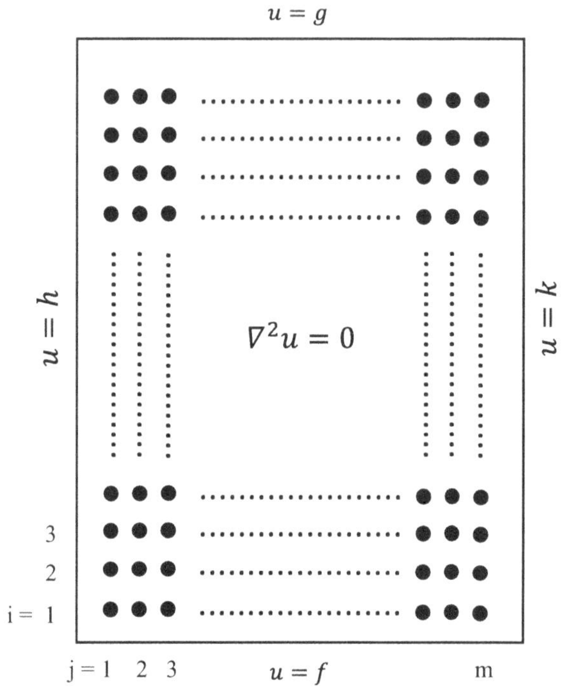

Fig. 9.5 Laplace equation on the discretized rectangular domain with given boundary conditions

Specifically, for $F_n(0, 1, 0)$, $F_m(-53.33, 10.66, -53.33)$, and $F_m(10.66, 2.66, 10.66)$, we will have

$$\lambda_i(n) = 2\cos\left(\frac{i\pi}{n+1}\right);\quad \varphi_i(n) = \sin\left(\frac{i(1{:}n)\pi}{n+1}\right),\ i = 1{:}n$$

And

$$\mu_j(m) = -53.33 + 21.32\cos\left(\frac{j\pi}{m+1}\right);\quad \varphi_j(m) = \sin\left(\frac{j(1{:}m)\pi}{m+1}\right),\ j = 1{:}m \tag{9.34}$$

And

$$\gamma_j(m) = 10.66 + 5.32\cos\left(\frac{j\pi}{m+1}\right);\quad \varphi_j(m) = \sin\left(\frac{j(1{:}m)\pi}{m+1}\right),\ j = 1{:}m$$

Suppose Λ and ϕ are the eigenvalues and eigenvectors of matrix M. Considering Eqs. (9.31) and (9.32), we will have

$$u = \sum_{l=1}^{mn} \frac{\phi_l \phi_l^t}{\Lambda_l} b;\quad \Lambda_l = \mu_j(m) + \lambda_i(n)\gamma_j(m);\ \phi_l = \varphi_i(n) \otimes \varphi_j(m) \tag{9.35}$$

Considering Eq. (9.34), we have:

$$\operatorname{norm}(\phi_l) = \frac{1}{2}\sqrt{(m+1)(n+1)} \tag{9.36}$$

Considering the relationship $(A \otimes B)(C \otimes D) = AC \otimes BD$ and $(A \otimes B)^t = A^t \otimes B^t$, one can write:

$$\phi_l \phi_l^t = [\varphi_i(n) \otimes \varphi_j(m)][\varphi_i(n) \otimes \varphi_j(m)]^t = [\varphi_i(n)\varphi_i^t(n)] \otimes [\varphi_j(m)\varphi_j^t(m)] \tag{9.37}$$

Using a block by block numbering from bottom to the top (Fig. 9.5), the vector b is obtained as follows:

$$f = \begin{bmatrix} f_1 \\ f_2 \\ \vdots \\ f_m \end{bmatrix};\; g = \begin{bmatrix} g_1 \\ g_2 \\ \vdots \\ g_m \end{bmatrix};\; b_i = \begin{bmatrix} h_i \\ 0 \\ \vdots \\ k_i \end{bmatrix};\; b = -\begin{bmatrix} b_1 + f \\ b_2 \\ \vdots \\ b_{n-1} \\ b_n + g \end{bmatrix} \tag{9.38}$$

Finally, by normalizing Eq. (9.37) and substituting in Eq. (9.35) we will have

$$u = \frac{4}{(m+1)(n+1)} \sum_{i=1}^{n}\sum_{j=1}^{m} \frac{[\varphi_i(n)\varphi_i^t(n)] \otimes [\varphi_j(m)\varphi_j^t(m)]}{\mu_j(m) + \lambda_i(n)\gamma_j(m)} b$$

$$u = \frac{4}{(m+1)(n+1)} \sum_{i=1}^{n}\sum_{j=1}^{m} \frac{\left[\sin\left(\frac{i(1:n)\pi}{n+1}\right)\left(\sin\left(\frac{i(1:n)\pi}{n+1}\right)\right)^t\right] \otimes \left[\sin\left(\frac{j(1:m)\pi}{m+1}\right)\left(\sin\left(\frac{j(1:m)\pi}{m+1}\right)\right)^t\right]}{-53.33 + 21.32\cos\left(\frac{j\pi}{m+1}\right) + 2\cos\left(\frac{i\pi}{n+1}\right)\left(10.66 + 5.32\cos\left(\frac{j\pi}{m+1}\right)\right)} b \tag{9.39}$$

Therefore, a closed-form mesh free solution is obtained for the Laplace equation on the rectangular domain. Similar formulation can be obtained using a finite difference method (FDM) wherein $F_n(0, 1, 0)$, $F_m(-4, 1, -4)$, and $F_m(1, 0, 1) = I$. Accordingly, updating Eqs. (9.34)–(9.39) for FDM results in:

$$u = \frac{4}{(m+1)(n+1)} \sum_{i=1}^{n} \sum_{j=1}^{m} \frac{\left[\sin\left(\frac{i(1:n)\pi}{n+1}\right)\left(\sin\left(\frac{i(1:n)\pi}{n+1}\right)\right)^t\right] \otimes \left[\sin\left(\frac{j(1:m)\pi}{m+1}\right)\left(\sin\left(\frac{j(1:m)\pi}{m+1}\right)\right)^t\right]}{-4 + 2\cos\left(\frac{j\pi}{m+1}\right) + 2\cos\left(\frac{i\pi}{n+1}\right)} b \tag{9.40}$$

While for both methods efficient closed-form formulas (Eqs. 9.39 and 9.40) with computational complexity ($O(N^2)$) of the same order are obtained, the convergence of the mesh free formulation is much faster due to the direct minimization of the functional and applying appropriate weight functions. In fact, FD formulation can be derived from a mesh free solution with specific assumptions and weight functions (resulting in $F_m(-4, 1, -4)$ instead of $F_m(-53.33, 10.66, -53.33)$ and $F_m(1, 0, 1) = I$ instead of $F_m(10.66, 2.66, 10.66)$); however, this specific solution will not be as fast as the golden standard solution. Furthermore, the proposed mesh free formulation is flexible and can be used with other standard weight and basis functions (for instance, exponential basis functions). Due to the use of a regular distribution of points in the domain, forming identical clouds, and a block by block numbering of the points, a coefficient matrix with desired decomposable pattern is achieved and solved using Kronecker product rules. The obtained efficient closed-form formulation is the advantage of the method over a conventional mesh free formulation. In short, the proposed method is as simple as a FDM (FDM formulated by Kronecker product rules) but as accurate as a mesh free solution.

9.4 Efficient Numerical Solution Using Conformal Mapping

In the previous sections it was indicated that the Laplace and Poisson's equations can be preserved from x-y plane to u-v plane and vice versa. Furthermore, the efficient mesh free solution for the rectangular domain was developed. Accordingly, instead of solving the more difficult problem in u-v domain above, we can solve the pre-developed rectangular domain. Consider the ellipses and rectangle domains in Fig. 9.6. Suppose solving the Laplace equation in the ellipses domain using the efficient solution in the rectangular domain is aimed. One should partition the ellipses in a way to achieve the desired points (vertical and horizontal lines) in the rectangle. Such unique partitioning is obtained through inverse mapping of the vertical and horizontal lines into the ellipses domain that is achieved using $w(z) = \sin(z)$. Therefore, to solve an arbitrary domain with an available rectangular domain transformation, first the rectangular domain is solved with a desirable mesh size as discussed in the previous section. Then, the obtained results; for example displacement field, is attributed to the corresponding nodes in the original domain obtained from mapping

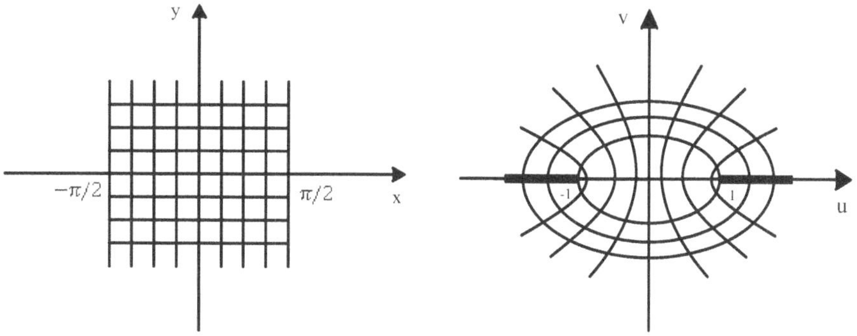

Fig. 9.6 Mapping of the Cartesian coordinate lines by $w(z) = \sin(z)$

of the nodes of the rectangular domain. Such mapping of nodes from the rectangular domain into the ellipses domain is shown in Fig. 9.6.

An important results from geometrical transformation and Kronecker product rules would be the following lemma:

For any arbitrary domain convertible to a rectangular domain there is a coordinate system in which the stiffness matrix of Laplace and Poisson's equations in a numerical method (MFM) can be written in the decomposable form $F_l(A_m, B_m, A_m)$.

Proof Suppose the matrix equation of the Laplace/Poisson's equation in an arbitrary domain is written as $Au = f$. Since Laplace/Poisson's equation is preserved (similar displacement (u) and force (f) vectors), the matrix equation in the rectangular domain will be as $A'u = F_l(A_m, B_m, A_m)u = f$ resulting in $Au = A'u$. If the relationship $Au = A'u$ holds for all u values, we will have $A = A'$. Since we can have numerous force vectors f, we can generate numerous displacement vectors u; this means $Au = A'u$ holds for all u values and therefore $A = A' = F_l(A_m, B_m, A_m)$.

There are a number of functions that transform a given domain directly into a rectangular one. However, in most cases one should employ combined transformations to achieve a rectangular domain. Specifically, Schwarz-Christoffel transformations can map any arbitrary polygon into a unit disc and then a transformation from the disc into the rectangle can be followed as shown in Fig. 9.7.

Schwarz-Christoffel transformations involve integral calculations that are not straightforward in many cases; however, these integrals are solved numerically in numerical methods [7, 8]:

$$w = A \int_0^z (S - z_1)^{-k_1} (S - z_2)^{-k_2} \cdots (S - z_n)^{-k_n} ds + B \tag{9.41}$$

The transformation maps the interior of the circle $|z| = 1$ onto the interior of a polygon, the vertices of the polygon are the images of the points z_j on the circle.

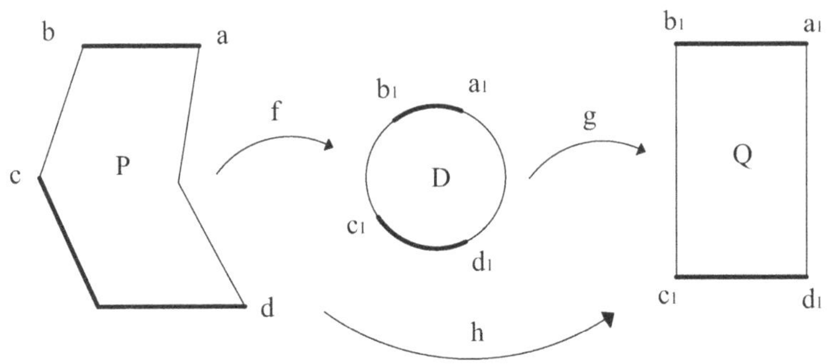

Fig. 9.7 Sequence of two simple mappings

Partitioning and numbering in our mesh free method is performed based on these transformations. Since defining a specific mesh in the polygon corresponding to the vertical and horizontal lines in the rectangle is difficult, an inverse approach is adopted. Accordingly, first rectangular domain is partitioned and numbered using our desirable approach in the mesh free method such that the stiffness matrix with the pattern $F_l(A_m, B_m, A_m)$ is constructed, then the nodes are mapped into the unit circle and polygon, respectively (Fig. 9.8). As shown in this figure, since the mapping is conformal, the angle (e.g. 90°) between lines in intersections is preserved in all three domains. Compared to the solution of the matrix equation $Mu = b$, numerical conformal mapping is an efficient procedure with the computational complexity of the order $n \log n$ [7], where n is the number of vertices of the polygon. Regarding $n \sim \sqrt{N}$ (N is degrees of freedom of the whole domain), we will have $\sqrt{N} \log \sqrt{N} \ll N$ that means the required effort for numerical conformal mapping is negligible against dominant computational complexity N^2 in the proposed method.

A summary of the method is presented via solving the following polygon. Consider the Laplace equation $\nabla^2 u(x, y) = 0$ in the pentagonal domain with the given boundary condition, Fig. 9.9.

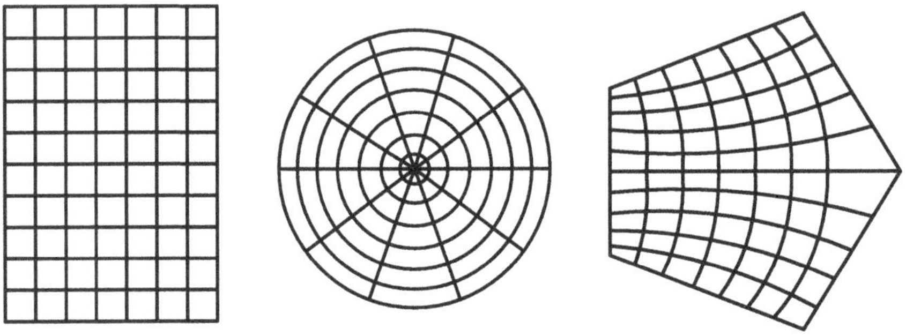

Fig. 9.8 Mapping of nodes into the unit circle and polygon in a conformal mapping

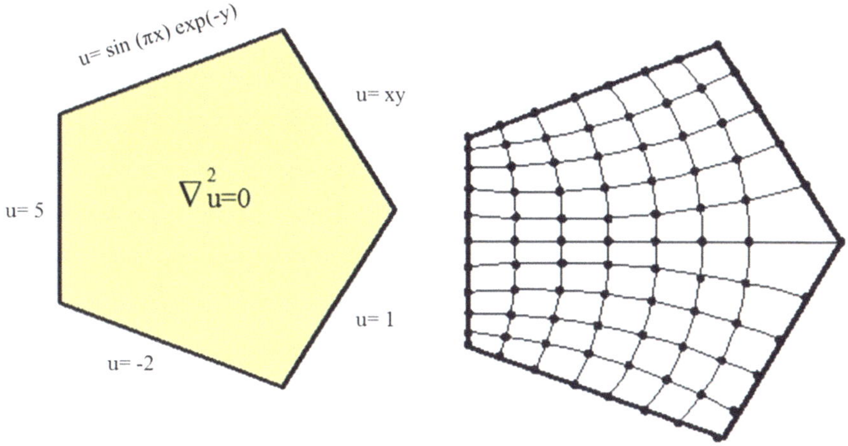

Fig. 9.9 Laplace equation in the pentagonal domain

First, suppose we need to solve the problem using a distribution of 88 nodes. A unit circle $|z| = 1$ including 11 sectors and 8 layers is considered. The rectangular image of the circle would also include 11 horizontal and 8 vertical lines. The only thing we need is the width and height of the rectangle that is obtained using the relationship $z = \ln w$ which is the mapping from circle to the rectangle. By substituting only three nodes of the circle into the equation above, three corresponding nodes in rectangle are obtained and dimensions of the rectangle are found. We do not even need to partition the circle and rectangle because we should only input the number of decided blocks (11×8) into the developed closed-form formula in Eq. (9.39). Accordingly, the displacement field in the rectangular domain is obtained. Finally, the displacement field is attributed to the corresponding points in the polygon (Fig. 9.10):

9.5 Examples

Example 1 Consider a uniform member subjected to two equal torques T with opposite directions at two of its ends with a cross section including eccentric circles, Fig. 9.11. Governing equations of the torsion problem can be expressed using Prandtl's stress function φ:

$$\nabla^2\varphi = -2G\alpha \quad \text{And } \varphi = 0 \text{ on } \Gamma$$

where G is the shear modulus of elasticity and α is the angle of twist per unit length.

Instead of solving the problem in the current form, the eccentric circles are mapped into a rectangle using two sequent transformations.

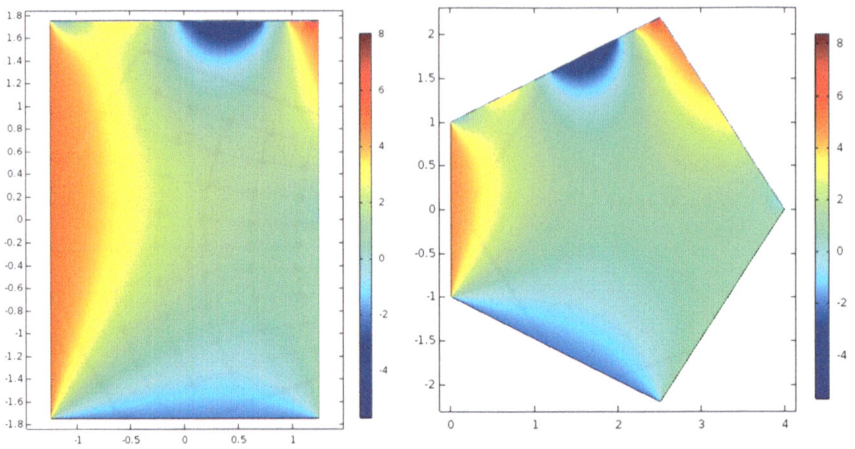

Fig. 9.10 Displacement field in the rectangular and pentagonal domain

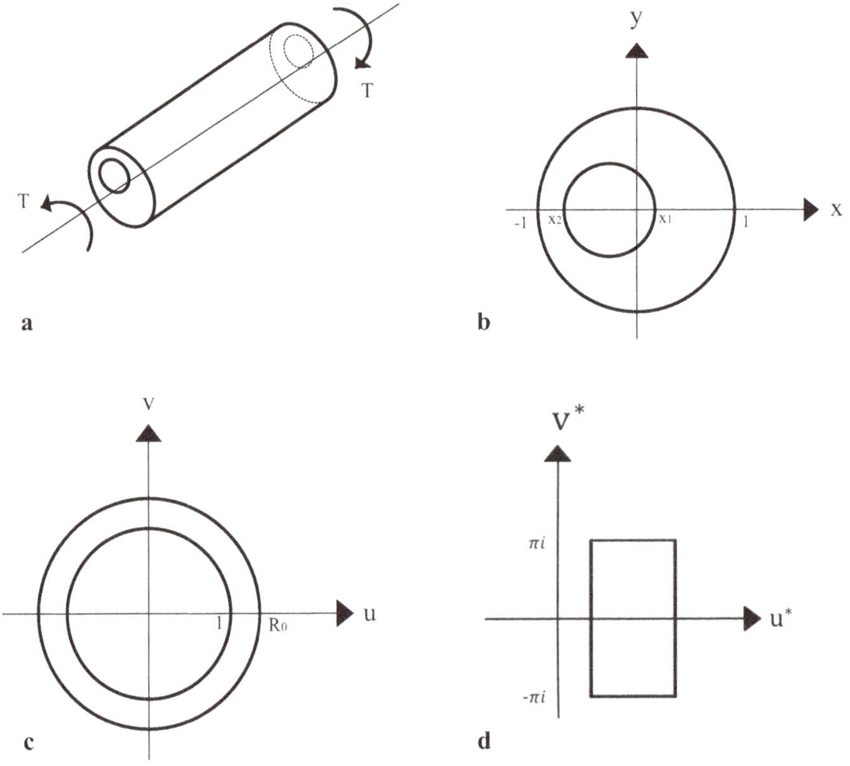

Fig. 9.11 The uniform member under equal torques (**a**) with the cross section including eccentric circles (**b**) the eccentric circles are mapped into concentric circles (**c**) and then a rectangle (**d**)

The following conformal mapping transforms eccentric circles into concentric circles

$$w = \frac{z - a}{az - 1}; \quad a = \frac{1 + x_1 x_2 + \sqrt{(1 - x_1^2)(1 - x_2^2)}}{x_1 + x_1}$$

$$R_0 = \frac{1 + x_1 x_2 + \sqrt{(1 - x_1^2)(1 - x_2^2)}}{x_1 - x_1} \; (a > 1 \text{ and } R_0 > 1 \text{ when } -1 < x_2 < x_1 < 1$$

Then, the concentric circles are mapped to a rectangle through $w = \exp(z)$. Combining the two transformation and utilizing the symbols w and w^* conformal mappings in (u, v) and (u^*, v^*) coordinate systems, respectively, result in:

$$w^* = g(w) = \exp(w) = \exp\left(\frac{z - a}{az - 1}\right)$$

Since we have the efficient solution of the Poisson's equation in the rectangle domain using Eq. (9.39), the inverse of the mapping above, $z = \frac{\ln(w^*) - a}{a \ln(w^*) - 1}$, defines the solution in the original domain (i.e. eccentric circles) of the problem. By considering $x_1 = 0.15$, $x_2 = -0.8$, and $\alpha = 0.26$ rad/m, the final result is obtained (Fig. 9.12).

Example 2 The head (u) of the water flowing into the soil with anisotropic material satisfies the following second order PDE:

$$k_x \frac{\partial^2 u}{\partial x^2} + k_y \frac{\partial^2 u}{\partial y^2} = 0$$

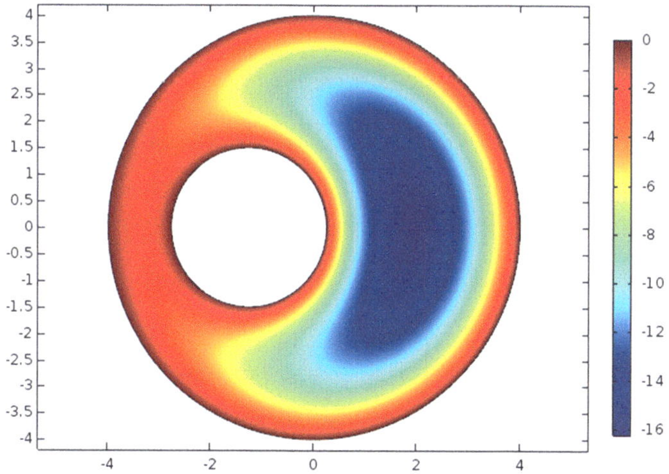

Fig. 9.12 Solution of a section including eccentric circles subjected to two equal torques

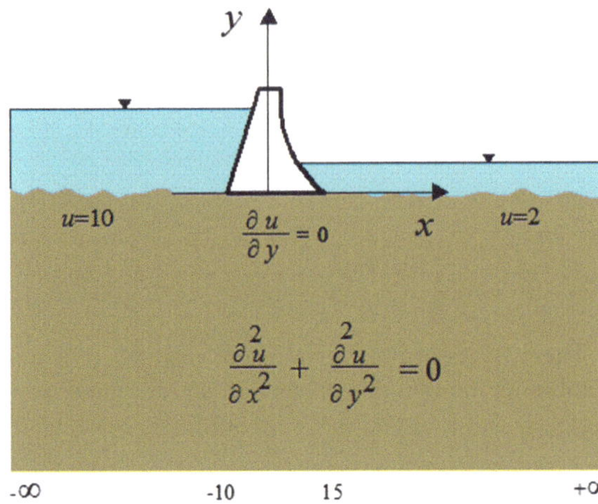

Fig. 9.13 The water flowing through the soil under the dam

where k_x and k_y are permeability coefficients in x and y directions, respectively.

Consider the soil with the given boundary conditions in Fig. 9.13 and $k_x = k_y = 1$. Suppose impermeable layers are far from the soil such that the soil can be considered as a semi-infinite plane. First, the mixed boundary condition on the upper layer is converted into the pure Dirichlet boundary condition.

$$-10 \le x \le 15 : u_{xx} + u_{yy} = 0 \overset{u_y=0 \to u_{yy}=0}{\to} u_{xx} = 0 \to u = Ax + B$$

$$\begin{cases} x = -10 \to u = 10 \\ x = 15 \to u = 2 \end{cases} \to \begin{cases} A = -8/25 \\ B = \frac{34}{5} \end{cases} \to u(x, 0) = \begin{cases} 10 & -\infty \le x \le -10 \\ \frac{34}{5} - \frac{8x}{25} & -10 \le x \le 15 \\ 2 & 15 \le x \le +\infty \end{cases}$$

The exact solution of the problem can be obtained through solving the following equation:

$$\Phi(\mathrm{x,y}) = \frac{1}{\pi} \int_{-\infty}^{\infty} \frac{y f(z)}{y^2 + (x - z)^2} dz = \frac{y}{\pi(x^2 + y^2)} * f(x)$$

where $*$ denotes convolution operation and $f(x)$ is the function for boundary conditions:

$$f(x) = \begin{cases} 10 & x < -10 \\ \frac{34}{5} - \frac{8x}{25} & -10 < x < 15 \\ 2 & x > 15 \end{cases}$$

Therefore,

$$\Phi(x, y) = \frac{10}{\pi}\tan^{-1}\left(\frac{y}{x+10}\right) + \int_{-10}^{15} \frac{y\left(\frac{34}{5} - \frac{8z}{25}\right)}{y^2 + (x-z)^2} dz - \frac{2}{\pi}\tan^{-1}\left(\frac{y}{x-15}\right) + 2$$

However, to solve the problem using a numerical method, the semi-infinite plane can be transformed into a unit circle and then into a rectangle using the conformal mappings $w = \frac{i+z}{i-z}$ and $w^* = \exp(w) = \exp\left(\frac{i+z}{i-z}\right)$, respectively. The solution of the differential equation on the semi-infinite domain in obtained using the inverse of the mapping w^*:

$$z = \frac{i(\ln(w^*) - 1)}{\ln(w^*) + 1}$$

The solution of the problem on the unit circle is shown in Fig. 9.14.

Example 3 Consider a fluid flow to the right in a semi-infinite region bounded by periodic walls. Using numerical conformal mapping between the half-plan ($u > 0$) and the semi-infinite region, the mesh grid in z plane is generated as shown in Fig. 9.15.

To find the potential flow in the original region, the region ABCDE in the w plane corresponding to the repetitive section A′B′C′D′E′ in the original domain is considered (Fig. 9.16). For a simple uniform (speed V_0) flow in the w plane, we will

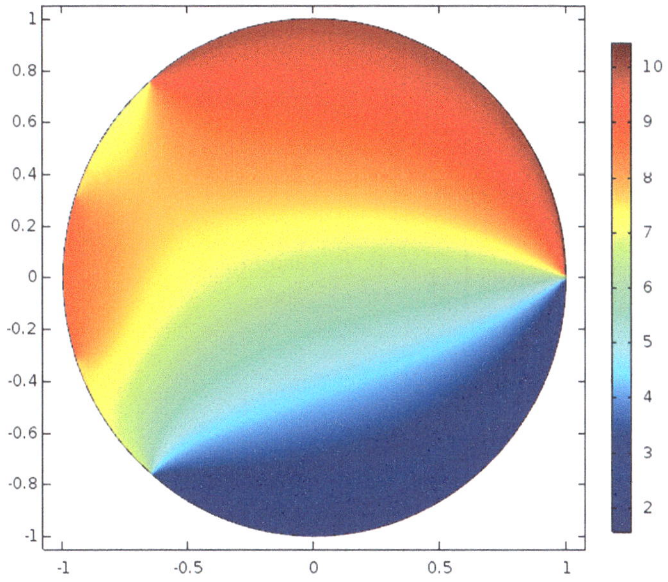

Fig. 9.14 Solution of a flow net for a seepage problem on the unit circle

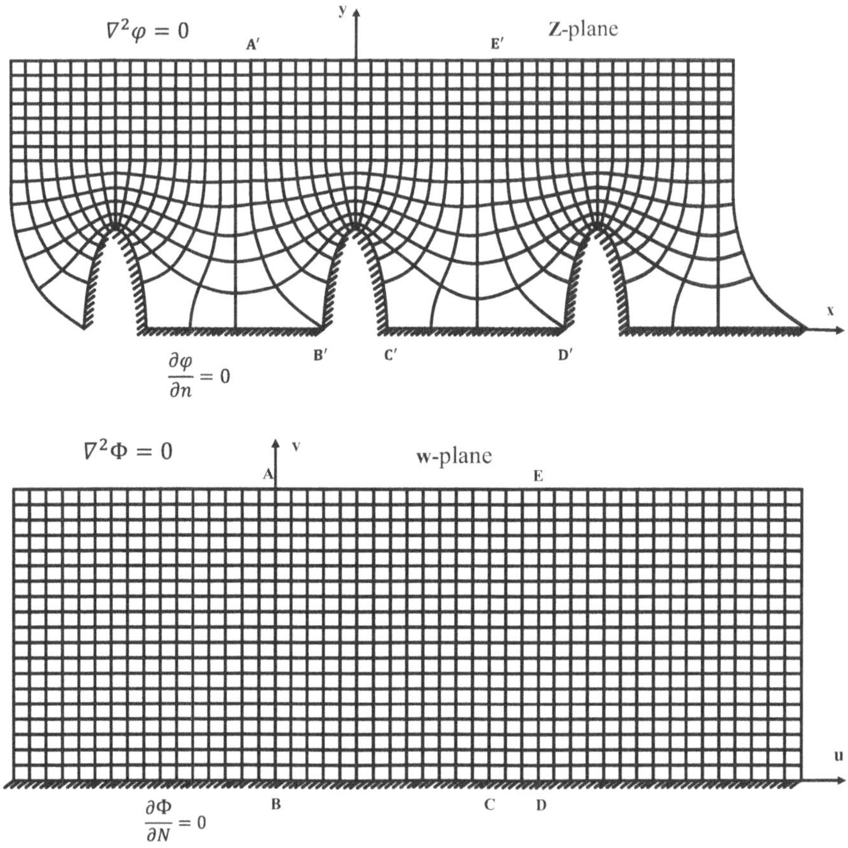

Fig. 9.15 Numerical conformal mapping between the half-plan (u > 0) and the semi-infinite region with periodic walls

have the following boundary conditions:

$$\begin{cases} \frac{\partial\Phi}{\partial N} = V_0 \ \Gamma_{AB} \text{ and } \Gamma_{DE} \\ \frac{\partial\Phi}{\partial N} = 0 \quad \Gamma_{BD} \text{ and } \Gamma_{AE} \end{cases}$$

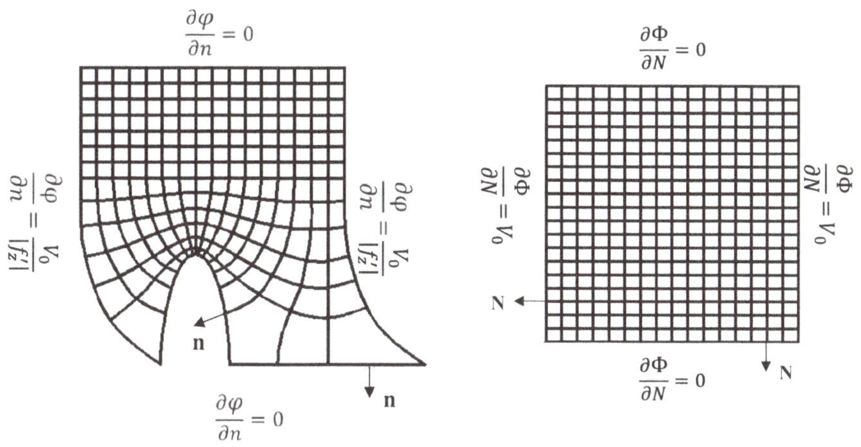

Fig. 9.16 A repetitive region in the original z domain and the corresponding rectangle in w plane

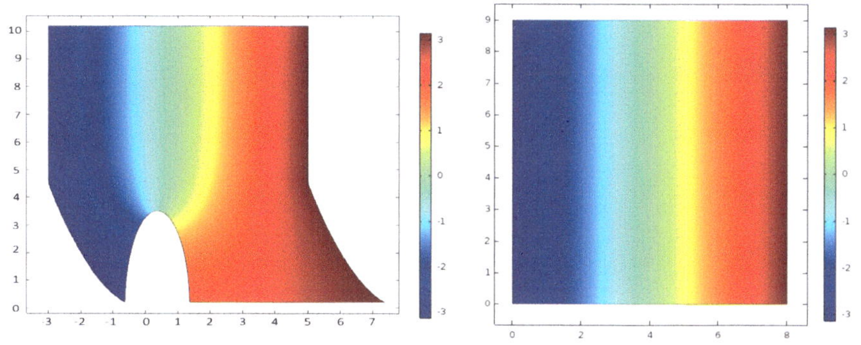

Fig. 9.17 The potential flow field in the rectangle and the corresponding original domain

The Neumann boundaries are converted to equivalent (the relationship between boundary and internal values) Dirichlet boundaries (For example for the points on the lower boundary, $\frac{\partial\Phi}{\partial N} = \frac{\Phi_1 - \Phi_0}{h} = 0 \rightarrow \Phi_0 = \Phi_1$) and using Eq. (9.39), the potential flow field in the rectangle and then the corresponding points in the original domain are obtained (Fig. 9.17).

References

1. Shojaei, I., Rahami, H., Kaveh, A.: A numerical solution for Laplace and Poisson's equations using geometrical transformation and graph products. Appl. Math. Modell. **40**(17–18), 7768–7783 (2016)

2. Zhang, X., Liu, X.H., Song, K.Z., Lu, M.W.: Least-squares collocation meshless method. Int. J. Numer. Methods Eng. **51**, 1089–1100 (2001)
3. Davie, A.M., Stothers, A.J.: Improved bound for complexity of matrix multiplication. Proc. R. Soc. Edinburgh: Section A: Math. **143**, 351–369 (2013)
4. Kaveh, A.: Optimal Analysis of Structures by Concepts of Symmetry and Regularity. Springer, GmbH, Wien-New York (2013). https://doi.org/10.1007/978-3-7091-1565-7
5. Kaveh, A., Shojaei, I., Rahami, H.: New developments in the optimal analysis of regular and near-regular structures: decomposition, graph products, force method. Acta Mech. **226**(3), 665–681 (2015)
6. Kaveh, A., Rahami, H.: Tri-diagonal and penta-diagonal block matrices for efficient eigen-solutions of problems in structural mechanics. Acta Mech. **192**, 77–87. (2007). http://www.springerlink.com/content/p62629m74120/?p=cc0dadc6e55a410daf169b171a7aca74&pi=0
7. Papamichael, N., Stylianopoulos, N.: Numerical Conformal Mapping: Domain Decomposition and the Mapping of Quadrilaterals. World Scientific, New Jersey (2010)
8. Driscoll, T.A., Trefethen, L.N.: Schwarz-Christoffel Mapping. Cambridge University Press (2002)

Chapter 10
Numerical Solution for System of Linear Equations Using Tridiagonal Matrix

10.1 Introduction

In this Chapter methods are developed for numerical solution of system of linear equations through taking advantages of the properties of repetitive tridiagonal matrices [1]. A system of linear equations is usually obtained in the final step of many science and engineering problems such as problems involving partial differential equations. In general, such systems can be solved utilizing direct or iterative methods. Direct methods, such as Gaussian elimination or LU decomposition, are not usually employed for problems with large and/or sparse matrices as these methods are computationally expensive and require storage of data and high speed computations. Therefore, several iterative methods [2–6] have been developed that can, in general, be classified as two groups of stationary and non-stationary approaches.

In the proposed solutions, the problem is first solved for repetitive tridiagonal matrices (i.e., system of linear equations) and a closed-from relationship is obtained. This relationship is then used for solving a general matrix through converting the matrix into a repetitive tridiagonal matrix and a remaining matrix that is moved to the right-hand side of the equation. The problem is solved iteratively by first utilizing an initial guess to define the vector on the right-hand side of the equation and then solving the problem using the closed-from relationship for repetitive tridiagonal matrices. This process is carried out iteratively until convergence is achieved. Computational complexity of the method is investigated and efficiency of the method is shown through several examples. As indicated in the examples, one of the advantages of the proposed method is its high rate of convergence in problems where the given matrix includes large off-diagonal entries. In such problems, a method like Jacobi will either have a low rate of convergence or be unable to converge.

A. Kaveh et al., *Swift Analysis of Civil Engineering Structures Using Graph Theory Methods*, Studies in Systems, Decision and Control 290,
https://doi.org/10.1007/978-3-030-45549-1_10

10.2 Jacobi Method for Solving System of Linear Equations $Au = C$

Consider matrix $\boldsymbol{A}$ below

$$\boldsymbol{A} = \begin{bmatrix} a_{11} & a_{12} & \cdots & a_{1n} \\ a_{21} & a_{22} & \cdots & a_{2n} \\ \vdots & \vdots & \ddots & \vdots \\ a_{n1} & a_{n2} & \cdots & a_{nn} \end{bmatrix} \tag{10.1}$$

Matrix $\boldsymbol{A}$ is decomposed into a diagonal matrix $\boldsymbol{D}$ and the remainder $\boldsymbol{R}$

$$\boldsymbol{A} = \begin{bmatrix} a_{11} & & & \\ & a_{22} & & \\ & & \ddots & \\ & & & a_{nn} \end{bmatrix} + \begin{bmatrix} 0 & a_{12} & \cdots & a_{1n} \\ a_{21} & 0 & \cdots & a_{2n} \\ \vdots & \vdots & \ddots & \vdots \\ a_{n1} & a_{n2} & \cdots & 0 \end{bmatrix} = \boldsymbol{D} + \boldsymbol{R} \tag{10.2}$$

We will have

$$(\boldsymbol{D} + \boldsymbol{R})\boldsymbol{u} = \mathbf{c} \tag{10.3}$$

And

$$\boldsymbol{D}\boldsymbol{u} = \mathbf{c} - \boldsymbol{R}\boldsymbol{u} \tag{10.4}$$

Setting an initial approximation as $\boldsymbol{u}^0 = \boldsymbol{D}^{-1}\mathbf{c}$ and obtaining a solution

$$\boldsymbol{u}^1 = \boldsymbol{D}^{-1}\left(\mathbf{c} - \boldsymbol{R}\boldsymbol{u}^0\right) \tag{10.5}$$

Repeating the procedure leads to an iterative algorithm from which the solution is obtained

$$\boldsymbol{u}^{n+1} = \boldsymbol{D}^{-1}\left(\mathbf{c} - \boldsymbol{R}\boldsymbol{u}^n\right) \quad \text{and } n = 0, 1, 2, \ldots \tag{10.6}$$

The condition for convergence of Jacobi method (and any other iterative method) is when the spectral radius of the iteration matrix ($\boldsymbol{D}^{-1}\boldsymbol{R}$) is less than 1. One sufficient condition for convergence of the method is that matrix $\boldsymbol{A}$ is diagonally dominant.

10.2.1 Closed-Form Solution of $Mx = f$ When M is a Tridiagonal Matrix

Eigenvalues and eigenvectors of a tridiagonal matrix, of dimension $N - 1$, of the form

$$M = \begin{bmatrix} b & c & & \cdots 0 \\ a & b & c & \vdots \\ & a & \ddots & c \\ \vdots & & a & b & c \\ 0 \cdots & & & a & b \end{bmatrix} \tag{10.7}$$

are calculated [7] using

$$\lambda_n = b + 2\sqrt{ac}\cos\frac{n\pi}{N} \quad and\ n = 1, 2, \ldots, N-1$$

$$v_j^n = \left(\sqrt{\frac{a}{c}}\right)^{j-1} \sin\frac{nj\pi}{N} \quad and\ j = 1, 2, \ldots, N-1$$

$$v^n = \left[v_1^n, v_2^n, \ldots, v_{N-1}^n\right]^t \tag{10.8}$$

And if $a = c$, we will have

$$\lambda_n = b + 2a\cos\frac{n\pi}{N} \quad and\ n = 1, 2, \ldots, N-1$$

$$v_j^n = \sin\frac{nj\pi}{N} \quad and\ j = 1, 2, \ldots, N-1$$

$$v^n = \left[v_1^n, v_2^n, \ldots, v_{N-1}^n\right]^t \tag{10.9}$$

Since the matrix is symmetric (of dimension $N - 1$), its eigenvectors $(\boldsymbol{v}^1, \boldsymbol{v}^2, \ldots, \boldsymbol{v}^{N-1})$ provide an orthogonal basis for $N - 1$ space. Therefore, we can expand $\boldsymbol{x}$ and $\boldsymbol{f}$ in terms of $(\boldsymbol{v}^1, \boldsymbol{v}^2, \ldots, \boldsymbol{v}^{N-1})$ basis:

$$\boldsymbol{x} = \sum_{i=1}^{N-1} x_i \boldsymbol{v}^i \text{and } \boldsymbol{f} = \sum_{i=1}^{N-1} b_i \boldsymbol{v}^i \tag{10.10}$$

where f_i's are known (i.e., can readily be computed)

$$f_i = \left(\boldsymbol{v}^i\right)^t \boldsymbol{f} \tag{10.11}$$

and x_i's are our unknowns that should be computed through substitution in $\boldsymbol{Mx} = \boldsymbol{f}$

$$\boldsymbol{M}\sum_{i=1}^{N-1} x_i \boldsymbol{v}^i = \sum_{i=1}^{N-1} f_i \boldsymbol{v}^i \tag{10.12}$$

where we can write

$$\boldsymbol{M}\sum_{i=1}^{N-1} x_i \boldsymbol{v}^i = \sum_{i=1}^{N-1} x_i \boldsymbol{M}\boldsymbol{v}^i = \sum_{i=1}^{N-1} x_i \lambda_i \boldsymbol{v}^i \tag{10.13}$$

We can now re-express Eq. (10.12) as

$$\sum_{i=1}^{N-1} \lambda_i x_i \boldsymbol{v}^i = \sum_{i=1}^{N-1} f_i \boldsymbol{v}^i \tag{10.14}$$

Because $\boldsymbol{v}^i$'s are linearly independent (they form a basis), we will have

$$\lambda_i x_i = f_i \rightarrow x_i = \frac{f_i}{\lambda_i} \quad \text{and } i = 1, 2, \ldots, N-1 \tag{10.15}$$

Therefore,

$$\boldsymbol{x} = \sum_{i=1}^{N-1} \frac{f_i}{\lambda_i}\boldsymbol{v}^i = \sum_{i=1}^{N-1} \boldsymbol{v}^i \frac{f_i}{\lambda_i} = \sum_{i=1}^{N-1} \frac{\boldsymbol{v}^i\left(\boldsymbol{v}^i\right)^t}{\lambda_i}\boldsymbol{f} \tag{10.16}$$

Finally,

$$\boldsymbol{x} = \sum_{i=1}^{N-1} \frac{\left[\sin\frac{i\pi}{N}, \sin\frac{2i\pi}{N}, \ldots, \sin\frac{(N-1)i\pi}{N}\right]^t \left[\sin\frac{i\pi}{N}, \sin\frac{2i\pi}{N}, \ldots, \sin\frac{(N-1)i\pi}{N}\right]}{b + 2a\cos\frac{i\pi}{N}}\boldsymbol{f} \tag{10.17}$$

10.2.2 A Numerical Method for the Solution of Au = c When A is an Arbitrary Matrix

Here, we want to get advantages of the closed form solution for tridiagonal matrices to develop an efficient numerical method for a linear system of equations $\boldsymbol{Au} = \boldsymbol{c}$ when $\boldsymbol{A}$ is arbitrary.

Consider matrix $\boldsymbol{A}$ below

$$\boldsymbol{A} = \begin{bmatrix} a_{11} & a_{12} & \cdots & a_{1n} \\ a_{21} & a_{22} & \cdots & a_{2n} \\ \vdots & \vdots & \ddots & \vdots \\ a_{n1} & a_{n2} & \cdots & a_{nn} \end{bmatrix} \tag{10.18}$$

Let us decompose matrix $\boldsymbol{A}$ into a tridiagonal matrix, $\boldsymbol{A}_1$, and the remaining terms, $\boldsymbol{A}_2$. b and a entries in tridiagonal matrix $\boldsymbol{A}_1$ are chosen to be, respectively, the average of diagonal entries (i.e., $a_{11}, \ldots, a_{nn}$) and the average of upper and lower neighboring-diagonals entries (i.e., $a_{12}, \ldots, a_{n-1,n}, \ldots, a_{21}, \ldots, a_{n,n-1}$).

$$\boldsymbol{A} = \begin{bmatrix} b & a & & \\ a & b & \ddots & \\ & \ddots & \ddots & a \\ & & a & b \end{bmatrix} + \begin{bmatrix} a_{11}-b & a_{12}-a & \cdots & a_{1n} \\ a_{21}-a & a_{22}-b & \cdots & a_{2n} \\ \vdots & \vdots & \ddots & \vdots \\ a_{n1} & a_{n2} & \cdots & a_{nn}-b \end{bmatrix} = \boldsymbol{A}_1 + \boldsymbol{A}_2 \tag{10.19}$$

We will have

$$(\boldsymbol{A}_1 + \boldsymbol{A}_2)\boldsymbol{u} = \mathbf{c} \tag{10.20}$$

And

$$\boldsymbol{A}_1\boldsymbol{u} = \mathbf{c} - \boldsymbol{A}_2\boldsymbol{u} \tag{10.21}$$

Replacing $\boldsymbol{A}_2$ by $\boldsymbol{A} - \boldsymbol{A}_1$

$$\boldsymbol{A}_1\boldsymbol{u} = \mathbf{c} - (\boldsymbol{A} - \boldsymbol{A}_1)\boldsymbol{u} \tag{10.22}$$

Let us initialize $\boldsymbol{u}$ using the following approximation

$$\boldsymbol{A}_1\boldsymbol{u}^0 = \boldsymbol{c} \tag{10.23}$$

Since $\boldsymbol{A}_1$ is tridiagonal, according to previous section we will have

$$\boldsymbol{u}^0 = \sum_{i=1}^{N-1} \frac{\left[\sin\frac{i\pi}{N}, \sin\frac{2i\pi}{N}, \ldots, \sin\frac{(N-1)i\pi}{N}\right]^t \left[\sin\frac{i\pi}{N}, \sin\frac{2i\pi}{N}, \ldots, \sin\frac{(N-1)i\pi}{N}\right]}{b + 2a\cos\frac{i\pi}{N}} \boldsymbol{c} \tag{10.24}$$

Now, we can improve the results through:

$$\boldsymbol{A}_1\boldsymbol{u}^1 = \boldsymbol{c} - (\boldsymbol{A} - \boldsymbol{A}_1)\boldsymbol{u}^0 \tag{10.25}$$

And in general we can use the following iterative equation to solve the problem

$$\boldsymbol{A}_1 \boldsymbol{u}^{n+1} = \boldsymbol{c} - (\boldsymbol{A} - \boldsymbol{A}_1)\boldsymbol{u}^n \tag{10.26}$$

Since $\boldsymbol{A}_1$ is tridiagonal, we can write the equation as follows

$$\boldsymbol{u}^{n+1} = \sum_{i=1}^{N-1} \frac{\left[\sin\frac{i\pi}{N}, \sin\frac{2i\pi}{N}, \ldots, \sin\frac{(N-1)i\pi}{N}\right]^t \left[\sin\frac{i\pi}{N}, \sin\frac{2i\pi}{N}, \ldots, \sin\frac{(N-1)i\pi}{N}\right]}{b + 2a\cos\frac{i\pi}{N}} \left[\boldsymbol{c} - (\boldsymbol{A} - \boldsymbol{A}_1)\boldsymbol{u}^n\right] \tag{10.27}$$

As such, an iterative formula for solving a general linear system of equations is obtained through properties of tridiagonal matrices. The condition for convergence of the method is when the spectral radius of the iteration matrix, $\boldsymbol{A}_1^{-1}(\boldsymbol{A} - \boldsymbol{A}_1)$, is less than 1.

We did not formally study the potential sufficient conditions for convergence of the proposed method. However, through several examples we have shown that in many cases that Jacobi method does not converge the proposed method converges.

10.2.3 Computational Complexity of the Method

For one iteration of the method, the dominant computational complexity is $O(n^2)$:

- $(\boldsymbol{A} - \boldsymbol{A}_1)\boldsymbol{u}^n = \boldsymbol{MV}$: Multiplication of a matrix ($\boldsymbol{M}$) and a vector ($\boldsymbol{V}$), of the complexity $O(n^2)$.

For sigma calculations we will have

- $\left[\sin\frac{i\pi}{N}, \sin\frac{2i\pi}{N}, \ldots, \sin\frac{(N-1)i\pi}{N}\right]\left[\boldsymbol{c} - (\boldsymbol{A} - \boldsymbol{A}_1)\boldsymbol{u}^n\right] = \boldsymbol{P}^t\boldsymbol{V}$: Multiplication of a row vector ($\boldsymbol{P}^t$) and a column vector ($\boldsymbol{V}$), of the complexity $O(n)$. The outcome here would be a scaler (S).
- $\left[\sin\frac{i\pi}{N}, \sin\frac{2i\pi}{N}, \ldots, \sin\frac{(N-1)i\pi}{N}\right]^t S = \boldsymbol{P}S$: Multiplication of a column vector ($\boldsymbol{P}$) and a scaler (S), of the complexity $O(n)$.
- n times calculation of the two steps above (because of sigma) leads to a computational complexity of $O(n^2)$.

Therefore, the dominant computational complexity of the method for one iteration is $O(n^2)$. If m iterations are required to achieve a desired accuracy, the computational complexity of the method would be $O(mn^2)$. However, since the required number of iterations to achieve convergence is much smaller than the dimension of matrix $\boldsymbol{A}$ (i.e., $m \ll n$), computational complexity of the method would be $O(n^2)$.

10.2.4 Examples

Example 1 Finite Difference Formulation of Diffusion Equation

A wide range of diffusion equations can be solved using analytical solutions (e.g., separation of variables, integral transformation, etc.). However, in more complicated cases numerical solution techniques are required to solve the problem. For instance, the problem

$$\begin{aligned}
&\alpha^2 u_{xx} = u_t, \ (0 < x < L, 0 < t < \infty)\\
&u(0, t) = p(t), \ (0 < t < \infty)\\
&u(L, t) = q(t), \ (0 < t < \infty)\\
&u(x, 0) = f(x), \ (0 < x < L)
\end{aligned}$$

is difficult to be solved using analytical solutions if $p(t)$ and $q(t)$ are not constant, but can readily be solved using a numerical solution like finite difference method.

Discretizing the problem using grid points and approximating $u_t(x, t)$ and $u_{xx}(x, t)$ as

$$u_t(x, t) \approx \frac{u(x, t + \Delta t) - u(x, t)}{\Delta t}$$

$$u_{xx}(x, t) \approx \frac{\frac{u(x+\Delta x, t) - u(x, t)}{\Delta x} - \frac{u(x, t) - u(x - \Delta x, t)}{\Delta x}}{\Delta x}$$

result in finite-difference approximation as follows

$$U_{j,k+1} = rU_{j-1,k} + (1 - 2r)U_{j,k} + rU_{j+1,k}$$

where

$$r = \alpha^2 \frac{\Delta t}{(\Delta x)^2}$$

Now using the finite-difference approximation we can calculate U (time $k + 1$) at a grid point as a linear combination of U's at the preceding time (k). As the equations are decoupled, the method is computationally efficient. However, the finite difference method is convergent and stable only if Δt and Δx satisfy the equation

$$r = \alpha^2 \frac{\Delta t}{(\Delta x)^2} \leq \frac{1}{2}$$

Satisfying this condition may result in high computational cost if we choose a small Δx, for the sake of accuracy, that in turn requires a very small Δt (i.e., many time steps). To remove such a restriction, the finite difference formulation can be modified.

Implicit Finite Difference Method: Crank–Nicolson Formulation
In construction of the equation for $U_{j,k+1}$ we have approximated $u_{xx}(x,t)$ using

$$u_{xx}(x,t) \approx \frac{\frac{u(x+\Delta x,t)-u(x,t)}{\Delta x} - \frac{u(x,t)-u(x-\Delta x,t)}{\Delta x}}{\Delta x}$$

However, this is not the only possible way of approximation of $u_{xx}(x,t)$. For instance, if we use some weighted average over the time interval, we will have

$$u_{xx}(x,t) \approx (1-\theta)\frac{u(x+\Delta x,t)-2u(x,t)+u(x-\Delta x,t)}{(\Delta x)^2} + \theta\frac{u(x+\Delta x,t+\Delta t)-2u(x,t+\Delta t)+u(x-\Delta x,t+\Delta t)}{(\Delta x)^2}$$

where θ is a number in interval [0, 1]. The finite difference formulation will be

$$\alpha^2\left[(1-\theta)\frac{U_{j-1,k}-2U_{j,k}+U_{j+1,k}}{(\Delta x)^2} + \theta\frac{U_{j-1,k+1}-2U_{j,k+1}+U_{j+1,k+1}}{(\Delta x)^2}\right] = \frac{U_{j,k+1}-U_{j,k}}{\Delta t}$$

which will reduce to the original equation for $\theta = 1$.

It can be shown that if $\theta \geq \frac{1}{2}$, the equation is both convergent and stable for all values of $r > 0$ (i.e., the condition in $r = \alpha^2\frac{\Delta t}{(\Delta x)^2} \leq \frac{1}{2}$ is discarded). If we set $\theta = \frac{1}{2}$, we will have

$$-rU_{j-1,k+1} + 2(1+r)U_{j,k+1} - rU_{j+1,k+1} = rU_{j-1,k} + 2(1-r)U_{j,k} + rU_{j+1,k}$$

$$j = 1, 2, \ldots, N-1 \quad \text{and } k = 0, 1, 2, \ldots$$

that is the Crank–Nicolson scheme. Expressing the equation in matrix form and moving boundary and U values from step k (i.e., $U_{j,k}$) to the right hand side of the matrix equation lead to

$$\begin{bmatrix} 2(1+r) & -r & & \cdots & 0 \\ -r & 2(1+r) & -r & & \vdots \\ & -r & \ddots & -r & \\ \vdots & & -r & 2(1+r) & -r \\ 0 & \cdots & & -r & 2(1+r) \end{bmatrix} \begin{bmatrix} U_{1,k+1} \\ U_{2,k+1} \\ \vdots \\ U_{N-2,k+1} \\ U_{N-1,k+1} \end{bmatrix}$$

$$= \begin{bmatrix} rp_{k+1} + rp_k + 2(1-r)U_{1,k} + rU_{2,k} & \\ rU_{1,k} + 2(1-r)U_{2,k} + rU_{3,k} & \\ \vdots & rU_{1,k} \\ rU_{N-3,k} + 2(1-r)rU_{N-2,k} + rU_{N-1,k} & \\ rU_{N-2,k} + 2(1-r)U_{N-1,k} + rq_{k+1} + rq_k & \end{bmatrix}$$

or in a compact form

$$\mathbf{AU}_{\mathrm{k+1}} = \mathbf{c}$$

Now starting with $k = 0$, we can solve the matrix equation for unknowns in the first time step (i.e., $U_{1,1}, U_{2,1}, \ldots, U_{N-1,1}$). Then setting $k = 1$ and solving the matrix equation, the next line of unknowns is found (i.e., $U_{1,2}, U_{2,2}, \ldots, U_{N-1,2}$). The procedure is repeated until unknowns in all time steps are calculated. This equation now can efficiently be solved using Eq. (10.27).

Example 2 Consider the following linear system of equations taken from:

$$\boldsymbol{A} = \begin{bmatrix} 10 & -1 & 2 & 0 \\ -1 & 11 & -1 & 3 \\ 2 & -1 & 10 & -1 \\ 0 & 3 & -1 & 8 \end{bmatrix}, \boldsymbol{x} = \begin{bmatrix} x_1 \\ x_2 \\ x_3 \\ x_4 \end{bmatrix}, \boldsymbol{b} = \begin{bmatrix} 6 \\ 25 \\ -11 \\ 15 \end{bmatrix}$$

The system of equations was solved using matrix inversion (i.e., the exact solution), the proposed method, and Jacobi method (Table 10.1). Matrix A_l in the proposed method and matrix D in Jacobi method are as follows:

$$\boldsymbol{A}_l = \begin{bmatrix} 9.75 & -1 & 0 & 0 \\ -1 & 9.75 & -1 & 0 \\ 0 & -1 & 9.75 & -1 \\ 0 & 0 & -1 & 9.75 \end{bmatrix} \boldsymbol{D} = \begin{bmatrix} 10 & 0 & 0 & 0 \\ 0 & 11 & 0 & 0 \\ 0 & 0 & 10 & 0 \\ 0 & 0 & 0 & 8 \end{bmatrix}$$

It is observed that the convergence rate of the proposed method is higher than that of the Jacobi method.

Table 10.1 Comparison between the exact solution, the proposed method, and the Jacobi method for solving a matrix equation of 4 unknowns

$x = A^{-1}b$	5 iterations	5 iterations
Exact	Proposed	Jacobi
1.0000	1.0003	0.9981
2.0000	2.0008	2.0023
−1.0000	−0.9995	−1.0019
1.0000	1.0008	1.0035
$\lVert Exact - Iterative \rVert$	0.0012	0.0050

Example 3 Black–Scholes equation

Black–Scholes equation is a partial differential equation in mathematical finance that describes the price of the option over time:

$$\frac{\partial V}{\partial t}+\frac{1}{2}\sigma^2 s^2\frac{\partial^2 V}{\partial s^2}+rs\frac{\partial V}{\partial s}-rV=0$$

Or in a more general form we can write:

$$\frac{\partial V}{\partial t}+a(s,t)\frac{\partial^2 V}{\partial s^2}+b(s,t)\frac{\partial V}{\partial s}+c(s,t)V=0$$

Using Crank-Nicholson finite difference method we will have:

$$\begin{aligned}&\frac{V_{n,j+1}-V_{n,j}}{\Delta t}+\frac{a_{n,j+1}}{2}\left(\frac{V_{n+1,j+1}-2V_{n,j+1}+V_{n-1,j+1}}{\Delta s^2}\right)\\&+\frac{a_{n,j}}{2}\left(\frac{V_{n+1,j}-2V_{n,j}+V_{n-1,j}}{\Delta s^2}\right)+\frac{b_{n,j+1}}{2}\left(\frac{V_{n+1,j+1}-V_{n-1,j+1}}{2\Delta s}\right)\\&+\frac{b_{n,j}}{2}\left(\frac{V_{n+1,j}-V_{n-1,j}}{2\Delta s}\right)+\frac{1}{2}c_{n,j+1}V_{n,j+1}+\frac{1}{2}c_{n,j}V_{n,j}=0\end{aligned}$$

We can rewrite the equation as

$$\begin{aligned}&A_{n,j+1}V_{n-1,j+1}+\left(1+B_{n,j+1}\right)V_{n,j+1}+C_{n,j+1}V_{n+1,j+1}\\&=-A_{n,j}V_{n-1,j}+\left(1+B_{n,j}\right)V_{n,j}-C_{n,j}V_{n+1,j}\end{aligned}$$

where

$$A_{n,j}=\frac{1}{2}v_1a_{n,j}-\frac{1}{4}v_2b_{n,j}$$

$$B_{n,j}=-v_1a_{n,j}+\frac{1}{2}\Delta tc_{n,j}$$

$$C_{n,j}=\frac{1}{2}v_1a_{n,j}+\frac{1}{4}v_2b_{n,j}$$

$$v_1=\frac{\Delta t}{\Delta s^2}\text{ and }v_2=\frac{\Delta t}{\Delta s}$$

These equations hold for $n=1,2,\ldots,N-1$. Boundary conditions provide two additional equations.

In a matrix of this form and for boundary conditions of $V(0,t)=0$ and $V(s_{\max},t)=s_{\max}-Ee^{-r(T-t)}$, we will have

$$\boldsymbol{M}_{j+1}^{\boldsymbol{L}}\boldsymbol{V}_{j+1}+\boldsymbol{r}_{j+1}^{\boldsymbol{L}}=\boldsymbol{M}_{j}^{\boldsymbol{R}}\boldsymbol{V}_{j}+\boldsymbol{r}_{j}^{\boldsymbol{R}}$$

$$V_{0,j} = 0,\ V_{N,j} = N\Delta s - Ee^{-r(j)\Delta t},\ V_{0,j+1} = 0,\ V_{N,j+1} = N\Delta s - Ee^{-r(j+1)\Delta t}$$

where

$$\boldsymbol{M}^{\boldsymbol{L}}_{j+1}\boldsymbol{V}_{j+1} + \boldsymbol{r}^{\boldsymbol{L}}_{j+1}$$

$$= \begin{bmatrix} 1+B_{1,j+1} & C_{1,j+1} & & & \\ A_{2,j+1} & 1+B_{2,j+1} & C_{2,j+1} & & \\ & A_{3,j+1} & \ddots & \ddots & \\ & & \ddots & 1+B_{N-2,j+1} & C_{N-2,j+1} \\ & & & A_{N-1,j+1} & 1+B_{N-1,j+1} \end{bmatrix} \begin{bmatrix} V_{1,j+1} \\ V_{2,j+1} \\ \vdots \\ V_{N-2,j+1} \\ V_{N-1,j+1} \end{bmatrix}$$

$$+ \begin{bmatrix} A_{1,j+1}V_{0,j+1} \\ 0 \\ \vdots \\ 0 \\ C_{N-1,j+1}V_{N,j+1} \end{bmatrix}$$

$$\boldsymbol{M}^{\boldsymbol{R}}_{j}\boldsymbol{V}_{j} + \boldsymbol{r}^{\boldsymbol{R}}_{j}$$

$$= \begin{bmatrix} 1-B_{1,j} & -C_{1,j} & & & \\ -A_{2,j} & 1-B_{2,j} & -C_{2,j} & & \\ & -A_{3,j} & \ddots & \ddots & \\ & & \ddots & 1-B_{N-2,j} & -C_{N-2,j} \\ & & & -A_{N-1,j} & 1-B_{N-1,j} \end{bmatrix} \begin{bmatrix} V_{1,j} \\ V_{2,j} \\ \vdots \\ V_{N-2,j} \\ V_{N-1,j} \end{bmatrix}$$

$$+ \begin{bmatrix} A_{1,j+1}V_{0,j+1} \\ 0 \\ \vdots \\ 0 \\ C_{N-1,j+1}V_{N,j+1} \end{bmatrix}$$

Therefore

$$\boldsymbol{M}^{\boldsymbol{L}}_{j+1}\boldsymbol{V}_{j+1} = \boldsymbol{M}^{\boldsymbol{R}}_{j}\boldsymbol{V}_{j} + \boldsymbol{r}^{\boldsymbol{R}}_{j} - \boldsymbol{r}^{\boldsymbol{L}}_{j+1}$$

The matrix and vectors on the right hand side of the equation are known, thus we can write the equation as

$$\boldsymbol{M}^{\boldsymbol{L}}_{j+1}\boldsymbol{V}_{j+1} = \boldsymbol{b}$$

Now, using the known matrix M^L_{j+1} and vector b, we can find vector V_{j+1}. The obtained vector is then used on the right hand side of the equation in the next iteration to find vector V in subsequent step.

Consider the following parameters for the problem:

$\sigma = 0.25$	Volatility of the stock
$r = 0.2$	Interest rate
$S_{max} = 20$	Maximum stock price
$S_{min} = 0$	Minimum stock price
$T = 1$	Maturation of contract
$E = 10$	Exercise price of the underlying
$M = 1600$	Number of time points
$N = 160$	Number of stock price points
$\Delta t = \frac{T}{M} = 0.000625$	Time step
$\Delta s = \frac{S_{max} - S_{min}}{N} = 0.125$	Price step.

Starting with $j = 0$, we will have

$$\boldsymbol{M}_1^L \boldsymbol{V}_1 = \left(\begin{bmatrix} 1.0000 & 0.0000 & & & & \\ 0.0000 & 1.0000 & 0.0000 & & & \\ & 0.0000 & \ddots & \ddots & & \\ & & \ddots & 0.5186 & 0.2358 & \\ & & & 0.2487 & 0.5125 & 0.2389 \\ & & & & 0.2519 & 0.5063 \end{bmatrix}_{159\times 159} \boldsymbol{V}_1 \right)$$

$$= \boldsymbol{b} = \begin{bmatrix} 0 \\ 0 \\ \vdots \\ 9.75 \\ 7.39 \end{bmatrix}_{159\times 1}$$

The linear system of equations was solved using Jacobi method, the proposed method, and matrix inversion (i.e., the exact solution). Matrix $\boldsymbol{A}_1$ in the proposed method and matrix $\boldsymbol{D}$ in Jacobi method are as follows:

$$\mathbf{A}_1 = \begin{bmatrix} 0.8339 & 0.0828 & & & & \\ 0.0828 & 0.8339 & 0.0828 & & & \\ & 0.0828 & \ddots & \ddots & & \\ & & \ddots & 0.8339 & 0.0828 & \\ & & & 0.0828 & 0.8339 & 0.0828 \\ & & & & 0.0828 & 0.8339 \end{bmatrix}$$

Table 10.2 Comparison between the exact solution, the proposed method, and the Jacobi method for solving a matrix equation of 159 unknowns

$V_1 = \left(M_1^L\right)^{-1} b$	8 iterations	8 iterations	34 iterations
Exact	Proposed	Jacobi	Jacobi
0.0000	0.0000	0.0000	0.0000
0.0000	0.0000	0.0000	0.0000
⋮	⋮	⋮	⋮
0.3762	0.3762	0.3762	0.3762
0.5012	0.5012	0.5012	0.5012
.6262	0.6262	0.6262	0.6262
⋮	⋮	⋮	⋮
9.5474	9.4974	7.0627	9.5040
9.8749	9.9234	7.9059	9.8395
9.6928	9.6620	8.5132	9.6717
$\lVert Exact - Iterative \rVert$	0.091402	5.8903	0.093887

$$\mathbf{D} = \begin{bmatrix} 1.0000 & & & & & \\ & 1.0000 & & & & \\ & & \ddots & & & \\ & & & 0.5186 & & \\ & & & & 0.5125 & \\ & & & & & 0.5063 \end{bmatrix}$$

Vector $\boldsymbol{V}_1$ of dimension 159×1 was obtained as (Table 10.2).

It can be observed that the difference between exact and proposed solutions (the norm of difference between the two vectors) is smaller than 0.1 after 8 iterations (Table 10.2; Fig. 10.1). To achieve almost same accuracy using a Jacobi method, we needed 34 iterations (Table 10.2; Fig. 10.1).

Now with $\boldsymbol{V}_1$ in hand, we can continue with the solution to calculate $\boldsymbol{V}_2$ in the next time step. The procedure is reaped until all unknowns are found ($j = 0, 1, \ldots, M = 1600$). This means that to solve the problem using a Jacobi method, we will need ~ $34 \times 1600 = 54{,}400$ iterations, whereas to solve the problem with the same accuracy using the proposed method we will need ~$8 \times 1600 = 12{,}800$ iterations.

Example 4 Another advantage of the proposed method over Jacobi method is its power in solving problems with large off-diagonal entries, as opposed to Jacobi method that is either slow or unable to converge in such cases.

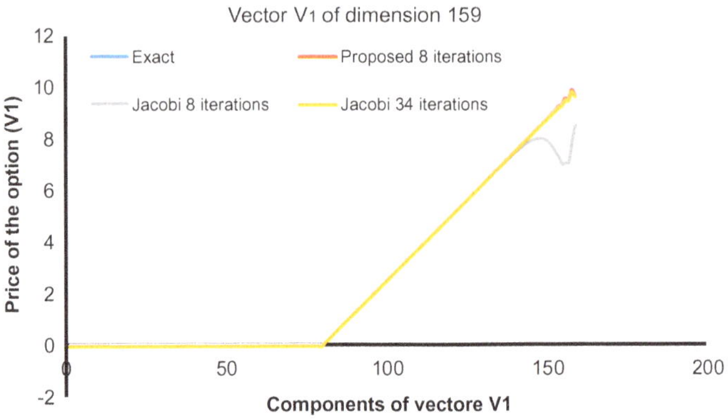

Fig. 10.1 Performance of the proposed method after 8 iterations is very close to the exact solution (norm of error is less than 0.1). Jacobi method needs ~34 iterations to become comparable with the proposed method

Consider a coefficient matrix $\boldsymbol{M}$ with relatively large off-diagonal entries above and below the main diagonal and consider same matrix $\boldsymbol{b}$ as in previous example:

$$\boldsymbol{M} = \begin{bmatrix} 2.0500 & 0.5000 & & & & \\ 0.5001 & 2.0500 & 0.5000 & & & \\ & 0.5002 & \ddots & \ddots & & \\ & & \ddots & 1.5686 & 0.7358 & \\ & & & 0.7487 & 1.5625 & 0.7389 \\ & & & & 0.7519 & 1.5563 \end{bmatrix}_{159\times 159} ; \ \boldsymbol{b} = \begin{bmatrix} 0 \\ 0 \\ \vdots \\ 9.75 \\ 7.39 \end{bmatrix}_{159\times 1}$$

It can be observed that the convergence rate of Jacobi method is much lower than the proposed method (Table 10.3; Fig. 10.2). Now, let's further increase the value of off-diagonal entries:

$$\boldsymbol{M} = \begin{bmatrix} 2.0500 & 4.0000 & & & & \\ 4.0001 & 2.0500 & 4.0000 & & & \\ & 4.0002 & \ddots & \ddots & & \\ & & \ddots & 1.5686 & 4.2358 & \\ & & & 4.2487 & 1.5625 & 4.2389 \\ & & & & 4.2519 & 1.5563 \end{bmatrix}_{159\times 159}$$

Table 10.3 Comparison between the exact solution, the proposed method, and the Jacobi method for a problem with relatively large off-diagonal entries

$V = M^{-1}b$	3 iterations	55 iterations
Exact	Proposed	Jacobi
0.0000	0.0000	0.0000
0.0000	0.0000	0.0000
⋮	⋮	⋮
0.7379	0.7379	0.7379
0.7789	0.7789	0.7789
0.8199	0.8199	0.8199
⋮	⋮	⋮
3.1328	3.1215	3.1402
3.2304	3.2404	3.2359
3.1907	3.1847	3.1936
$\|Exact - Iterative\|$	0.0217	0.0238

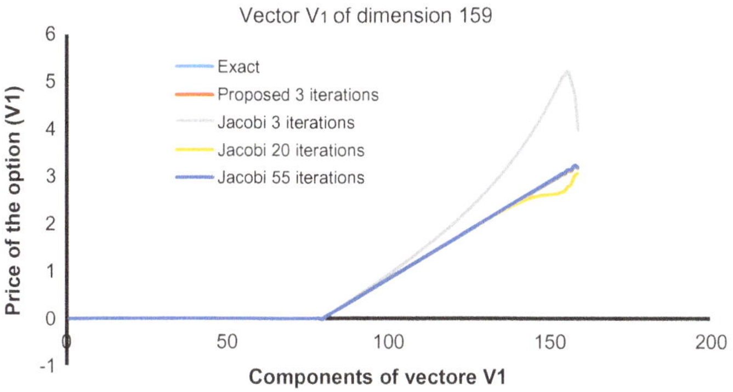

Fig. 10.2 In a problem with relatively large off-diagonal entries the norm of error for the proposed solution is ~0.02 after 3 iterations. The Jacobi method needs ~55 iterations to achieve same accuracy

It can be observed that while in the proposed method the error after 15 iterations is less than 0.06, the Jacobi method cannot converge (Table 10.4).

Table 10.4 Comparison between the exact solution, the proposed method, and the Jacobi method for a problem with dominant off-diagonal entries

$V = M^{-1}b$	15 iterations	55 iterations
Exact	Proposed	Jacobi
0.9948	1.0002	0
−0.5098	−0.5117	0
−0.7335	−0.7383	0
⋮	⋮	⋮
0.8424	0.8380	4.9508e+24
−0.7934	−0.7886	6.6326e+24
−0.4439	−0.4413	3.4968e+25
⋮	⋮	⋮
1.6268	1.6300	6.1527e+40
1.7270	1.7315	4.4047e+40
0.0330	0.0266	2.3087e+40
$\|Exact - Iterative\|$	0.0569	Diverged

The norm of error for the proposed solution is ~0.05 after 15 iterations. The Jacobi method cannot converge

References

1. Shojaei, I., Rahami, H.: A new numerical solution for linear system of equations. Appl Math Comput, Submitted for Publication (2020)
2. Tian, Zh, Tian, M., Zhang, Y., Wen, P.: An iteration method for solving the linear system Ax = b. Comput. Math Appl. **75**(8), 2710–2722 (2018)
3. Pratapa, PhP, Suryanarayana, Ph, Pask, J.E.: Anderson acceleration of the Jacobi iterative method: an efficient alternative to Krylov methods for large, sparse linear systems. J. Comput. Phys. **306**(1), 43–54 (2016)
4. Nguyen, N.C., Fernandez, P., Freun, R., Peraire, J.: Accelerated residual methods for the iterative solution of systems of equations. SIAM J. Sci. Comput. **40**(5), A3157–A3179 (2018)
5. Bohacek J, Kharicha A, Ludwig A, Wu M, Holzmann T, Karimi-Sibakib E. (2019) A GPU solver for symmetric positive-definite matrices vs. traditional codes, Comput Math Appl. 78(9):2933–2943
6. Lemita, S., Guebbai, H., Aissaoui, M.Z.: Generalized Jacobi method for linear bounded operators system. Comput. Appl. Math. **37**(3), 3967–3980 (2018)
7. Yueh, WCh.: Eigenvalues of several tridiagonal matrices. Appl Math E-Notes **5**, 66–74 (2005)

The manufacturer's authorised representative in the EU is Springer Nature Customer Service Centre GmbH, Europaplatz 3, 69115 Heidelberg, Germany. If you have any concerns regarding our products, please contact ProductSafety@springernature.com

Printed and bound by CPI Group (UK) Ltd, Croydon, CR0 4YY
30/09/2026
02225648-0010